GROWTH AND IMPERFECTIONS OF METALLIC CRYSTALS

ROST I NESOVERSHENSTVA METALLICHESKIKH KRISTALLOV

РОСТ И НЕСОВЕРШЕНСТВА МЕТАЛЛИЧЕСКИХ КРИСТАЛЛОВ

ISBN 978-1-4899-4745-1   ISBN 978-1-4899-4743-7 (eBook)
DOI 10.1007/978-1-4899-4743-7

The original Russian text was published by Naukova Dumka in 1966 in
Kiev for the Scientific Councils of Solid State Physics of the Academies
of Sciences of the USSR and the Ukrainian SSR and the Institute of Metal
Physics of the Academy of Sciences of the Ukrainian SSR and has been
corrected by the editor for this edition.

Library of Congress Catalog Card Number 68-13058

© 1968 Springer Science+Business Media New York
Originally published by Consultants Bureau in 1968.
Softcover reprint of the hardcover 1st edition 1968

A Division of Plenum Publishing Corporation
227 West 17 Street, New York, N. Y. 10011
All rights reserved

No part of this publication may be reproduced in any
form without written permission from the publisher

FOREWORD

The vigorous development of new fields of science and technology and intensive research into the theory of strength, the electron structure of metals, and other questions of solid state physics demand the production of metal single crystals with special structures and properties. The solution of this extremely important and difficult problem is impossible by means of ordinary sampling tests, which inevitably involve errors. This question calls for a deep understanding of the mechanism and kinetics of the atomic processes associated with the growth of crystals, the manner in which imperfections of various types develop, etc.

In recent years many Soviet and other scientists have begun showing considerable interest in these problems, and following their investigations substantial progress has been made in the problem in hand. This is especially true of questions relating to crystallization from solution and from the vapor phase, in which crystal-growth theory has received experimental support and already serves as a reliable basis for the practical growth of crystals. As regards crystallization from the melt, especially of metals, there is unfortunately only a meager array of information on both the formation of the crystals and the development of dislocations and other defects in them. This is to some extent due to the nontransparency of metals, their relatively high melting points, and other factors impeding the acquisition of even indirect evidence on growth processes. It may well be that this is why metals have received scant attention. We may nevertheless now hope that the application of modern experimental techniques and the united efforts of experimental and theoretical physicists will soon result in corresponding advances in this direction also, bringing us closer to the solution of one of the main problems in solid state physics: conscious control of growth processes and the production of metal crystals with assigned structure and properties.

This end should in some measure be served by the publication of this volume, which contains the results of the principal investigations in this field carried out in the Soviet Union during the last two or three years.

The first part of the book contains theoretical and experimental papers on the mechanism and kinetics of crystal growth from the melt and in the solid phase, and also methods of growing single crystals.

The second part is concerned with imperfections in crystals. The modern views on the formation of dislocation structure are presented together with methods of studying such structures and some corresponding experimental results. In addition to metals, some data relating to defects in semiconducting crystals are presented; the latter are of general importance, especially as regards various methods of studying stress fields around dislocations in silicon, etc.

The last part deals with the theory underlying the generation of crystallization centers in solutions and the laws of crystallization in certain alloys at high cooling rates.

FOREWORD

The material in this collection was presented to the All-Union Conference on the Growth of Metal Crystals, called by the Scientific Councils on Solid State Physics of the Academies of Sciences of the USSR and UkrainianSSR in Kiev in June 1965.

D. E. Ovsienko

CONTENTS

SECTION I

GROWTH MECHANISM AND THE GROWTH OF CRYSTALS

SECTION II

CRYSTAL IMPERFECTIONS AND THEIR STUDY

SECTION III

GENERATION OF CRYSTALLIZATION CENTERS AND
THE EFFECT OF HIGH COOLING RATES

SECTION I

GROWTH MECHANISM AND THE GROWTH
OF CRYSTALS

SOME QUESTIONS RELATING TO THE THEORY OF THE NONSTATIONARY GROWTH OF CRYSTALS BY THE MECHANISM BASED ON THE FORMATION OF TWO-DIMENSIONAL NUCLEI

B. Ya. Lyubov

*Institute of Metal Science and the Physics of Metals of the Central
Scientific-Research Institute of Ferrous Metallurgy*

The theory of the growth of crystals from the liquid phase by the formation of two-dimensional nuclei constitutes one particular question in the general problem of the kinetics of phase transformations. It is well known that phase transformations of the first kind (to which we shall confine our attention) take place by the generation and growth of centers of a new phase. Particles of the new phase with less than critical dimensions we shall call seeds or nuclei, and larger particles we shall call centers. In general, both the rate of generation of centers (the number of nuclei becoming centers per unit time) and the velocity of their growth are functions of time t and temperature [1]. However, these quantities, which characterize the kinetics of phase transformations, are usually regarded as independent of time. Thus, in essence, the existence of a nonstationary period in the process under consideration is ignored; the existence of such a period requires establishment in each individual case. In order to analyze the kinetics of an isothermal phase transformation, it is convenient to describe the state of the "metastable original phase—new phase" system at a given instant t by a function $Z(n, t)$ representing the distribution of the nuclei and centers with respect to the number of atoms n contained in these. Knowing $Z(n, t)$, an exhaustive representation of the kinetics of the phase transformation may be obtained. During the growth of a center by the formation of two-dimensional nuclei from the liquid phase, a decisive factor in the kinetics of the process is the rate of growth over the face of the nucleus of a solid phase having the form of a layer with a thickness of the order of the interatomic distance in the crystal. Here, $Z(n, t)$ will indicate the distribution of the two-dimensional new-phase formations with respect to dimensions at the given instant t. If $Z(n, t)$ is known, then the rate of generation of two-dimensional centers is given as a "flux" in dimensional space in terms of the value of $n = n^*$. In this paper we shall attempt to construct a theory of the nonstationary growth of crystals, allowing for the time-dependence of the characteristics of the process.

General Theory of the Kinetics

of Isothermal Phase Transformation

In a space of dimensions characterized by the number of atoms n in a new-phase formation, let there be an axis on which three successive points correspond to $n-1$, n, and $n+1$. It is easy to see that if $P_+(n)$ and $P_-(n)$ are the probabilities that an atom will either be joined to the center or torn away from it per unit time, then

$$\frac{\partial Z(n, t)}{\partial t} = -Z(n, t)[P_+(n) + P_-(n)] + Z(n-1, t)P_+(n-1) + Z(n+1, t)P_-(n+1). \qquad (1)$$

If we take $n \gg 1$ and expand the functions in (1) in series, we obtain

$$\frac{\partial Z}{\partial t} = \frac{1}{2}\frac{\partial^2}{\partial n^2}[Z(P_+ + P_-)] + \frac{\partial}{\partial n}[Z(P_- - P_+)]. \qquad (2)$$

It should be noted that the absence of any dependence of P_- and P_+ on Z implies that the growth of each center is regarded as being independent of the presence of others. From quite general considerations it follows that [2]:

$$P_{\pm} = \tilde{n}\omega e^{-\frac{U}{kT} \mp \frac{1}{2kT}\frac{d\Delta F_n}{dn}}, \qquad (3)$$

where $\tilde{n}$ is the number of atoms situated directly on the surface of a center containing n atoms; ω is the oscillation frequency of the atoms in the direction of the surface of the center; U is the activation energy of the process; ΔF_n is the change in the free energy of the system on formation of a center containing n atoms. From expressions (2) and (3) we obtain

$$\frac{\partial Z}{\partial \theta} = \frac{\partial^2}{\partial n^2}\left[\tilde{n}\cosh\left(\frac{1}{2kT}\frac{d\Delta F_n}{dn}\right)Z\right] + 2\frac{\partial}{\partial n}\left[\tilde{n}\sinh\left(\frac{1}{2kT}\frac{d\Delta F_n}{dn}\right)Z\right], \qquad (4)$$

$$\theta = \omega e^{-\frac{U}{kT}}t,$$

or on condition

$$\frac{1}{2kT}\frac{d\Delta F_n}{dn} \ll 1,$$

$$\frac{\partial Z}{\partial \theta} = \frac{\partial^2}{\partial n^2}(\tilde{n}Z) + \frac{\partial}{\partial n}\left[\frac{\tilde{n}}{kT}\frac{d\Delta F_n}{dn}Z\right]. \qquad (5)$$

Equation (5) differs somewhat from the well-known Zel'dovich equation [3] which replaces the term $\frac{\partial^2}{\partial n^2}(\tilde{n}Z)$ by $\frac{\partial}{\partial n}\left(\tilde{n}\frac{\partial Z}{\partial n}\right)$. The difference is due to the more gradual series expansion of the functions of Eq. (1) in our case. For $\frac{1}{P_{\pm n}}\frac{\partial P_{\pm}(n)}{\partial n} \ll 1$ both considerations lead to the same equations.

In order to make the problem specific, Eq. (5) must be supplemented with boundary and initial conditions. As initial conditions we must take

$$Z(n, 0) = 0, \qquad (6)$$

i.e., the absence of new-phase formations at t = 0.

The boundary conditions of the problem may be formulated in accordance with the model of Becker and Döring [4]. According to this, as $n \to 0$, $Z(n, t) \to b(n)$, where b(n) is the

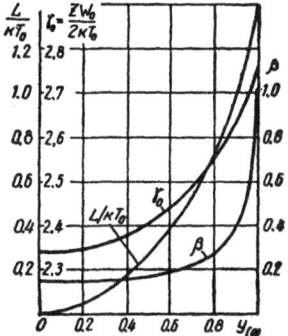

Fig. 1. Results of the calculation of phase equilibrium.

For all values of parameter γ and β, Eq. (5) has a root $y = 0$. However, the disordered phase to which this solution corresponds is only stable or metastable [i.e., $(\partial^2 F/\partial y^2) > 0$] for temperatures above T'_c, determined from the relation

$$\frac{4kT'_c}{zw_0} = 1 - \beta. \tag{6}$$

It is not difficult to show that for $\beta < 1/7$ there is a phase transformation of the second kind at a temperature of T'_c. For $\beta > 1/7$, Eq. (5) has three solutions in the temperature range $T'_c < T < T''_c$ (the temperature T''_c is the temperature corresponding to the loss of stability in the ordered phase). In this case, there is a phase transformation of the first kind. At the transformation temperature T_0 an ordered phase with a degree of long-range order $y_{(0)}$ is in equilibrium with the disordered phase ($y = 0$). The values of $y_{(0)}$ and T_0 are determined by solving Eq. (5) together with the equation

$$\frac{\gamma_0}{2}(1 + \beta y_{(0)}^2)(1 - y_{(0)}^2) - \frac{\gamma_0}{2} + (1 + y_{(0)}) \ln(1 + y_{(0)}) + (1 - y_{(0)}) \ln(1 - y_{(0)}) = 0, \tag{7}$$

corresponding to the condition of equality between the free energies of the two phases.

The results of the solution of Eqs. (5) and (7) for various values of parameter β appear in Fig. 1. In the same figure is a curve characterizing the ratio L/kT_0 (L is the heat of transformation per atom)

$$\frac{L}{kT_0} = \frac{\gamma_0}{2} y_{(0)}^2 [1 - \beta + \beta y_{(0)}^2]. \tag{8}$$

It follows from Eqs. (5), (7), and (8) that in the limiting cases:

for $y_{(0)} \ll 1$,

$$\beta \simeq \frac{1}{7}\left(1 + \frac{24}{35} y_{(0)}^2\right), \qquad \gamma_0 \simeq \frac{7}{3}\left(1 + \frac{4}{35} y_{(0)}^2\right), \qquad \frac{L}{kT_0} \simeq y_{(0)}^2; \tag{9}$$

for $y_{(0)} = 1 - \varepsilon$ ($\varepsilon \ll 1$),

$$\beta \simeq -\frac{\ln \varepsilon}{4 \ln 2}, \qquad \gamma_0 \to 4 \ln 2, \qquad \frac{L}{kT_0} \to 2 \ln 2. \tag{10}$$

In the model considered, the maximal value of L/kT_0 equals $2 \ln 2 \approx 1.4$.

Let us now use the model for describing the state of the boundary of the ordered and disordered phases. In the present case the phase boundary is characterized by a gradual change in the degree of long-range order. Let us consider a boundary parallel to the (100) plane of a fcc lattice. Let us divide the system into atomic layers parallel to this plane. Each layer contains N atoms. Any α point has $z_1 (=4)$ closest β points in the same layer and $z_2 (=1)$ each in the layers above and below it. Any layer n is characterized by a degree of long-range order y_n and a corresponding interaction energy between nearest neighbors $w_n = w_0(1 + \beta y_n^2)$. The interaction energy of the closest atoms in neighboring layers we shall take as being equal to the arithmetic-mean energy in the corresponding layers [for example, the interaction energy of the atoms in the n-th and $(n + 1)$-th layers equals $(w_n + w_{n+1})/2$].

Considering the interaction between nearest neighbors, the configuration part of the free energy F_K of the inhomogeneous system considered may be written on the Bragg–Williams approximation in the form equilibrium distribution of nuclei with respect to size. For $n = n_0$, $Z(n_0, t) = 0$. Thus we assume that formations of the new phase equal to or larger than n_0 are absent from the system. In addition to this we assume that each atom constitutes the formation of a new phase and that the total number of atoms does not vary during the process, i.e., we neglect the loss of atoms belonging to the old phase as the process continues. This scheme is very artificial and may only be regarded as an extremely rough approximation to the true state of affairs. Let us accept it as a first approximation in a slightly changed form. The boundary conditions given for the problem strictly relate to variable limiting values of n. Actually, at any moment there is a maximum size of the new-phase center to which the condition $Z[n_0(t), t] = 0$ should refer. If we suppose that at the initial instant $Z(0, 0) = b(0)$, then as time progresses $Z(n, t) \equiv b(n)$ for larger and larger dimensions of the center, until a stationary state is reached over the whole range of n values considered. This state cannot be realized if we consider the vanishing of old-phase atoms during the process and the collision of centers.

Kinetic Theory of the Growth of New-Phase Centers
by the Formation of Two-Dimensional Nuclei

Let us consider the face of a growing crystal on which a new-phase region of size l is deposited (Fig. 1). In this case (a = interatomic distance):

$$\Delta F_n = -\Delta F_0 l^2 a + 4\sigma l a, \tag{7}$$

$$\tilde{n} = 4\frac{l}{a}, \qquad n = \frac{l^2}{a^2}, \qquad dn = 2\frac{l\,dl}{a^2},$$

and according to Eq. (4),

$$\frac{\partial Z_1}{\partial \theta} = a^3 \frac{\partial}{\partial l} \frac{1}{l} \frac{\partial}{\partial l} \left\{ \cosh\left[\frac{a^3}{2kT}\left(-\Delta F_0 + 2\frac{\sigma}{l}\right)\right] Z_1 \right\} + 4a \frac{\partial}{\partial l} \left\{ \sinh\left[\frac{a^3}{2kT}\left(-\Delta F_0 + 2\frac{\sigma}{l}\right)\right] Z_1 \right\}. \tag{8}$$

Here, ΔF_0 is the change in free energy associated with the transformation of unit volume of the coexisting phases, σ is the boundary (interphase) tension, and

$$Z(n)\,dn = Z_1(l)\,dl.$$

Let us introduce the dimensionless variables

$$x = \frac{l}{l_k}, \qquad l_k = \frac{2\sigma}{\Delta F_0}, \qquad \tau = \frac{\theta a^3}{l_k^3}.$$

Then, if

$$\left| \frac{a^3 \Delta F_0}{kT}\left(-1 + \frac{1}{x}\right)\right| \ll 1,$$

$$\frac{\partial Z_1}{\partial \tau} = \frac{\partial}{\partial x}\left[\frac{1}{x} \frac{\partial Z_1}{\partial x} - B\left(1 - \frac{1}{x}\right) Z_1 \right], \tag{9}$$

$$B = 2\frac{l_k^2}{a^2} \frac{a^3 \Delta F_0}{kT} = 2\frac{\Delta F_{\max}}{kT}, \tag{10}$$

where $\Delta F_{\max}$ is the maximum value of function (7).

The corresponding expression for the "flux" has the form

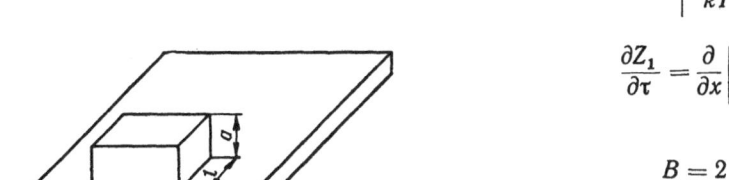

Fig. 1. Face of a growing crystal carrying a two-dimensional formation of a new phase.

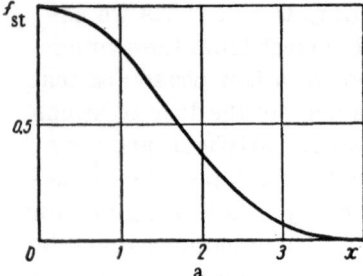

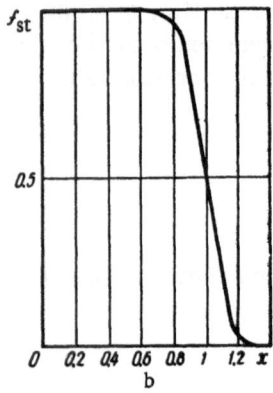

Fig. 2. Stationary value of function f:
a) $B = 1$; b) $B = 100$.

$$I(x, \tau) = -\frac{a^3\omega e^{-\frac{U}{kT}}}{l_k^2 x}\left[\frac{\partial Z_1}{\partial x} + B(1-x)Z_1\right]. \quad (10a)$$

The equilibrium distribution $b_1(x)$ corresponds to the condition

$$I \equiv 0.$$

$$b_1(x) = \alpha e^{B\left(\frac{x^2}{2} - x\right)}, \quad (10b)$$

where α is the constant factor.

If at the initial instant there were N atoms in the system

$$Z(0)\Delta n = Z_1(0)\Delta l. \quad (10c)$$

For $\Delta n = 1$, $\Delta l = a$,

$$Z_1(0) = \frac{N}{a}. \quad (11)$$

Thus

$$b_1(x) = \frac{N}{a}e^{B\left(\frac{x^2}{2} - x\right)} \quad (12)$$

Let us introduce a new function $f(x, \tau) = Z_1(x, \tau)/b_1(x)$; then

$$I = -\frac{a^3\omega}{l_k^2 x}e^{-\frac{U}{kT}}b_1\frac{\partial f}{\partial x} = -N\frac{a^2}{l_k^2}\frac{\omega}{x}e^{-\frac{U}{kT}+\frac{B}{2}(x^2-2x)}\frac{\partial f}{\partial x}, \quad (13)$$

$$x\frac{\partial f}{\partial \tau} = \frac{\partial^2 f}{\partial x^2} - \left[\frac{1}{x} + B(1-x)\right]\frac{\partial f}{\partial x}. \quad (14)$$

According to the foregoing,

$$f(x, 0) = 0, \quad f(0, \tau) = 1, \quad f(\infty, \tau) = 0. \quad (14a)$$

Certain features of the problem in hand may be observed by analyzing the stationary process. It should be noted that the replacement of the requirement $f(x_0, \tau) = 0$, where x_0 corresponds to n_0/n^*, by the condition $f(\infty, \tau) = 0$, is only possible for a very rapid fall in $Z_1(l, t)$ with increasing l. The solution of Eq. (14) in the stationary case ($\partial f/\partial \tau \equiv 0$) (Fig. 2a, b) is

$$f_{st}(x) = \frac{e^{-\frac{B}{2}(x-1)^2} + \sqrt{\frac{\pi B}{2}}\,\mathrm{erfc}\left[\sqrt{\frac{B}{2}}(x-1)\right]}{e^{-\frac{B}{2}} + \sqrt{\frac{\pi B}{2}}\left(\mathrm{erf}\sqrt{\frac{B}{2}} + 1\right)}, \quad (15)$$

where

$$\mathrm{erfc}\,y = \frac{2}{\sqrt{\pi}}\int_y^\infty e^{-\xi^2}d\xi = 1 - \mathrm{erf}\,y.$$

Correspondingly,

$$\frac{\partial f_{st}}{\partial x} = - \frac{Bxe^{-\frac{B}{2}(x-1)^2}}{e^{-\frac{B}{2}} + \sqrt{\frac{\pi B}{2}}\left(\text{erf}\sqrt{\frac{B}{2}} + 1\right)} . \tag{16}$$

Hence, according to Eq. (13), for x = 1,

$$I_{st} = N\frac{a^2}{l_k^2}\frac{\omega e^{-\frac{U}{kT}-\frac{B}{2}}B}{e^{-\frac{B}{2}} + \sqrt{\frac{\pi B}{2}}\left(\text{erf}\sqrt{\frac{B}{2}} + 1\right)} = N\omega e^{-\frac{U}{kT}}\frac{2\frac{a^3\Delta F_0}{kT}e^{-\frac{\Delta F_{max}}{kT}}}{e^{-\frac{B}{2}} + \sqrt{\pi\frac{\Delta F_{max}}{kT}}\left(\text{erf}\sqrt{\frac{\Delta F_{max}}{kT}} + 1\right)}, \tag{17}$$

I_{st} is the "flux" of new-phase formations passing through the critical size.

If we neglect the tangential growth of the centers, then the rate of their growth in a direction perpendicular to the face of the crystal is

$$v_n = aI_{st} . \tag{18}$$

The latter assumption is clearly valid for centers not very far from the critical dimensions. For crystal-face dimensions L(N = L^2/a^2) much larger than l_k it is clearly impermissible to neglect the tangential growth. In order to include this factor in the computing scheme, let us calculate the total volume of transformed phase

$$W = a\int_0^{l_{max}} Z_1(l, t)\, l^2 dl, \tag{19}$$

where l_{max} is the maximum size of a center in the system.

In dimensionless variables, for $f(x, \tau) \equiv f_{st}(x)$,

$$W = Nl_k^3\int_0^{x_{max}} f(x, \tau) e^{B\left(\frac{x^2}{2}-x\right)}x^2 dx = \frac{Nl_k^3 e^{-\frac{B}{2}}\left\{\frac{x_{max}^3}{3} + \sqrt{\frac{\pi B}{2}}\int_0^{x_{max}} e^{\frac{B(x-1)^2}{2}}\text{erfc}\left[\sqrt{\frac{B}{2}}(x-1)\right]dx\right\}}{e^{-\frac{B}{2}} + \sqrt{\frac{\pi B}{2}}\left(\text{erf}\sqrt{\frac{B}{2}} + 1\right)} . \tag{20}$$

For very large x_{max} the first term is much larger than the second, and

$$W \cong N\frac{l_k^3 e^{-\frac{B}{2}}\frac{x_{max}^3}{3}}{e^{-\frac{B}{2}} + \sqrt{\frac{\pi B}{2}}\left(\text{erf}\sqrt{\frac{B}{2}} + 1\right)} . \tag{21}$$

It follows from expression (9) that the equation for the directional growth of the center is

$$\frac{dx_0}{d\tau} = B\left(1 - \frac{1}{x_0}\right) . \tag{22}$$

For large x_0,

$$x_{max} \cong B\tau. \tag{23}$$

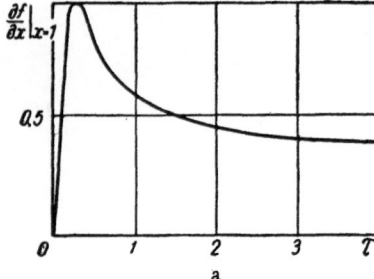

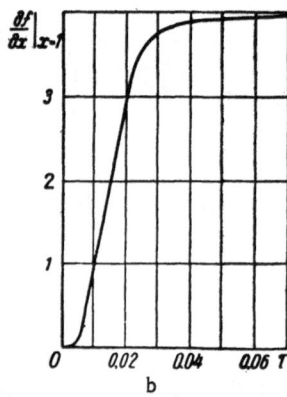

Fig. 3. Variation of $\left.\dfrac{\partial f}{\partial x}\right|_{x=1}$ with τ:
a) B = 1; b) B = 100.

Finally, the relative proportion of transformed material is

$$\frac{W}{V_0} = \frac{e^{-\frac{B}{2}} \dfrac{(B\tau)^3}{3}\left(\dfrac{l_k}{a}\right)^3}{e^{-\frac{B}{2}} + \sqrt{\dfrac{\pi \bar{B}}{2}}\left(\operatorname{erf}\sqrt{\dfrac{\bar{B}}{2}} + 1\right)}, \qquad (24)$$

$$V_0 = L^2 a, \qquad N = \frac{L^2}{a^2}.$$

If we regard the process of the overgrowth of the face as finished, for example, as $W/V_0 = \eta = 0.9$, we can easily find the corresponding τ_η and

$$v_n = \frac{a}{t_\eta}. \qquad (25)$$

In the case of the nonstationary process,

$$I_{\text{nonst}}(x=1) = N\,\frac{a^2}{l_k^2}\,\omega e^{-\frac{U}{kT}-\frac{B}{2}}\left.\frac{\partial f}{\partial x}\right|_{x=1}. \qquad (26)$$

On solving Eq. (14) by numerical methods, we obtain the relation between $\left.\dfrac{\partial f}{\partial x}\right|_{x=1}$ and τ for B = 1, 10, 20, 50, and 100 (Fig. 3a, b). By using this relationship it is not difficult to find the "expectation time" for the appearance of the first center of critical dimensions, $\tau_{\exp}$, from the relation

$$\int_0^{t_{\exp}} I_{\text{nonst}}(x=1)\,dt = 1 = N\,\frac{l_k}{a}\,e^{-\frac{B}{2}}\int_0^{\tau_{\exp}}\left.\frac{\partial f}{\partial x}\right|_{x=1}d\tau. \qquad (27)$$

Neglecting the tangential growth of the two-dimensional nuclei, we find

$$v = \frac{a}{t_{\exp}}. \qquad (28)$$

If we consider the period of the nonstationary process τ_n as being the time required to reach a state in which $f(x, \tau)$ differs from $f_{\text{st}}(x)$ by no more than 1%, then from the numerical solution of Eq. (14) we may find the values of τ_n for various B (Table 1). In the present case the condition of the necessity of taking the nonstationary properties of the process into account is $\tau_{\exp} \ll \tau_n$. If $\tau_{\exp} \gg \tau_n$, then the crystal-growth process in our system may be regarded as stationary. In order to estimate the value of $N = N_0$ requiring the nonstationary state to be taken into account, let us put $\tau_{\exp} \cong \tau_n$. Then from relation (27),

$$N_0 = \frac{1}{\dfrac{l_k}{a}\,e^{-\frac{B}{2}}\displaystyle\int_0^{\tau_n}\left.\frac{\partial f}{\partial x}\right|_{x=1}d\tau} = \frac{\left(\dfrac{a^3 \Delta F_0}{kT}\right)^{1/2}}{\sqrt{\dfrac{\bar{B}}{2}}\,e^{-\frac{B}{2}}\displaystyle\int_0^{\tau_n}\left.\frac{\partial f}{\partial x}\right|_{x=1}d\tau}. \qquad (29)$$

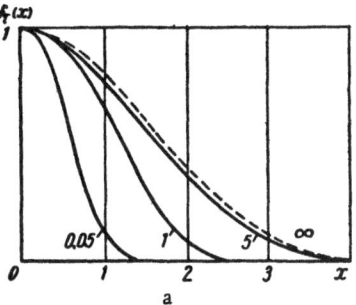

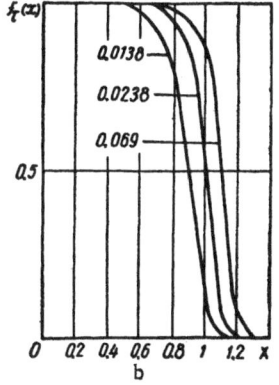

Fig. 4. Function $f(x, \tau) = f_\tau(x)$:
a) B = 1; b) B = 100. Figures on
the curves indicate the corresponding value of τ.

Table 1

B	τ_n
1	4.9
10	0.54
20	0.38
50	0.16
100	0.067
1000	0.010
10000	0.0014
15000	0.00069

Table 2

$\dfrac{N_0 \sqrt{kT}}{V a^3 \overline{\Delta F_0}}$	B
1	1
120	10
$3 \cdot 10^4$	20
$8 \cdot 10^{12}$	50
$3 \cdot 10^{21}$	100

Direct calculation gives the values of $N_0\sqrt{kT}/\sqrt{a^3\Delta F_0}$ as a function of B (Table 2). Knowing the specific characteristics of the system, there is no difficulty in finding crystal dimensions L_0 and l_k from Eq. (29). If $L_0 \gg l_k$, the tangential growth of the two-dimensional nuclei
must be taken into account in the computing scheme. To this end we must use formula (20) to
estimate τ_η for the given η and compare this quantity with τ_n. It should be noted that the use
of formula (20) for estimating the time for the growth of the first center on the face is very
arbitrary, since it is strictly valid for the beginning of the process only. In order to calculate
τ_η for $\eta = 0.9$, expression (20) must be put in the form

$$\frac{W}{V_0} = 0.9 = \left(\frac{l_k}{a}\right)^3 \int_0^{x_{max}(\tau_\eta)} f(x, \tau_\eta)\, e^{\frac{B}{2}(x^2 - 2x)} x^2 dx. \tag{30}$$

The form of the function $f(x, \tau)$ for B = 1 and B = 100 is shown in Fig. 4a, b; $x_{max}(\tau)$
may arbitrarily be identified with the value of x corresponding to $f(x, \tau) = 0$. The rate of crystal growth

$$v_n = \frac{a}{t_\eta} = \frac{a^4}{l_k^3 \tau_\eta}\, \omega e^{-\frac{U}{kT}}. \tag{31}$$

For numerical calculations based on the scheme described, we must first establish the value
of σ, which is directly associated with the structure of the surface separating the phases [5].

In conclusion, it should be mentioned that the numerical solution of Eq. (14) was obtained
in the "Stal'proekt" Institute.

Conclusions

The theory here described enables us to estimate the period of the nonstationary process of crystal growth based on the formation of two-dimensional nuclei and to study the kinetic aspects of this growth in the nonstationary stage. Further development of the theory should proceed along the lines of taking the interaction of the two-dimensional new-phase formations during growth into account and refining the boundary conditions of the problem. It is important to note that the nonstationary aspects of the process must be taken into account when there is a rapid variation in the conditions under which the phase transformation is taking place. Thus, in the case of crystallization with rapid cooling, this factor may play a substantial part.

Literature Cited

1. Lyubov, B. Ya., and Roitburd, A. L., In: "Problems of Metal Science and the Physics of Metals," Vol. V, p. 91. Metallurgizdat, Moscow (1958).
2. Lyubov, B. Ya., In: "Problems of Metal Science and the Physics of Metals," Vol. V, p. 294. Metallurgizdat, Moscow (1958).
3. Zel'dovich, Ya. B., Zh. Éksperim. i Teor. Fiz., 12:525 (1942).
4. Becker, R., and Döring, W., Ann. Physik, 24:719 (1935).
5. Cahn, I. W., Hillig, W. B., and Sears, G. W., Acta Met., 12:1421 (1964).

ON THE "BLURRING" (DIFFUSE NATURE) OF
THE CRYSTAL—MELT BOUNDARY

D. E. Temkin

*Institute of Metal Science and the Physics of Metals of the Central
Scientific-Research Institute of Ferrous Metallurgy*

The question of the structure of the bounding surface between the solid phase and the melt is perhaps the most important in studying the kinetics of the motion of this boundary during crystallization or melting [1]; so far, however, it has not really been satisfactorily solved. Only individual attempts at studying this process are to be found in the literature.

Roitburd and Cahn [2, 3] gave a phenomenological description of the state of a plane boundary separating the phases in a crystalline medium. The boundary was considered as a transitional region in which there was a variation in a certain parameter y from the value y' characterizing one phase to y" characterizing the other. It was found that metastable states of the interface (i.e., states corresponding to minima of the free energy of the two-phase system in question) only existed in a definite range of deviations from the equilibrium temperature. This range was the narrower, the more "blurred" the boundary. These facts are very important when studying the kinetics of the motion of the phase boundary. Cahn suggested [3] that, for those deviations from equilibrium temperature at which metastable equilibrium states existed, the boundary moved in accordance with the mechanism of layer-like growth. For deviations from equilibrium greater than critical, when there are no metastable states, the boundary may be displaced continuously in the direction of its normal, without requiring the formation of steps for such motion.

This phenomenological approach does not enable us to give a complete description of the state of the boundary between the solid and liquid phases, since the meaning of the parameter introduced into the consideration remains uncertain. Hence, a coherent statistical analysis of the state of the boundary is required. However, the difficulties of such an analysis are exacerbated in the case of a crystal—melt boundary by the fact that at the present time there is no coherent statistical theory for the crystal—melt transformation.

Frequently, the model of a rough crystal face [6] is used to describe the crystal—melt boundary [4, 5]. The use of this model is equivalent to the assumption that there is no "blurring" of the boundary. In fact, with this model, we may be quite certain as to which of the phases in the transitional region a given atom belongs for a given configuration of the boundary. For a "blurred" boundary, however, this concept loses its meaning. It is clear that this model gives a good description of the state of a crystal face when the crystal is in contact with a vapor, but is very rough when considering crystal—melt interfaces.

In this paper we shall attempt to estimate diffuse or "blurred" nature of the interfaces by means of a very simple model of the crystal−melt transition. Melting is often considered as a process of "disordering" [7, 8]. If, following [7] and considering a fcc lattice, we introduce two sublattices α and β (both equivalent), then in a state of complete ordering (crystal at absolute zero) the atoms occupy all the lattice points of one of these sublattices. On raising the temperature, the number of atoms lying at lattice points of the second sublattice (formed by the centers of the cube edges) increases and the degree of long-range order falls. In the completely disordered phase (liquid) the number of atoms at the points of the two sublattices is the same.

Let the system contain N atoms, of which N_α lie at the lattice points of the α sublattice. Using the Bragg−Williams approximation [9] and considering only interaction between nearest neighbors, we may write the free energy F of the partly disordered system in the form

$$\frac{F(N, N_\alpha, T)}{NkT} = \frac{F_0(N, T)}{NkT} + \frac{zw}{kT} \frac{N_\alpha}{N} \left(1 - \frac{N_\alpha}{N}\right) + \ln\left[\frac{N!}{N_\alpha!(N - N_\alpha)!}\right]^2, \tag{1}$$

where $F_0(N, T)$ is the free energy of the completely ordered state; w is the interaction energy of two neighboring atoms, one of which lies in an α and the other in a β point; and z is the number of β points closest to the α point (in the case under consideration, z = 6).

Let us introduce the degree of long-range order

$$y = 2\frac{N_\alpha}{N} - 1. \tag{2}$$

In the completely ordered state y = 1 or y = −1 (all the atoms lie at β points); in the completely disordered state $N_\alpha = N/2$ and y = 0.

It is well known [9] that in a system described by expression (1), at a temperature T_c' given by the relation $4kT_c'/zw = 1$, there is a phase transformation of the second kind. It is also known that, allowing for the dependence of the interaction energy w on the distance between the lattice points (or in essence on the occupied volume) leads, under certain conditions, to a phase transformation of the first kind [7, 10] taking place with a change in the degree of long-range order and the volume. However, allowing for the dependence of w on volume leads to such serious complication in the description of the surface of separation that there is no point in considering it for the very coarse model under consideration. We shall therefore consider that the energy w does not depend on volume, but on the degree of long-range order. This assumption will not lead to any change in the qualitative picture as compared with that in which w depends on volume. The simplest form of the dependence of w on y is

$$w = w_0(1 + \beta y^2), \tag{3}$$

since, in view of the equivalence of the α and β lattices, we must satisfy the condition w(y) = w(−y). Here, $w_0(>0)$ and β are certain parameters. Considering (2) and (3), expression (1) may be written as

$$\frac{F(N, y, T) - F_0(N, T)}{NkT} = \frac{\gamma}{2}(1 + \beta y^2)(1 - y^2) - 2\ln 2 + (1 + y)\ln(1 + y) + (1 - y)\ln(1 - y). \tag{4}$$

where $\gamma = zw_0/kT$.

The equilibrium values of the degree of long-range order y are determined from the condition

$$\frac{\partial F}{\partial y} = 0 \quad \text{or} \quad \gamma y[1 - \beta + 2\beta y^2] = \ln\frac{1 + y}{1 - y}. \tag{5}$$

$$\frac{F_\kappa}{NkT} = \sum_{n=-\infty}^{\infty} \left\{ f\left(y_n\right) - \frac{\gamma}{2}\frac{z_2}{z}\left(1 + \beta y_n^2\right) y_n \left(y_{n-1} - 2y_n + y_{n+1}\right)\right\}, \tag{11}$$

where

$$f\left(y_n\right) = \frac{\gamma}{2}\left(1 + \beta y_n^2\right)\left(1 - y_n^2\right) + \left(1 + y_n\right)\ln\frac{1+y_n}{2} + \left(1 - y_n\right)\ln\frac{1-y_n}{2} \tag{12}$$

is $(1/kT)$ times the configuration free energy (for one particle) of the homogeneous phase characterized by a degree of long-range order y_n [see relation (4)]. The equilibrium values of y_n correspond to a minimum of free energy and are determined from the equation $\partial F_K/\partial y_n = 0$ or

$$\left(2 + 3\beta y_n^2\right)\left(y_{n-1} - 2y_n + y_{n+1}\right) + \beta\left(y_{n-1}^3 - 2y_n^3 + y_{n+1}^3\right) - \frac{2z}{\gamma z_2}\frac{df\left(y_n\right)}{dy_n} = 0, \tag{13}$$

$$y_{-\infty} = 0, \qquad y_{+\infty} = y_{(0)}.$$

Here we assume that at $n = -\infty$ we have the liquid (disordered) phase, and at $n = \infty$ the ordered phase, characterized by a degree of long-range order $y_{(0)}$. Since $y = 0$ and $y_{(0)}$ corresponds to equilibrium values of the degree of long-range order in the phases under consideration, we have $f'(0) = f'(y_{(0)})$, and $f''(0)$ and $f''(y_{(0)})$ are positive.

If we take relation (13) into account, expression (11) may be written in the form

$$\frac{F_\kappa}{NkT} = \sum_{n=-\infty}^{\infty} \left\{ f\left(y_n\right) - \frac{1}{2} y_n \frac{df\left(y_n\right)}{dy_n} + \frac{\gamma z_2}{2z}\beta y_n^3\left(y_{n-1} - 2y_n + y_{n+1}\right)\right\}. \tag{14}$$

Let us endeavor to analyze a number of limiting cases in which the nonlinear difference equation (13) may be linearized or replaced by a differential equation.

1. Behavior of the Solution a Long Way from the Boundary (for Large $|n|$). It is clear that, for fairly large n, y_n differs little from $y_{(0)}$ and y_{-n} from 0. Remembering this and introducing the new quantities

$$\begin{aligned} \delta_n &= y_{(0)} - y_n \quad (\text{for } n > 0), \\ \varepsilon_n &= y_n \quad\quad\;\; (\text{for } n < 0). \end{aligned} \tag{15}$$

We may simplify Eq. (13) (for large $|n|$), retaining only terms linear in ε_n and δ_n

$$\varepsilon_{n+1} - \left(2 + \frac{f''(0)}{2\mu}\right)\varepsilon_n + \varepsilon_{n-1} = 0 \qquad (\text{for } n < 0), \tag{16}$$

$$\delta_{n+1} - \left[2 + \frac{f''\left(y_{(0)}\right)}{2\mu\left(1 + 3\beta y_{(0)}^2\right)}\right]\delta_n + \delta_{n-1} = 0 \quad (\text{for } n > 0).$$

Here we have considered that

$$f'\left(\varepsilon_n\right) \simeq f'(0) + \varepsilon_n f''(0) = \varepsilon_n f''(0), \qquad f'\left(y_{(0)} - \delta_n\right) \simeq -\delta_n f''\left(y_{(0)}\right),$$

and denoted $\mu = \gamma z_2/2z$.

Solutions of Eq. (16) satisfying the conditions $\varepsilon_{-\infty} = 0$ and $\delta_\infty = 0$ have the form

$$\varepsilon_{-n} = \frac{b_1}{\nu^n}, \qquad \delta_n = b_2 \varrho^n, \tag{17}$$

where b_1 and b_2 are constants and ν and ρ are determined from the system of relations

$$\nu = 1 + \frac{f''(0)}{4\mu} + \sqrt{\left(1 + \frac{f''(0)}{4\mu}\right)^2 - 1}, \quad \varrho = 1 + \frac{f''(y_{(0)})}{4\mu(1 + 3\beta y_{(0)}^2)} - \sqrt{\left[1 + \frac{f''(y_{(0)})}{4\mu(1 + 3\beta y_{(0)}^2)}\right]^2 - 1}. \quad (18)$$

It is clear that the closer the quantities ν and ρ are to unity, the more "blurred" will the boundary be.

2. Analysis of a Severely "Blurred" Boundary. In this case, instead of y_n we introduce a function $y(x)$ of a continuous variable x (in atomic planes parallel to the interface x = na, where a is the interplane distance), and replace the difference equation (13) by a differential equation

$$(2 + 3\beta y^2)\frac{d^2 y}{dx^2} + \beta \frac{d^2(y^3)}{dx^2} - \frac{1}{\mu a^2}\frac{df}{dy} = 0,$$

$$y(-\infty) = 0, \qquad y(\infty) = y_{(0)}. \tag{19}$$

This equation may be written in the form

$$\frac{d}{dy}\left[(1 + 3\beta y^2)\left(\frac{dy}{dx}\right)^2\right] - \frac{1}{\mu a^2}\frac{df}{dy} = 0, \tag{20}$$

after integration

$$\frac{dy}{dx} = \frac{1}{a\sqrt{\mu}}\sqrt{\frac{f(y) - f(0)}{1 + 3\beta y^2}}. \tag{21}$$

Here we have remembered that for y = 0, dy/dx = 0. Integrating this equation, we obtain

$$x = A + a\sqrt{\mu}\int \sqrt{\frac{1 + 3\beta y^2}{f(y) - f(0)}}\,dy, \tag{21a}$$

where A is a constant of integration determining the position of the boundary.

The effective width of the boundary l (in numbers of interplane distances) may be determined as

$$l = y_{(0)}\Big/a\left(\frac{dy}{dx}\right)_{\max}. \tag{22}$$

The calculated widths of the boundary for various values of $y_{(0)}$ are given in Fig. 2. Using relation (9), it is quite easy to show that for $y_{(0)} \ll 1$, $l \simeq 3\sqrt{35}/4\,y_0^2$. In the other limiting case in which $y_{(0)} = 1$, we have $\lim\limits_{y_0=1} l = 1/\sqrt{2}$. Thus we see that with increasing $y_{(0)}$ (L/kT$_0$ increasing in parallel with this) the width of the interface becomes smaller, i.e., the boundary becomes sharper. For $y_{(0)} \ll 1$, according to relations (9) and (12), we have

$$f(y) - f(0) \simeq \frac{y_{(0)}^4}{15}\,y^2\left(1 - \frac{y^2}{y_{(0)}^2}\right)^2,$$

which after substitution in Eq. (21) and integration gives

$$x = A + \frac{a\sqrt{70\frac{z_2}{z}}}{2y_{(0)}^2}\ln\frac{\dfrac{y}{y_{(0)}}}{\sqrt{\dfrac{\left(1 - \dfrac{y^2}{y_{(0)}^2}\right)}{2}}}.$$

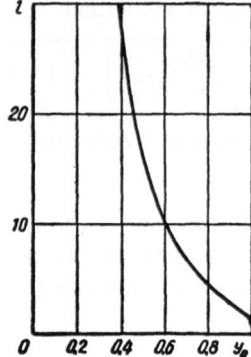

Fig. 2. Width of the boundary for various values of the degree of long-range order $y_{(0)}$ of the ordered phase.

3. Analysis of a Sharp Interface. We see from the graph in Fig. 2 that the boundary will be sharp for $y_{(0)} \simeq 1$. In the case of a sharp boundary we may suppose that $\varepsilon_0 \ll 1$ and $\delta_1 \ll 1$ [in the notation of (15)]. Then, on requiring that relations (17) should describe ε_0 and δ_1, we find

$$\varepsilon_{-n} = \frac{\varepsilon_0}{\nu^n} \quad (n \geqslant 0), \qquad \delta_n = \delta_1 \varrho^{n-1} \quad (n \geqslant 1). \qquad (23)$$

In order to determine the unknown quantities ε_0 and δ_1, i.e., in order to "match" the solution of (23), we use relations (13) for n = 0 and n = 1:

$$\left.\begin{array}{l} (2 + 3\beta\varepsilon_0^2)\left(\dfrac{\varepsilon_0}{\nu} - 2\varepsilon_0 + y_{(0)} - \delta_1\right) + \\[2mm] \quad + \beta\left[\dfrac{\varepsilon_0^3}{\nu^3} - 2\varepsilon_0^3 + (y_{(0)} - \delta_1)^3\right] - \dfrac{f'(\varepsilon_0)}{\mu} = 0, \\[3mm] [2 + 3\beta(y_{(0)} - \delta_1)^2][\varepsilon_0 - 2(y_{(0)} - \delta_1) + y_{(0)} - \delta_1\varrho] + \\[2mm] \quad + \beta[\varepsilon_0^3 - 2(y_{(0)} - \delta_1)^3 + (y_{(0)} - \delta_1\varrho)^3] - \dfrac{f'(y_{(0)} - \delta_1)}{\mu} = 0 \end{array}\right\} \qquad (24)$$

Let us consider the limiting solution of this set of equations at $\beta \to \infty$. It is not difficult to show that as $\beta \to \infty$ we have $\nu \to \infty$, $\rho \to 0$, $\delta_1 \to 0$, while ε_0 tends to a nonzero limit given by the equation

$$3\varepsilon_0^2(1 - 2\varepsilon_0) + (1 - 2\varepsilon_0^3) - \frac{f'(\varepsilon_0)}{\beta\mu} = 0.$$

Considering that at $\beta \to \infty$

$$f'(\varepsilon_0) \simeq \gamma\beta\varepsilon_0(1 - 2\varepsilon_0^2),$$

we have for ε_0 the equation

$$1 - 12\varepsilon_0 + 3\varepsilon_0^2 + 16\varepsilon_0^3 = 0,$$

a necessary solution of which is $\varepsilon_0 = 0.0860$. Thus, for $\beta \to \infty$, the solution of Eq. (13) is

$$y_n = \begin{cases} 0 & (n < 0) \\ 0.0860 & (n = 0) \\ 1 & (n > 0). \end{cases} \qquad (25)$$

However, Eq. (13) has a solution of another type, tending to a limit differing from (25) as $\beta \to \infty$. In order to find this limit, we suppose that $\varepsilon_0 \ll 1$, $\delta_2 \ll 1$, and that y_1 differs considerably from both 0 and 1. In this case, relations (17) take the form

$$\varepsilon_{-n} = \frac{\varepsilon_0}{\nu^n} \quad (n \geqslant 0), \qquad \delta_n = \delta_2 \varrho^{n-2} \quad (n \geqslant 2). \qquad (26)$$

In order to determine ε_0, δ_2, and y_1, we must use relations (13) for n = 0, 1, and 2. As $\beta \to \infty$, $\delta_2 \to 0$, and the system of equations for ε_0 and y_1 takes the form

$$\left.\begin{array}{l} 3\varepsilon_0^2 y_1 + y_1^3 + 16\varepsilon_0^3 - 12\varepsilon_0 = 0 \\[2mm] 16y_1^3 + 3(1 + \varepsilon_0)y_1^2 - 12y_1 + 1 + \varepsilon_0^3 = 0 \end{array}\right\} \qquad (27)$$

An approximate solution of this system is $\varepsilon_0 \simeq 0.032$ and $y_1 \simeq 0.73$. Thus, as $\beta \to \infty$, the solution of the second type tends to the limit

$$y_n = \begin{cases} 0 & (n \leqslant -1) \\ 0.032 & (n = 0) \\ 0.73 & (n = 1) \\ 1 & (n > 1). \end{cases} \tag{28}$$

Solutions (25) and (28) describe different states of the phase boundary. Solution (25) (solution of the first type) corresponds to a minimum of the free energy of the system F_K, while (28) (solution of the second type) describes a saddle point on the surface F_K. Solutions of the first and second types differ outwardly in that, in the solution of the second type, there is an intermediate layer, the state of which is close to the maximum of function $f(y)$. Thus, as $\beta \to \infty$, the function $f(y)$ is maximum at $y = 1/\sqrt{2}$; a state close to this is given by y_1 in expression (28).

Although the existence of two types of solutions has been proved for $\beta \to \infty$ (and for any finite γ), we may suppose that analogous solutions exist at the phase−equilibrium temperature for any β greater than $1/7$ (for $\beta > 1/7$ there is a phase transformation of the first kind). By analogy with the results of the phenomenological description [2, 3] we may suppose that Eq.(13) only has the solutions indicated in a certain region around the equilibrium temperature and none outside this. Clearly, the critical deviations from equilibrium at which solutions of the two types coincide depend on the width of the boundary (or β), being the smaller, the greater the width (or the smaller β). An additional basis for this supposition is provided by the analysis of a boundary consisting of a single intermediate layer.

It is clear from all that has been said that for suitably large values of β the boundary may, to a fair accuracy, be considered as consisting of a single intermediate layer. Let us suppose that in this layer $y = y_1$ while $y_0 = y_{-1} = \ldots = 0$ and $y_2 = y_3 = \ldots = y_{(0)}$. Then, according to relation (11), the free energy equals (only considering terms with $n = 0, 1, 2$ containing y_1):

$$\frac{F_\kappa}{NkT} = f(0) + f(y_{(0)}) + f(y_1) - \mu(1 + \beta y_1^2)(y_{(0)} - 2y_1) - \mu(1 + \beta y_{(0)}^2) y_{(0)}(y_1 - y_{(0)}). \tag{29}$$

In order to determine the equilibrium value of y_1 we have the equation

$$\frac{d}{dy_1}\left[\frac{F_\kappa}{NkT}\right] = \frac{1}{\mu} \ln \frac{1 + y_1}{1 - y_1} + y_1\left[4 - \frac{2z}{z_2}(1 - \beta) - 3\beta y_{(0)} y_1 + \left(8 - \frac{4z}{z_2}\right)\beta y_1^2\right] - y_{(0)}(2 + \beta y_{(0)}^2) = 0. \tag{30}$$

Equation (30) has the following properties analogous to the properties of the equation describing the roughness of the single-layer boundary [4]:

1. For β smaller than a certain value β_0 and any temperatures (any γ), this equation has one root. The value of β_0 lies between 0.5 and 1, but has not been determined exactly. This case is of no interest to us, since, for these values of β, it is insufficient merely to consider one intermediate layer and a multilayer boundary must be taken into account.

2. For $\beta > \beta_0$, Eq. (30) has three roots in a certain temperature range including the phase−equilibrium temperature T_0. If we place the roots in ascending order, then the first and last correspond to minimum F_K and the middle one to a maximum.

3. For $\beta > \beta_0$ and deviations from T_0 greater than some critical value, Eq. (30) has one root instead of three. In the case of supercooling, the largest root is retained, and in the case of superheating, the smallest. This may be regarded as the "crystallization" of the transitional layer in the case of the supercooling of the system and as "melting" in the case of super-

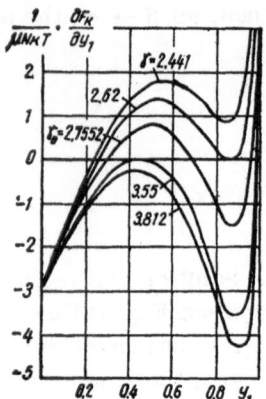

Fig. 3. Left-hand side of Eq. (30) as a function of the degree of long-range order y_1 in the intermediate layer for various temperatures.

heating. If under these conditions, considering, for example, supercooled systems, we remove the limitations according to which for $n \leq 0$ there is a disordered and for $n > 1$ an ordered phase, and expand the transitional region (i.e., consider a boundary consisting of two, three, etc., layers), then we find that the "crystallization" propagates behind the boundary of the disordered phase. This indicates that under these conditions, i.e., for supercoolings greater than critical, Eq. (13) has no solutions.

By way of illustration we have carried out a calculation for the single-layer boundary for $\beta = 1$ and various $\gamma = zw_0/kT$. The results of calculating the function $d/dy_1 [F_K/\mu NkT]$ from formula (30) are reflected in Fig. 3. For $\beta = 1$, phase equilibrium corresponds to $\gamma_0 = 2.7552$ and $y_{(0)} = 0.9906$. For $\gamma < \gamma_0$ the system is in the superheated and for $\gamma > \gamma_0$ in the supercooled state. We see from Fig. 3 that critical supercooling is given by $\gamma \simeq 3.55$ ($y_{(0)} \simeq 0.998$) and critical superheating by $\gamma \simeq 2.62$ ($y_{(0)} \simeq 0.987$). Using these values, let us find the critical supercooling $\Delta T'$ and the critical superheating $\Delta T''$: $\Delta T'/T_0 \simeq 0.3$, $\Delta T''/T_0 \simeq 0.05$. We notice that the critical supercooling is much greater than the superheating.

As already mentioned, for $\beta \rightarrow \infty$, the solutions of Eq. (13) tend to (25) and (28) independently of temperature. Hence the value of the critical deviations from equilibrium increase without limit as β rises.

Conclusions

The crystal—melt boundary is considered as a transitional region between the ordered and disordered phases, in which the degree of long-range order gradually changes from zero (disordered phase of melt) to some finite value $y_{(0)}$ (order phase or crystal).

The width of the boundary l increases as $y_{(0)}$ falls (L/kT_0 falls in parallel with $y_{(0)}$).

For $y_{(0)} = 1$ ($\beta = \infty$) we have proved the existence of two states of the boundary: one corresponds to a minimum of the free energy F_K and the other to a saddle point on the F_K surface.

We have considered the possible existence of these two states of the boundary for deviations from equilibrium smaller than critical and the absence of metastable states for deviations greater than critical. This may be confirmed on the basis of the one-layer boundary.

Literature Cited

1. Cahn, J. W., Hillig, W. B., and Sears, G. W., Acta Met., 12(12): 1421 (1964).
2. Roitburd, A. L., Crystallization and Phase Transformations. Minsk (1962). See also Kristallografiya, 7(2): 291 (1962).
3. Cahn, J. W., Acta Met., 8(8): 554 (1960).
4. Jackson, K., Liquid Metals and Their Solidification [Russian translation], p. 200. Metallurgizdat, Moscow (1962).
5. Temkin, D. E., Mechanism and Kinetics of Crystallization. p. 86. Minsk (1964).
6. Barton, W., Cabrera, H., and Frank, F., Elementary Processes of Crystal Growth [Russian translation]. p. 11. IL, Moscow (1959).

7. Lennard-Jones, J. E., and Devonshire, A. F., Proc. Roy. Soc., A170:464 (1939).

8. Frank, F. C., Proc. Roy. Soc., A170:182 (1939).

9. Krivoglaz, M. A., and Smirnov, A. A., Theory of Ordered Alloys. Fizmatgiz, Moscow (1958).

10. Ross, A. W., and Haar, D. ter., Physica, 25:343 (1959).

ON THE THEORY OF THE NORMAL GROWTH
OF CRYSTALS

V. T. Borisov

*Institute of Metal Science and the Physics of Metals of the Central
Scientific-Research Institute of Ferrous Metallurgy*

The crystallization of metals takes place for a very slight supercooling at the phase boundary. In order to describe the process taking place for a small deviation from the equilibrium state we may make use of thermodynamic theory. The growth law thus obtained has a linear character

$$V = L\Delta\mu = K\Delta T \tag{1}$$

($\Delta\mu$ is the difference in the chemical potentials of the phases, V is the velocity, and ΔT is the supercooling) and qualitatively differs from the relations characterizing the nuclear or spiral growth of a crystal. This relates to both the classical forms of these theories, which are concerned with ideal boundaries, and modernized versions [1], which include effective values of surface tension and other parameters in order to take account of the atomically rough structure of the crystal—melt boundary. Equation (1) indicates the existence, at slight supercoolings, of a particular crystallization mechanism known [2, 3, 4] as the "normal" type. The thermodynamic derivation of (1) does not reveal the details of this phenomenon; however, on the basis of general considerations of the thermodynamics of irreversible processes we may suppose that the contribution of the normal growth mechanism to the total crystallization rate will be considerable if there are substantial fluctuations in the volumes of the co-existing phases in the neighborhood of equilibrium.

The statistical consideration presented in [3] directly relates the normal growth mechanism to fluctuations of the phase-separation boundary at equilibrium. It follows from the resultant expression for the velocity

$$V = a\nu e^{-Q/kT} [\beta_0'\Delta\mu/kT + \Delta\beta(\Delta T)], \tag{2}$$

where a is the lattice parameter, ν is the oscillation frequency, and Q is the activation energy, that, for a slight supercooling the normal mechanism predominates if the density of the points of growth β_0' is sufficiently great. In the contrary case, when $\beta_0' \ll 1$, growth is determined by tangential mechanisms [term $\Delta\beta(\Delta T)$], except for the immediate neighborhood of the melting point. The quantity β_0' represents the probability of the formation, at a given point of the crystal face, of an atomic configuration favorable toward the attachment of a new particle to the

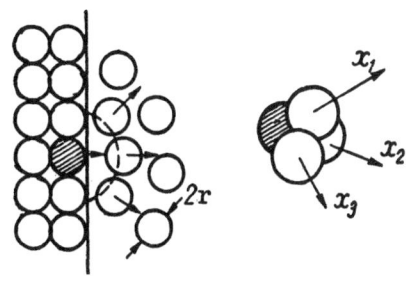

Fig. 1

crystal (probability of a microscopic fluctuation of relief) and is expressed by the ratio of the corresponding statistical sums

$$\beta_0' = \int\limits_{\gamma'} e^{-H(X,P)/kT_0} dXdP \Big/ \int\limits_{\Gamma'} e^{-H(X,P)/kT_0} dXdP, \qquad (3)$$

where X, P are sets of coordinates and momenta, Γ' is the total phase volume, and γ' is the volume corresponding to the favorable configurations.

The density of the points of growth β_0' is given for a particle in transit from the liquid to the crystal; it is related to the analogous quantity β_0'' for a particle passing in the opposite direction by

$$\beta_0' = \beta_0'' e^{-2q/3kT_0}, \qquad (4)$$

where q and T_0 are the heat of fusion and the melting point, respectively. The quantity β_0'' is described by an expression analogous to the ratio (3) and may be calculated on the following simplifying assumptions.

The models of the liquid and crystal will initially be regarded as oscillators. For a selected particle (Fig. 1) belonging to the solid phase to be able to pass into the liquid, the nearest particles of liquid must be removed (there are considered to be three of these) to distances x_1, x_2, and x_3, so that the volume $\pi r^2 (x_1 + x_2 + x_3)$ freed exceeds a certain critical value Δv sufficient for the transition to take place. If $\rho(x_i)$ is the probability of the displacement of particle i to a distance x_i, then the probability of a favorable volume developing is

$$\iiint\limits_{x_1+x_2+x_3 \geqslant \delta} \varrho(x_1)\varrho(x_2)\varrho(x_3)\, dx_1\, dx_2\, dx_3,$$

where $\delta = \Delta v/\pi r^2$. Bearing in mind similar expressions for the energy of each particle in the neighborhood of the equilibrium position

$$H(X,P) = U_0 + m\omega^2 (x^2 + y^2 + z^2)/2 + (p_x^2 + p_y^2 + p_z^2)/2m, \qquad (5)$$

where m and ω are the mass and mean oscillation frequency of the particle and U_0 is a constant, we find, in accordance with (3),

$$\beta_0'' = \iiint\limits_{x_1+x_2+x_3 \geqslant \delta} e^{-m\omega^2(x_1^2+x_2^2+x_3^2)/2kT_0} dx_1 dx_2 dx_3 \Big/ \left(\int\limits_{-\infty}^{\infty} e^{-m\omega^2 x^2/2kT_0} dx \right)^3. \qquad (6)$$

Expression (6) may be brought to the form $\beta_0'' = \varphi(\xi \delta/a)$, where $\xi = (ma^2\omega^2/2kT_0)^{1/2}$. The Morse potential $U = U_0[1 - \exp(-\alpha x)]^2$, $U_0 = 2H/zN$ (where H is the heat of sublimation, z is the coordination number, and N Avogadro's number) has the expansion $U = U_0\alpha^2 x^2$ in the neighborhood of equilibrium. Comparison with expression (5) gives $m\omega^2 = 2U_0\alpha^2$. This enables us to express the constant ξ in terms of the potential parameter α and by means of relations $D = z^2N(kT)^2/32h\alpha^2H$ [5] and $\varkappa mD = a^2kT$ [2] in terms of the diffusion coefficient in the liquid D or the kinematic viscosity $\varkappa$:

$$\xi = \alpha a(2H/zRT_0)^{1/2} = a(zkT_0/48hD)^{1/2} = (\varkappa Mz/48hN)^{1/2},$$

$$M = mN. \qquad (7)$$

Table 1

Substance	$10^3 \varkappa \dfrac{cm^2}{sec}$	q/kT_0	ξ	β_0'	β_0' exp	
					I	II
Mercury	1.4	1.2	6.6	0.19	0.03	0.86
Cadmium	3.0	1.0	7.3	0.22	0.04	0.69
Potassium	6.5	0.9	6.3	0.24	0.15	8.9
Rubidium	4.4	0.8	7.7	0.34	0.03	0.55
Gallium	3.3	2.2	6.0	0.10	—	—
Salol	78	7.5	51	0.0007	—	—

For $z = 10$, $(z/48hN)^{1/2} = 7.2 \; g^{1/2} \cdot cm \cdot sec^{-1/2}$, and thus

$$\beta_\theta' = e^{-2q/3kT_0} \, \varphi \left(7.2 \, \frac{\delta}{a} \, (\varkappa M)^{1/2} \right). \tag{8}$$

Results of calculating the density of the points of growth for a number of substances are presented in Table 1.

The ratio $\delta/a = \Delta v / \pi r^2 a$ is probably close to the relative change in volume on solidification. In the calculations the value $\delta/a = 0.03$ was taken for all substances.

Expression (8) leads to high values of the density of the points of growth for substances such as metals having a small q/kT_0 ratio and low viscosity. For substances with a large relative heat of fusion and high viscosity, such as Salol, the rate of normal crystal growth is negligibly small. Both these factors introduce a considerable contribution into the reduction in β_0'.

The last columns of Table 1 present experimental data regarding the density β_0' [6]. The values in columns I and II were calculated on the assumption that the activation energy Q for the passage of a particle from the liquid into the solid phase equals, in the first case, zero, and in the second the activation energy of diffusion. The theoretical values of β_0' do not contradict the experimental. We may also conclude that the activation energy for crystallization is probably lower than that relating to diffusion in the liquid.

Relation (8) enables us to estimate the order of magnitude of the fluctuations in surface relief due to the passage of particles from phase to phase. Since $(x_1 + x_2 + x_3)/a \simeq 0.03$ corresponds to experimental data, fluctuations associated with a variation of about 1% in interatomic distances may be regarded as favorable.

The arrangement of particles near the phase boundary depends on the atomic structure of the face, especially on its crystallographic direction and the extent of its blurred or diffuse nature. The analysis of the influence exerted by these factors, which determine the value and anisotropy of the density of growth points and the velocity of normal growth, is an important problem in future development of the theory.

Below we shall consider questions relating to the structure and stability of the interface between the liquid and the crystal. The change ΔF in the free energy of the system associated with the formation of a loose face instead of a plane one, for the case in which a complex of n particles belonging to the solid phase is "adsorbed" on an ideal face containing n_0 atomic sites, is given by the expression

$$f = \frac{\Delta F}{n_0 kT} = -\zeta \frac{x}{1-x} + \theta \frac{x}{1-x^2} + \frac{x}{1-x} \ln x + \ln(1-x), \tag{9}$$

where $\zeta = \dfrac{\Delta \mu}{kT}$, $\theta \simeq \dfrac{2q}{3kT}$, $x = \dfrac{n}{n+n_0}$ (see [4, 7] for the derivation and assumptions made).

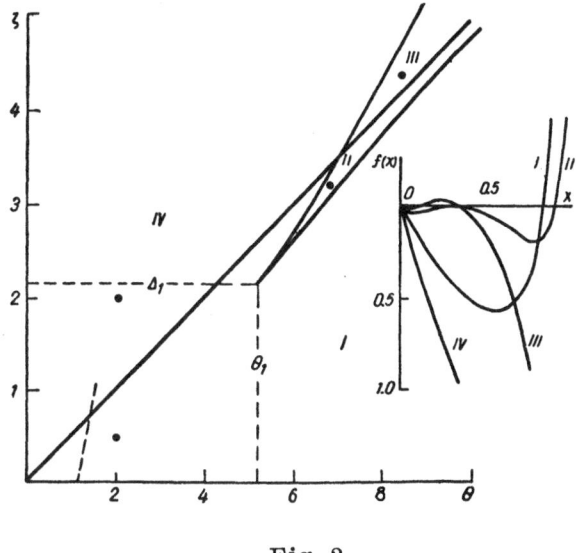

Fig. 2

There are two types of singular points of potential $f(x, \theta, \zeta)$ associated with the breakdown of stability. First of all, these include the set of values of θ, ζ such that (for an appropriate variation of x) an unlimited fall of potential becomes possible. It follows from expression (9) that the region in question forms part of the plane

$$\zeta > \theta/2. \tag{10}$$

Then a breakdown of stability is possible at those extremal points of the function $f(x, \theta, \zeta)$ where, in addition to $\nu f/\nu x = 0$ the condition $\nu^2 f/\nu x^2 = 0$ is satisfied. The geometrical locus of such values of θ, ζ is determined by eliminating x from the equations

$$\zeta = \frac{(1+x)(1+x^2)}{2x(1-x)} + \ln x, \qquad \theta = \frac{(1+x)^3}{2x(1-x)}. \tag{11}$$

Analysis shows the existence of two branches of the function $\zeta(\theta)$ defined by these equations, touching at the cusp $x_1 = 0.268$, $\theta_1 = 5.196$, $\zeta_1 = 2.147$ (Fig. 2). The branches have asymptotes $\zeta = \theta$ ($x \to 0$) and $\zeta = \theta/2$ ($x \to 1$), and the plane is divided into four regions. The first two correspond to a stable boundary, two different states being possible in region II. In region III the transitional zone may only exist as a metastable formation, stable with respect to small variations of the parameter x. In region IV the rough boundaries of the crystal are labile. These conclusions are drawn on the basis of a study of the arrangement and number of roots of the function $\nu f(x, \theta, \zeta)/\nu x$ and are illustrated by the graphs of the functions $f(x)$ (Fig. 2), plotted for the four points marked on the plane.

On reducing the temperature, i.e., on increasing the rate of crystal growth, ζ and θ rise, varying along the path schematically indicated by a broken line. Beyond the intersection with the straight line $\zeta = \theta/2$ the stability of the boundary is broken. This occurs, however, for a considerable supercooling $\Delta T \simeq T_0/3$, which in the case of metals ($K \sim 10$ cm/sec · deg) corresponds to velocities of the order of 10^3 cm/sec. Such conditions are only possible during the growth of a crystal in a severely supercooled melt, and then only at the initial instant.

It was suggested in [1] that continuous growth (apparently identified with normal growth) is only realized when the energy barrier vanishes for the motion of the boundary as a whole, i.e., essentially in the region of the unstable boundary. This point of view differs substantially from that developed in this paper (and those cited above), according to which normal growth takes place for a slight supercooling. It should be noted that when the liquid−crystal system is labile thermodynamic equations of the type of (1) and (13) are not applicable and the kinetic aspects of the crystallization of an unstable system require special consideration.

In the above discussion, as in other treatments of the same type such as [4] and [7], we have studied a motionless phase-separation boundary. The question of the stationary structure of a growing face should be considered on the basis of the thermodynamics of irreversible processes. Let a certain system A be characterized by a thermodynamic potential $Z(T, p, x, \xi)$ depending on temperature T, pressure p, and the continuous internal parameters x, ξ varying with time. In the present case, x characterizes the state of the crystal face and ξ determines its position in space. In order to find the irreversible part of the change of entropy, let us enclose the system in a thermostat A_0 containing a large number of particles and insulated from the external medium. The state of the thermostat is characterized by the potential $Z_0(T, p)$ and we may suppose that the temperature does not change on exchanging energy with the system under consideration. If x, ξ change by dx, dξ, the entropy of the system A changes by ds and that of the thermostat by ds_0. The total change $ds^* = ds + ds_0$ is due solely to irreversible processes in system A, since the system $A + A_0$ is isolated; $du_0 = Tds_0 - pdv_0$, where u_0 and v_0 are the energy and volume of the thermostat. Hence, $ds_0 = (du_0 + pdv_0)/T = dH_0/T = -dH/T$ (the heat contents H_0 and H relate to systems A_0 and A, respectively). Thus the change in entropy associated with processes in A may be expressed in terms of the characteristics of this system: $ds^* = ds - dH/T = -dZ(T, p, x, \xi)/T$. Since $Tds^*/dt = -(\nu Z/\nu x)dx/dt - (\nu Z/\nu \xi)d\xi/dt$, the kinetic equations for x and ξ have the form

$$\frac{d\xi}{dt} = -L_{11}\frac{\nu Z}{\nu \xi} - L_{12}\frac{\nu Z}{\nu x}, \qquad \frac{dx}{dt} = -L_{21}\frac{\nu Z}{\nu \xi} - L_{22}\frac{\nu Z}{\nu x}. \tag{12}$$

If the velocity $d\xi/dt = V$ is kept constant, there ensues a stationary condition characterized by a minimum in the increment ds^*/dt, which is expressed by the vanishing of the velocity dx/dt. If x and ξ are variables describing the roughness and position of the face, then $Z = f(x) - \xi\zeta$, where $f(x)$ is given by the previous expression (9), and the kinetic equations

$$V = \frac{L_{11}L_{22} - L_{12}^2}{L_{22}} \cdot \frac{\Delta\mu}{kT} = K\Delta T, \qquad \frac{\nu Z}{\nu x} - \frac{L_{12}}{L_{22}}\frac{\nu Z}{\nu \xi} = 0 \tag{13}$$

describe the kinetics of the displacement and structure of the growing face. The second of these may be written in the form

$$\frac{\nu}{\nu x}\left[f(x) + \frac{L_{12}}{L_{22}}\frac{\Delta\mu}{kT}\, x\right] = 0. \tag{14}$$

The stationary structure of the phase-separation boundary during the growth of the crystal is determined, therefore, by a minimum of the function $f(x) + xL_{12}\Delta\mu/L_{22}kT$, which differs from the free energy used to describe the motionless boundary by the presence of an extra term associated with the cross-interaction between the process of changing the structure of the face and the growth process. Formally the role of the extra term reduces to a reduction in $\zeta = \Delta\mu/kT$ by a quantity $(1 - x)\zeta L_{12}/L_{22}$, which is equivalent to a displacement along the axis of ζ in Fig. 2 and leads either to an increase of stability (reduction in the degree of roughness), if $L_{12} > 0$, or to a reduction in stability if $L_{12} < 0$. The first case is the more likely. Independently of the sign and magnitude of L_{12} we may note that our foregoing consideration based on a motionless boundary corresponds to the neglect of the "cross phenomenon" in question and in a qualitative respect remains valid for $V \neq 0$. The roughness of the face does not vanish during growth of the crystal, and varies continuously with varying growth rate.

Literature Cited

1. Cahn, J. W., Hillig, W. B., and Sears, G. W., Acta Met., 12(12) : 1421 (1964).
2. Jäntsch, O., Z. Krist., No. 108, p. 185 (1956).
3. Borisov, V. T., Dokl. Akad. Nauk SSSR, 151(6) : 1311 (1963).
4. Borisov, V. T., Dukhin, A. I., and Matveev, Yu. E., Problems of Metal Science and the Physics of Metals. No. 8, p. 269 (1964).
5. Swalen, R. A., Acta Met., 7(11) : 736 (1959).
6. Borisov, V. T., and Dukhin, A. I., Problems of Metal Science and the Physics of Metals. No. 7, p. 363 (1962).
7. Temkin, D. E., In: Mechanism and Kinetics of Crystallization. p. 55. Minsk (1964).

ON THE KINETICS OF THE GROWTH OF
A NEW-PHASE NUCLEUS

B. I. Birman and B. L. Timan

All-Union Scientific-Research Institute
of Single Crystals

The kinetic aspects underlying the growth of the nucleus of a new phase have been studied in a number of papers. In the majority of cases isothermal conditions of growth have been considered, although there have been some papers, such as [1], in which both isothermal and non-isothermal conditions have been treated, using the thermodynamics of irreversible processes.

In this paper we shall adopt a rather different approach to this problem and thereby further elucidate certain laws.

In order to study the principal kinetic aspects underlying the growth of a new-phase nucleus, let us consider a simplified scheme of the growth of a spherical nucleus (or to be specific, a crystal) developing in the center of a spherical vessel filled with the mother phase (a melt of the crystallizing material). In this case we may suppose that the physical system under consideration consists of three parts: the nucleus of the new phase, the melt, and the surrounding medium.

The change of entropy of the whole system per unit time resulting from the irreversible process may be written in the form

$$\frac{dS}{dt} = \frac{dS_n}{dt} + \frac{dS_s}{dt} + \frac{dS_m}{dt} + \frac{dS_{s.m.}}{dt},$$ (1)

where S_n, S_s, S_m, and $S_{s.m.}$ are, respectively, the entropies of the nucleus, the surface of phase separation, the melt, and the surrounding medium.

Let us find an expression for each of these terms.

For this purpose we introduce the entropy of unit volume, which we denote s. Then, for the entropy of the nucleus and the melt, we may write

$$S_n = \int_0^\varrho s_n 4\pi r^2 dr, \qquad S_m = \int_\varrho^R s_m 4\pi r^2 dr.$$

Considering that the radius of the nucleus ρ varies in time, we obtain

$$\frac{dS_n}{dt} = s_n(\varrho)\, 4\pi\varrho^2 \frac{d\varrho}{dt} + \int_0^\varrho \frac{ds_n}{dt}\, 4\pi r^2 dr,$$

$$\frac{dS_m}{dt} = -\, s_m(\varrho)\, 4\pi\varrho^2 \frac{d\varrho}{dt} + \int_\varrho^R \frac{ds_m}{dt}\, 4\pi r^2 dr. \tag{2}$$

Using the well-known thermodynamic relation

$$ds = \frac{1}{T}\, C_p dT - \beta dp - \frac{\mu}{T}\, dN, \tag{3}$$

where ds is the change in the entropy of unit volume of the system, C_p is the specific heat at constant pressure referred to unit volume, β is the coefficient of volume expansion, μ is the chemical potential, dN is the increment in the number of particles, and T, p, and V are, respectively, the temperature, pressure, and volume.

Remembering formula (3), we may write expression (2) in the form

$$\frac{dS_n}{dt} = S_n(\varrho)\, 4\pi\varrho^2 \frac{d\varrho}{dt} + \int_0^\varrho \frac{C_p'}{T}\frac{dT}{dt}\, 4\pi r^2 dr - \int_0^\varrho \beta_n \frac{dp}{dt}\, 4\pi r^2 dr - \frac{\mu_n(\varrho)}{T}\frac{dN_n}{dt},$$

$$\frac{dS_m}{dt} = -S_m(\varrho)\, 4\pi\varrho^2 \frac{d\varrho}{dt} + \int_\varrho^R \frac{C_p''}{T}\frac{dT}{dt}\, 4\pi r^2 dr - \int_\varrho^R \beta_m \frac{dp}{dt}\, 4\pi r^2 dr - \frac{\mu_m(\varrho)}{T}\frac{dN_m}{dt}, \tag{4}$$

where C_p' and C_p'' are, respectively, the specific heats of the nucleus and melt per unit volume.

The integrals in the expressions (4) are associated with a definite type of temperature distribution in the system. For simplicity of subsequent calculations, we assume that the temperature in the nucleus is the same everywhere, while in the melt it is distributed in accordance with the law

$$T(r) = \frac{1}{(R-\varrho)}\left[RT_R\left(1 - \frac{R}{r}\right) + \varrho T_\Phi\left(\frac{R}{r} - 1\right)\right], \tag{5}$$

where T_R and T_Φ are, respectively, the temperatures at the outer boundary of the melt and the phase-separation boundary.

This kind of temperature distribution corresponds to a stationary solution of the equation of thermal conductivity (heat-conduction equation) and is a fairly good approximation on satisfying the condition vR/$a \ll 1$ (where v is the rate of growth and a is the diffusivity).

Differentiating the temperature with respect to time, allowing for the variation of ρ, dividing by the temperature, and neglecting the comparatively weak temperature dependence of the specific heat, we determine the values of

$$\int_0^\varrho \frac{C_p'}{T}\frac{dT}{dt}\, 4\pi r^2 dr \quad \text{and} \quad \int_\varrho^R \frac{C_p'}{T}\frac{dT}{dt}\, 4\pi r^2 dr.$$

Subsequently we shall neglect the comparatively small variations of pressure in the melt arising in the course of crystallization.

The change of pressure in the nucleus on changing its dimensions is, in view of the curvature of the surface,

$$\frac{dp}{dt} = -\frac{2\sigma}{\varrho^2}\frac{d\varrho}{dt},\tag{6}$$

where σ is the surface-tension coefficient at the phase-separation boundary. For the rate of change of the number of particles in the nucleus and in the melt, we have the relation

$$\frac{dN_n}{dt} = -\frac{dN_m}{dt} = \frac{4\pi\varrho^2}{V_0}\frac{d\varrho}{dt},$$

where V_0 is the volume associated with one particle.

Let us determine the rate of change of entropy of the phase boundary. It is well known that the expression for the change in the entropy of the surface may be written in the form

$$dS_s = \frac{dE_s}{T} - \frac{\sigma}{T}d\zeta,\tag{7}$$

where dE_S is the increment in surface energy and $d\zeta$ is the change in the surface of the nucleus.

Using this expression, we may write

$$\frac{dS_s}{dt} = \frac{1}{T_\Phi}\frac{dE_s}{dt} - \frac{\sigma}{T_\Phi}8\pi\varrho\frac{d\varrho}{dt}.\tag{8}$$

Since it is assumed that the external surface of the melt and the temperature on it remain constant, the change of entropy there is zero. The rate of change of the entropy of the surrounding medium may be determined in terms of the heat flux q_R from the melt to the surrounding medium by using the formula

$$\frac{dS_{s,m.}}{dt} = -4\pi R^2 \frac{q_R}{T_R}.\tag{9}$$

Here we shall assume that the temperature of the surrounding medium coincides with that of the outer boundary of the melt.

After certain transformations, and after allowing for the foregoing formulas, we obtain an expression for the rate of change of entropy of the whole physical system in the following form:

$$\frac{dS}{dt} = \frac{C_p'}{1+\varepsilon}\frac{4\pi}{3}\varrho^3\frac{d\varepsilon}{dt} + C_p'4\pi\left[\frac{d\varepsilon}{dt} + \frac{\varepsilon R}{\varrho(R-\varrho)}\frac{d\varrho}{dt}\right]\times$$

$$\times\left[\frac{\varrho(R+2\varrho)(R-\varrho)}{6} - \varepsilon\frac{\varrho^2(R-\varrho)}{3} + \varepsilon^2\frac{\varrho^3}{(R-\varrho)^2}\left(\frac{R^3}{R-\varrho}\ln\frac{R}{\varrho} - \frac{11R^2-7R\varrho+2\varrho^2}{6}\right)\right]\tag{10}$$

$$+\frac{1}{T_\Phi}\frac{dE_s}{dt} + \frac{4\pi\varrho^2}{T_\Phi}\frac{d\varrho}{dt}\left[(\Phi_m-\Phi_n) - Q - \frac{2\sigma(1-\beta_n T_\Phi)}{\varrho}\right] + \frac{q_R 4\pi R^2}{T_R},$$

where $\varepsilon = (T_\Phi - T_R)/T_R$; Q is the latent heat of fusion per unit volume, and Φ_m and Φ_n are, respectively, the thermodynamic potentials per unit volume of the melt and the nucleus.

The rate of growth of the nucleus, the rate of change of temperature at the crystallization front, and the heat flux at the boundary may be interrelated by the following relation arising from the law of energy conservation:

$$T_R C_p'\frac{4\pi}{3}\varrho^3\frac{d\varepsilon}{dt} + \frac{dE_s}{dt} + T_R C_p'4\pi\frac{\varrho(R+r\varrho)(R-\varrho)}{6}\left[\frac{d\varepsilon}{dt} + \frac{\varepsilon R}{\varrho(R-\varrho)}\frac{d\varrho}{dt}\right] = 4\pi\varrho^2 Q\frac{d\varrho}{dt} - q_R\cdot 4\pi R^2.\tag{11}$$

We introduce the surface energy referred to unit area, calling this u_s. Then we may write

$$\frac{dE_s}{dt} = u_s 8\pi\varrho \frac{d\varrho}{dt} + 4\pi\varrho^2 C_s T_R \frac{d\varepsilon}{dt}, \tag{12}$$

where $C_s = du_s/dT$.

For the present formulation of the problem, in which the temperature at the outer boundary remains constant, we may consider that $T_R = T_0$, where T_0 is the original temperature in the melt before the formation of the nucleus.

After using relations (11) and (12), expanding $\Delta\Phi = (\Phi_m - \Phi_n)$ in series, near the equilibrium crystallization temperature, confining attention to terms of the first order of smallness in ε, neglecting βT_Φ in comparison with unity, and making a few transformations, we obtain an expression for the rate of change of the entropy of the physical system under consideration, referred to unit surface of the phase boundary:

$$\frac{dS}{dt} = \frac{\Delta\Phi_0}{T_\Phi}\left(1 - \frac{\varrho^*}{\varrho}\right)\frac{d\varrho}{dt} - \varepsilon\left\{\frac{C_p' R}{3} + \frac{\left(C_p'\frac{T_0}{T_\Phi} - C_p'\right)\varrho}{3} + C_s\frac{T_0}{T_\Phi}\right\}\frac{d\varepsilon}{dt}, \tag{13}$$

where

$$\varrho^* = \frac{2\sigma}{\Delta\Phi_0}, \quad \Delta\Phi_0 = \frac{Q(T_K - T_0)}{T_K}.$$

In the limiting case of an isothermal process for $dT_\Phi/dt = 0$, the expression for the rate of change of entropy of the system coincides with that obtained in [1]. In the case of a nonisothermal process, however, the rate of change of entropy contains not only a term proportional to the rate of growth of the new-phase nucleus, but also a term proportional to the rate of temperature change at the crystallization front.

According to the thermodynamics of irreversible processes, for a comparatively slight deviation of the system from the state of thermodynamic equilibrium, the rate of change of entropy associated with the irreversibility of the process may be put in the form

$$\frac{dS}{dt} = \frac{1}{T}\sum J_i X_i, \tag{14}$$

$$J_i = \sum L_{ik} X_k, \tag{15}$$

where X_i and J_i are the generalized forces and corresponding fluxes, and L_{ik} are the kinetic coefficients.

Comparing expressions (14) and (13), we determine the forces producing an irreversible change in the state of the physical system under consideration and the fluxes corresponding to these:

$$X_1 = \frac{(T_K - T_0)}{T_K}\left(1 - \frac{\varrho^*}{\varrho}\right), \tag{16}$$

$$X_2 = -\frac{(T_\Phi - T_0)}{T_0^2}T_\Phi \cong -\frac{(T_\Phi - T_0)}{T_0}, \tag{17}$$

$$J_1 = Q\frac{d\varrho}{dt}, \tag{18}$$

$$J_2 = \frac{dT_\Phi}{dt}\left[\frac{C_p^{'}R}{3} + \frac{\left(C_p^{'}\frac{T_0}{T_\Phi} - C_p^{'}\right)\varrho}{3} + C_s\frac{T_0}{T_\Phi}\right] \cong \frac{C_p^{'}R}{3}\frac{dT_\Phi}{dt}. \tag{19}$$

According to relation (15) we may write

$$Q\frac{d\varrho}{dt} = L_{11}\frac{(T_K - T_0)}{T_K}\left(1 - \frac{\varrho^*}{\varrho}\right) - L_{12}\frac{(T_\Phi - T_0)}{T_0}, \tag{20}$$

$$\frac{C_p^{'}R}{3}\frac{dT_\Phi}{dt} = L_{21}\frac{(T_K - T_0)}{T_K}\left(1 - \frac{\varrho^*}{\varrho}\right) - L_{12}\frac{(T_\Phi - T_0)}{T_0}. \tag{21}$$

This system of equations enables us in principle to solve the problem of the kinetics of the growth of a new-phase nucleus under the conditions in question. We may eliminate the quantity T_Φ from the equations and thus obtain an equation for determining the time variation of the radius of the nucleus in the form

$$Q\frac{d^2\varrho}{dt^2} = \frac{L_{11}(T_K - T_0)}{T_0^2}\frac{\varrho^*}{\varrho^2}\frac{d\varrho}{dt} - \frac{3L_{22}Q}{C_p^{'}RT_0^2}\frac{d\varrho}{dt} + \frac{3(L_{22}L_{11} - L_{12}L_{21})}{C_p^{'}RT_0^4}\left(1 - \frac{\varrho^*}{\varrho}\right). \tag{22}$$

This expression is nonlinear and cannot be solved in general form.

However, for $\rho \gg \rho^*$, Eqs. (20) and (21) simplify considerably and their solution is written as

$$T_\Phi = T_0 + \frac{L_{21}}{L_{22}}\frac{T_0}{T_K}(T_K - T_0)(1 - e^{-\frac{3L_{22}}{C_p^{'}RT_0}t}), \tag{23}$$

$$\varrho = \varrho_0 + \frac{(L_{22}L_{11} - L_{12}^2)}{L_{22}Q}\frac{(T_K - T_0)}{T_K}t + \frac{L_{12}^2 C_p^{'}R(T_K - T_0)}{3L_{22}^2 QT_K}(1 - e^{-\frac{3L_{22}}{C_p^{'}RT_0}}). \tag{24}$$

We see from expression (23) that the supercooling at the crystallization front under the conditions in question (at the boundary $T_0 = $ const) becomes smaller, and after a time $\tau > C_p^{'}RT_0/3L_{22}$ approaches a constant value $T_\Phi - T_0 = (T_0 - T_K)\left(1 - \frac{L_{21}}{L_{22}}\frac{T_0}{T_K}\right)$, differing from zero (if $T_0 \neq T_K$).

It follows from expression (24) that, following the reduction in supercooling, the growth rate also falls off with time and after the same period τ tends to a constant value

$$\frac{d\varrho}{dt} = \frac{(L_{22}L_{11} - L_{12}^2)}{L_{22}Q}\frac{(T_K - T_0)}{T_K}.$$

Determination of the kinetic coefficients involves additional consideration of the kinetics of the molecular processes taking place during the growth of the nucleus. The kinetic coefficient L_{11} was determined in [1] on the basis of the transitional complex considered in the theory of reaction velocities. The cross kinetic coefficient L_{12} occurs here for the first time, and its determination constitutes the subject of an independent investigation. The kinetic coefficient L_{22} involves the diffusivity of the melt and may be written in the form

$$L_{22} = \frac{T_0}{R}C_p^{'}a.$$

In view of this we may estimate the time τ required for the establishment of stationary supercooling at the crystallization front and a constant growth rate as $\tau \approx R^2/3a$. We see from this expression that τ is associated with the thermal inertia of the melt. Putting $R = 10^{-1}$ m, $a \approx 10^{-3}$ m^2/h, for example, we obtain $\tau \approx 1$ h.

Literature Cited

1.   Lyubov, B. Ya., Problems of Metal Science and the Physics of Metals, Vol. 5. p. 294. Metallurgizdat, Moscow (1958).

STUDY OF THE GROWTH MECHANISM
OF CERTAIN METAL CRYSTALS
FROM THE MELT

G. A. Alfintsev and D. E. Ovsienko

Institute of Metal Physics of the Academy of Sciences
of the Ukrainian SSR

According to existing opinions, the growth of crystals may take place by way of three different mechanisms: (1) by the formation of two-dimensional nuclei and their subsequent growth; (2) by a dislocation mechanism, i.e., by the attachment of particles (atoms, molecules) to steps formed by the outcrops of screw dislocations; (3) by the disordered attachment of particles to the surface of separation; in the third case, as distinct from the first and second (which involve layer-by-layer growth), the crystallization front moves homogeneously in a direction normal to itself.

Each mechanism has its own functional dependence of the growth rate (v) on the supercooling at the phase-separation boundary (ΔT). In the case of the growth of crystals by the two-dimensional nucleus mechanism, this relationship has the form [1-3]:

$$v = K_1 e^{-\frac{K_2}{T(\Delta T)}}, \tag{1}$$

for the dislocation mechanism [4, 5]:

$$v = K_3 (\Delta T)^2, \tag{2}$$

and for the case of normal growth [4, 6-10]:

$$v = K_4 \Delta T. \tag{3}$$

The quantities K_1, K_2, K_3, and K_4 may be regarded as constants for a small range of supercoolings. The realization of one or other of the mechanisms is determined by the structure of the crystal—melt boundary surface. The mechanism of the two-dimensional nucleus is characteristic of atomically smooth, low-index boundaries, the dislocation mechanism of surfaces containing steps formed by the outcrops of screw dislocations, and the normal mechanism of rough surfaces [23]. It was assumed in [26] that the latter case was realized for smaller supercoolings than those corresponding to the layer-by-layer mechanisms.

However, the authors of [12, 13] expressed another point of view regarding the conditions for the development of one or other of the mechanisms. This theory was based on the assumption [11, 12] that the crystal–melt interface was "diffuse" (blurred), i.e., that there was a gradual transition from the properties of one phase to the other. The thickness of the transitional layer may vary from one (sharp boundary) to several lattice parameters. Using this assumption, the authors came to the conclusion that, independently of the diffuseness of the interface, the mechanisms of layer-by-layer growth (nucleus-forming and dislocation) should take place for small motive forces, i.e., for slight supercoolings, and the mechanism of normal growth for large forces. Here the mechanism of normal growth is not associated with the roughness of the surface of separation. In accordance with the theory, a parameter g characterizing the diffuseness of the boundary and a coefficient β reflecting the distinction between the transfer of atoms during crystallization and self-diffusion in the liquid, were introduced into formulas (1)-(3). The authors of [13] show that the main conclusions of their theory agree with certain experimental data, but the theory was not submitted to any special experimental verification.

It should be mentioned that, despite a certain amount of success in the theory of crystal growth from the vapor phase and from solution, the mechanism of growth from the melt remains little understood, largely because there have not been enough experimental investigations into this subject. This relates particularly to the growth of metallic crystals, which have only been treated in a few papers. Let us briefly consider the results of some of these.

V. T. Borisov and A. I. Dukhin [14-16] studied the temperature dependence of the growth rate of Hg, K, and Cd crystals. Varying the growth rate over wide limits, they found no marked supercoolings at the crystallization front. Even for the maximum growth rates the measured supercoolings were close to the measuring error.

In the view of the authors, for such slight supercoolings neither nucleus formation nor the dislocation mechanism can produce such high growth rates; these are only possible by way of the normal-growth mechanism resulting from a high density of growth points (rough surface of separation). However, if these data are analyzed on the basis of Kane's theory, the observed kinetics may also be explained as a result of layer-by-layer mechanisms.

The authors of [17] (which was published after our own work had been concluded) studied the kinetics of the growth of tin crystals at low growth rates, using an original method of thermal waves. Analyzing the resultant data from the point of view of different theories, the authors came to the conclusion that the experimental dependence of v on ΔT was best described by the expression

$$v = 50\,(\Delta T)^2 \ \text{cm/sec,} \tag{4}$$

which corresponds to the dislocation mechanism of growth. Using the resultant data, they estimated the parameter g, which proved to be equal to roughly 0.002 (for $\beta = 600$) and the value of the limiting supercooling up to which (according to Kane) layer-by-layer mechanisms should operate. This value proved to be 0.3°C. The authors of [13] also indicated the possibility of layer-by-layer growth for such metals as Bi, Sb, Al, and Zn, in which the boundary of the crystallization front and anisotropy of the growth rate were observed. Data relating to the growth kinetics, however, are required in addition to these indications in order to provide reliable proof.

Thus, on the basis of available experimental data we cannot yet give preference to one particular growth theory or another, or draw conclusions regarding the growth characteristics of different metals. Further research will be required for this.

In our own investigations we studied the kinetics of the growth of various pure metal crystals (gallium, bismuth, and tin) and the variation in the macrostructure of the crystallization front in order to obtain information relating to the growth mechanism. In addition to this, for gallium we determined the effect of deformation and small additions of silver on the rate of growth. These metals were mainly chosen for methodical reasons: low melting points, facilitating the use of reliable thermostating methods and methods of temperature measurement, and the possibility of observing the phase boundary directly. Moreover, according to theory, different mechanisms may be involved in the growth of these metals [23].

Method

The work was carried out with a special apparatus. The metal was poured into a glass cuvette with plane-parallel walls. The heating, cooling, and thermostating of one part of the cuvette were effected either by an ultrathermostat (for gallium, and partly for tin) or by a small flat furnace. The other part of the cuvette was at room temperature. The metal in the latter was always in the solid state and served as a "seed." The orientation of the boundary was known and could be varied by means of the seed crystal.

The growth rate was measured in a metallographic microscope by observing the motion of the crystallization front. For this purpose we used polarized light, giving good contrast between the liquid and solid phases of the metals and thus enabling changes in the structure of the front to be followed during the growth process. This method had some advantage over the decantation method, in which the liquid tends to adhere and so hide the details of surface relief [19, 20].

The temperature at the crystallization front was measured with a fine thermocouple (diameter 50 μ). The part of the thermocouple situated in the metal was insulated with a thin layer (1-2 μ) of heat-resistant lacquer. The error in measuring the temperature was ±0.03°C.

We used gallium of the Gl−0 and Gl−00 types, tin of the OVCh-000 type, bismuth of the V-000 type, and high-purity silver. The proportion of the nominal element was over 99.999 wt.% in each case.

Results and Discussion

Gallium. On studying the growth of gallium crystals we found that the same face had different growth rates for the same bath temperature. The growth rate was in fact extremely sensitive to deformation of the growing crystal. Conditions were therefore chosen so as to avoid shaking or vibrating the specimen and thus producing deformation. It was later found that when the front met the thermocouple there was a change in growth rate. However, this only occurred in the growth of undeformed crystals of pure gallium, so that in studying these crystals we had to dispense with the thermocouple and take the supercooling at the crystallization front as the difference between the melting point and the bath temperature. This proved to be justified, since special experiments showed that the bath temperature practically coincided with the temperature of the front in the velocity range between 0 and $2 \cdot 10^{-6}$ m/sec.

If the experiments with pure gallium were carried out under conditions excluding deformation, the crystals usually failed to grow, even for relatively large supercoolings. Thus, for $\Delta T = 0.48°C$, there was no movement of the (001) face over a period of 1 h, although measurements were made to an accuracy of ±2μ . For $\Delta T = 0.76°C$ this face grew at a rate of $1.56 \cdot 10^{-6}$ m/sec. On reducing the supercooling to 0.45°C, growth again stopped for 3 h. On once more increasing the supercooling, growth again accelerated. Analogous data were obtained for the (111) face, the threshold of supercooling below which no growth occurred this time being 1.10°C.

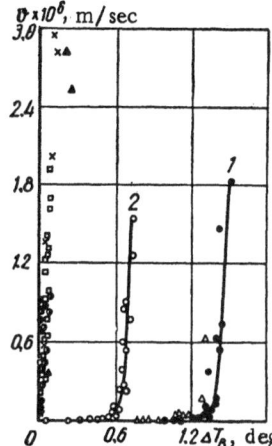

Fig. 1. The v(ΔT_B) relationship for pure gallium and Ga−0.01 wt.% Ag alloy. Undeformed crystals: 1) Ga, GL-0, Δ−Ga, GL-00, (111) face; 2) Ga, (001) face; × − alloy, (111) face. Deformed crystals: □ − Ga, (111) face; ○ − Ga, (001) face; △ − alloy, (111) face.

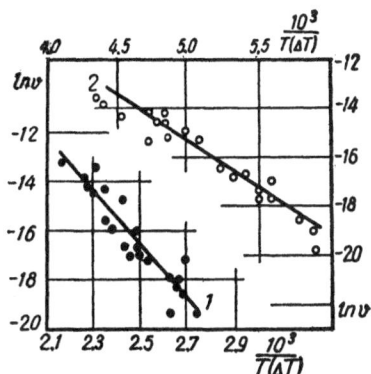

Fig. 2. Relation between ln v and $\frac{1}{T(\Delta T)}$ according to the data of curves 1 and 2 in Fig. 1.

The relation between v and ΔT for such crystals is shown in Fig. 1 [curve 1 for the (001) and curve 2 for the (111) face]. The same data are shown in a plot of ln v vs. $1/T(\Delta T)$ in Fig. 2. We see from Fig. 2 that the experimental points lie on a straight line, which agrees with the requirements of the growth theory based on the mechanism of two-dimensional nucleation. The experimentally obtained temperature dependence of the growth rate for the (001) face is described by the following equation:

$$v = 70e^{-\frac{3900}{T(\Delta T)}} \quad \text{m/sec;} \qquad (5)$$

for the (111) face

$$v = 5400e^{-\frac{11000}{T(\Delta T)}} \quad \text{m/sec,} \qquad (6)$$

which are similar in form to formula (1). It should be mentioned that the data obtained for the two types of gallium were almost identical (Fig. 1).

The existence of threshold supercoolings (critical supercooling) below which the growth rate is negligibly low and the character of the temperature/growth-rate relationship strongly suggests that the growth of the (001) and (111) faces of gallium crystals in the supercooling range studies is effected by way of the formation of two-dimensional nuclei.

We also studied the effect of deformation on the above crystals. Deformation was created by lightly bending the crystal or pricking the growing face with a fine glass capillary. The effect of deformation is illustrated by Fig. 3, which shows the displacements of the (001) face in time for various supercoolings of the bath. We see from the figure that deformation (the instant of applying which is shown by an arrow) leads to a sharp rise in growth rate. Thus crystals 1 to 4 hardly grew at all for supercoolings below 0.48°C, but after deformation they grew at $2.9 \cdot 10^{-6}$, $6.1 \cdot 10^{-6}$, $8.7 \cdot 10^{-6}$, and $12.4 \cdot 10^{-6}$ m/sec, respectively. For a supercooling of 0.63°C, crystal 5 grew before deformation at $1.28 \cdot 10^{-6}$ m/sec, but after deformation at $2 \cdot 10^{-4}$ m/sec, i.e., 160 times faster. Analogous results were obtained for the (111) face. We see from Fig. 1 that the deformed crystals grow at a considerable rate even for supercoolings of less than 0.05°C. The temperature dependence of growth rate in this range of supercoolings is almost the same for the (001) and (111) faces of the deformed crystals, but differs greatly for the undeformed ones.

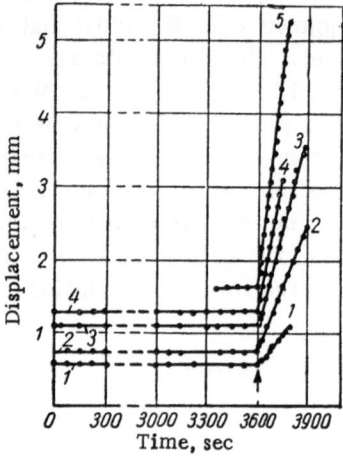

Fig. 3. Effect of deformation on the growth rate of the (001) face for various ΔT_B. Arrow indicates the instant of deformation. Crystal 1) $\Delta T_B = 0.13°C$; 2) $\Delta T_B = 0.23°C$; 3) $\Delta T_B = 0.35°C$; 4) $\Delta T_B = 0.43°C$; 5) $\Delta T_B = 0.63°C$.

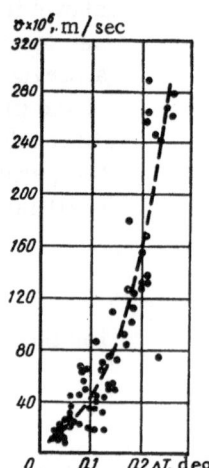

Fig. 4. Relation between v and ΔT for the (111) face of deformed pure Ga crystals.

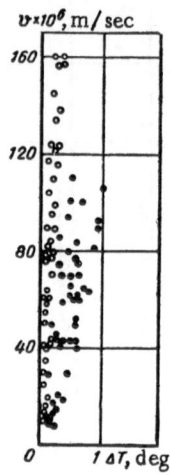

Fig. 5. Relation between v and ΔT for the (001) face: $\bigcirc$ — deformed pure Ga crystals; $\bullet$ — Ga−0.01 wt.% Ag alloy.

Figures 4 and 5 present v−ΔT graphs for high growth rates of the (111) and (001) faces of deformed crystals. Owing to the great scatter of the experimental points and the relatively narrow range of supercoolings, it is very difficult here to draw any definite conclusions regarding the character of the temperature dependence and hence the mechanism of growth. However, analysis of the data for the (111) face (Fig. 4) shows that the variation of v with ΔT corresponds better to expression (2) than (1) and (3). This relationship may be written approximately in the form

$$v \simeq 4.2 \cdot 10^{-3} (\Delta T)^2 \quad \text{m/sec.} \tag{7}$$

This gives some grounds for supposing that the growth of the deformed crystals in the velocity range under consideration is based on the dislocation mechanism. Apparently plastic deformation creates defects of the screw-dislocation type in the gallium crystals, and the presence of these leads to a change in the growth mechanism, the two-dimensional nucleation being replaced by the dislocation type.

The fact that after deformation the crystals grow for quite a long time (a matter of hours) at a constant rate indicates the absence of any impurity−adsorption effect on the growth. This may also serve as an additional proof of the fact that the threshold supercoolings and low growth rates of the undeformed crystals are due to certain aspects of the growth mechanism rather than being associated with impurity adsorption.

We also studied the effect of the addition of silver on the growth of gallium crystals, the experiments being conducted under the same conditions as for pure gallium. The effect of adding 0.01 wt.% Ag on the growth rate of the (111) face is shown in Fig. 1. It follows from this figure that the presence of even a small amount of silver leads to the removal of the "threshold" and to a considerable increase in the growth rate as compared with the growth of undeformed crystals of pure gallium. We see from Fig. 5, however, that the growth rate of the alloys is rather lower than that of a pure gallium crystal after deformation. On raising the amount of

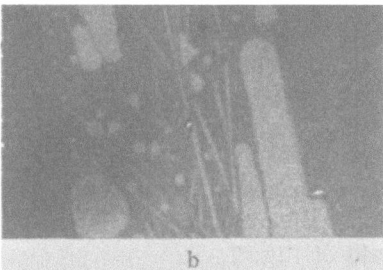

Fig. 6. Capture of impurity-rich liquid by the crystal. Dark parts correspond to the solid phase, light parts to the liquid. ×100.

silver in the gallium to 1%, the fall in the growth rate becomes more substantial. It is a characteristic fact that deformation has no effect on the growth rate of the alloy crystals (see Fig. 1). This fact, together with the similarity between the temperature dependence of the growth rate of the alloy crystals and that of the deformed crystals of pure gallium, gives grounds for supposing that the silver impurity acts in the same way as deformation, by creating defects of the screw−dislocation type. The validity of this principle is supported by the fact that small additions of impurities produce a great increase in the dislocation density of metals [27, 28].

The reduction in the growth rate of the alloy crystals as compared with that of deformed crystals of pure gallium may well be due to the fact that, as growth continues, the silver builds up in front of the crystallization front and impedes the transport of the principal material to the growing crystal.

The enrichment of the melt with impurity leads to loss of stability in the plane crystallization front, and this creates conditions for the nonuniform capture of impurity by the crystal. This phenomenon was observed in the Ga−0.01 wt.% Ag alloy for a growth rate of $3.5 \cdot 10^{-5}$ m/sec and bath temperature 28.17°C. The (111) face was at first smooth and grew at a constant rate. After some time a projection appeared in the form of a step; this grew in a tangential direction, passing around the layer of liquid adjacent to the crystallization front. This resulted in the formation of a closed island of liquid (Fig. 6a). After this the boundary again became plane and moved at the previous velocity, while the liquid island captured by the crystal diminished at a considerably slower rate. Subsequently the growth of the solid phase of the principal material was accompanied by the precipitation of a new phase in needle form (Fig. 6b).

It is still hard to say, on the basis of our experiments with the alloy, whether we can regard these laws as common to other impurities, or whether they represent a special property of silver, since, in this case, the nonuniform capture of the impurity is accompanied by the precipitation of a new phase, which should intensify the effect of the impurity in forming dislocations. Further investigations will be required in order to solve this problem.

Bismuth and Tin. Data relating to the kinetics of the growth of bismuth and tin are presented in Fig. 7. We see from the graph that the relation between the growth rate of the (100) face of the bismuth crystals and the supercooling at the crystallization front is nonlinear. The relation cannot be described by an exponential expression either. It follows from the ln v vs. ln ΔT graph (Fig. 8) plotted from the data of Fig. 7 that the relationship may be expressed in the following form:

$$v = 1.23 \cdot 10^{-4} (\Delta T)^{1.7} \quad \text{m/sec.} \tag{8}$$

In general this expression agrees with formula (2), which comes from the theory based on the dislocation mechanism of growth. This agreement may be regarded as quite satisfactory if we consider the possibility of a reduction in the power index as a result of thermal interaction between the spirals [5].

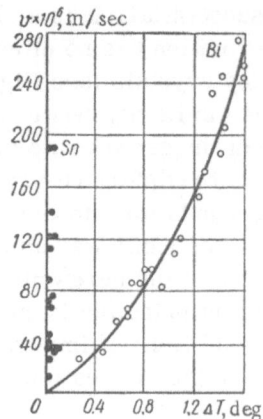

Fig. 7. Relation between v and ΔT for Bi on the (100) face and Sn.

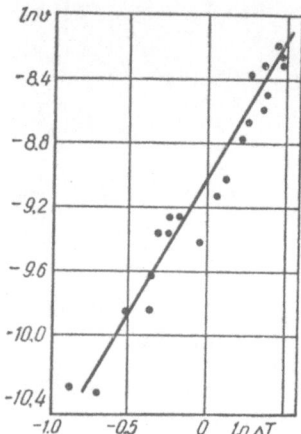

Fig. 8. Relation between ln v and ln (ΔT) for Bi, taken from the data of Fig. 7.

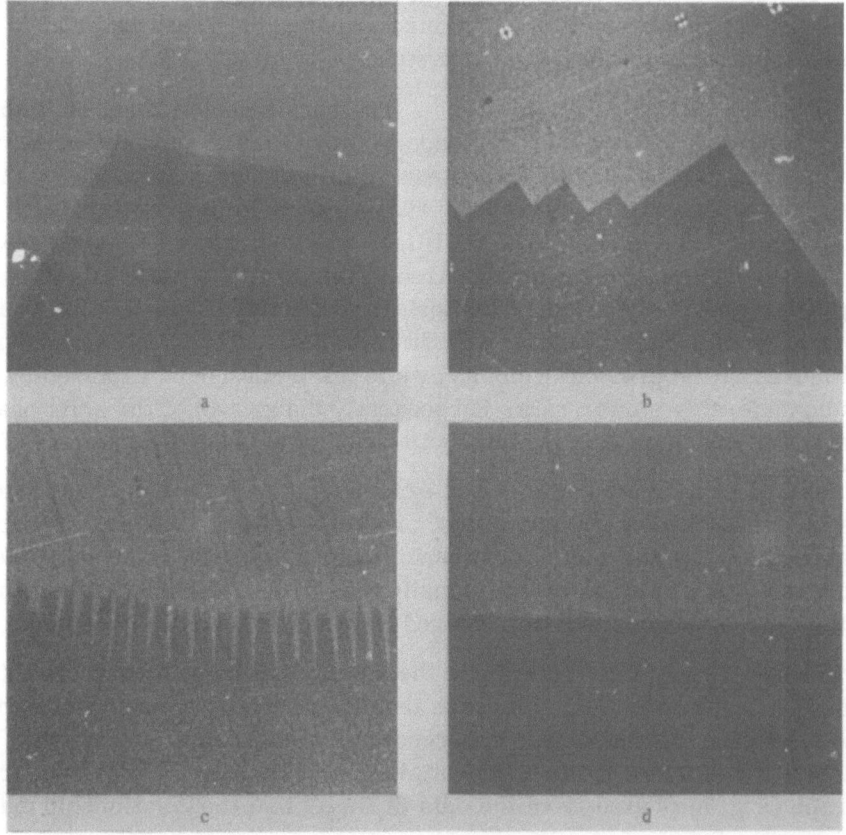

Fig. 9. Photographs of the macrostructure of the crystallization front. Dark parts refer to the crystal and light parts to the melt. ×100. a) Ga, ΔT = 1°C; b) Bi, ΔT = 1°C; c) Sn, ΔT < 0.03°C; d) Sn, conditions close to equilibrium. Pictures a–c correspond to a supercooled melt and d to a superheated one.

We see from Fig. 7 that for tin, even at high growth rates ($2 \cdot 10^{-4}$ m/sec), the super-coolings at the front are comparable with measuring error, never exceeding 0.06°C. It is therefore not possible to establish the character of the temperature dependence of growth rate and hence pass judgement on the growth mechanism. We note that the relationship between v and ΔT calculated from formula (4) falls within the experimental values of Fig. 7. In view of the large measuring errors, however, this cannot be regarded as confirmation of the tin−crystal growth mechanism proposed in [17], although it by no means contradicts this. Despite the inadequacy of the data presented as regards judging the growth mechanism, they do reveal a distinguishing feature in the kinetics of tin−crystal growth as compared with bismuth and gallium, in that the growth rate of the tin crystals is much greater than that of the other metals for the same supercoolings at the front.

Additional data regarding the special features in the growth of tin crystals were obtained by studying the macrostructure of the crystallization front. Whereas the crystallization front of gallium and bismuth has a strict crystallographic facing, in the case of tin this is always absent. Characteristic shapes of the crystallization front appear in Fig. 9. The shape of the crystal−melt interface associated with the growth of bismuth and gallium (showing distinct crystal faces as in Figs. 9a, b) is preserved up to quite large supercoolings at the front (up to 2°C). The well-faced form of the crystallization front associated with these metals may serve as an additional argument regarding the layer-by-layer form of growth in these metals [13, 23, 24]. For tin, even under conditions close to equilibrium, the interphase boundary is even, though not plane (Fig. 9d), while for slight supercoolings (less than 0.03°C) the crystallization front loses stability and decomposes, with the formation of needles projecting into the melt (Fig. 9c). The absence of sharp faces from the crystallization front of tin in our experiments may well be due to the effects of the normal-growth mechanism.

It is interesting to estimate the critical supercooling at which, according to Kane's theory, the normal-growth mechanism should appear. This can only be done for gallium on the basis of the data presented; for bismuth and tin only very coarse estimates of the "diffuseness" parameter g and the coefficient β can be made. In estimating g and β for the different metals we used various relations taken from [13]. For gallium these were:

$$g = \frac{50kL\Delta T_\Gamma}{\pi\sigma^2 aV_m},$$

(9)

$$g/\beta = \frac{L^2 D (\Delta T)^2}{4\pi RT^3\sigma V_m v},$$

(10)

where L is the heat of fusion; D is the self-diffusion coefficient in the liquid; σ is the interphase surface energy; V_m is the molar volume; a is the lattice parameter; and ΔT_Γ is the threshold supercooling.

Metal	Face	D, cm²/sec	σ [25], ergs/cm²	a, cm	g	β
Ga	(001) (111)	$1.63 \cdot 10^{-5}$	56	$7.64 \cdot 10^{-8}$ $5.9 \cdot 10^{-8}$	0.02 0.06	0.14 0.98
Bi	(100)	$1 \cdot 10^{-5}$	55	$6.6 \cdot 10^{-8}$	>0.017	>0.13
Sn	—	$2.5 \cdot 10^{-5}$	59	$5.8 \cdot 10^{-8}$	$<1 \cdot 10^{-6}$	>0.23

In the case of bismuth we used expression (10) and the relation

$$g = \frac{aL(\Delta T)^*}{V_m \sigma T_0}, \tag{11}$$

where $(\Delta T)^*$ is the critical supercooling of the onset of normal growth, estimated as being equal to 1.6°C, although it may be much higher.

For tin the calculations were based on the formulas

$$\beta = \frac{vaRT^2}{DL\Delta T}, \tag{12}$$

$$g\beta = \frac{va^2RT}{\pi D\sigma V_m} \tag{13}$$

by substituting the experimental values v = 1.9 · 10^{-4} m/sec, ΔT = 0.06°C in (12) and v < 1·10^{-6} m/sec in (13), on the assumption that the mechanism of normal growth is realized under these conditions.

The resultant values of g and β are given in the table, which also shows the values of the principal constants used in the calculations.

Under the assumptions and approximations made, the diffuseness parameters for gallium and bismuth were of the same order, but differed greatly from that corresponding to tin. The thickness of the diffuse separation boundary for Ga and Bi was of the order of two lattice parameters, while for Sn it was almost twice this.

Using the value of g given for gallium, we estimated the critical supercooling above which only the normal-growth mechanism should operate. This proved to be 12°C for the (111) face. For such supercoolings the growth rate of gallium crystals is too large, and the experimental proof of the validity of this estimate, serving as confirmation of Kane's theory, constitutes a very difficult problem.

Conclusions

We have studied the growth kinetics of gallium, bismuth, and tin, together with the structure of their crystallization fronts. We have also obtained some information regarding the effects of deformation and small traces of silver on the growth of gallium crystals. The results enable us to draw certain conclusions regarding possible growth mechanisms and indicate the distinguishing features in the growth of the metals in question.

The growth of very pure gallium crystals, taking place under specific conditions and at low supercoolings, is characterized by an exponential relationship between the growth rate and the supercooling, and by a threshold supercooling below which the growth rate changes from a measurable to a vanishingly small quantity. These indications, together with the existence of a crystallization front incorporating sharp crystal faces, strongly suggest that gallium crystals grow by a mechanism based on the formation of two-dimensional nuclei. This is also supported by the fact that deformation influences the growth of the crystals. Deformation results in the development of defects and removes the supercooling threshold, sharply increasing the growth rate and changing the character of the temperature-dependence of growth rate from an exponential to a quadratic form; this may be regarded as a change from the nucleation to the dislocation mechanism of growth.

Addition of silver has a marked effect. The presence of even 0.01% Ag leads to an increase in growth rate and the elimination of the threshold supercooling, as in the case of deformation. However, the growth rate in this case is lower than that of deformed crystals of pure gallium grown at the same supercoolings. This behavior of the impurities may result from the effects of two opposing processes. On the one hand, the nonuniform capture of impurities results in the formation of dislocations, facilitating crystal growth, and on the other hand the buildup of impurities in front of the crystallization front impedes access of the principal element (gallium) and thus retards the growth of the defect-containing crystal.

As in the case of deformed gallium crystals, the kinetics of the growth of bismuth crystals is characterized by an approximately quadratic relation between the growth rate and the supercooling, which tends to prove the predominance of the dislocation mechanism of growth. The crystallization front also contains sharp crystal faces in this case. In the growth of bismuth crystals, no signs of two-dimensional nucleation could be seen; this was probably because of the difficulty of creating the conditions required for this purpose.

The results obtained for tin led to no definite conclusions regarding the growth mechanism, only indicating certain peculiarities in the growth of tin crystals as compared with gallium and bismuth. For the same supercoolings at the front, the growth rate of tin crystals is several orders greater than that of the other metals. The crystallization front has no sharp crystal faces, even under conditions close to equilibrium. For supercoolings smaller than 0.03°C, the front breaks down into a large number of needles projecting into the melt.

These differences in crystal growth are undoubtedly related to the nature of the metal, manifest in the structure and properties of the surface of separation.

Literature Cited

1. Volmer, M., and Marder, M., Z. Phys. Chem., 154(A) : 97 (1931).
2. Stranskii, I. N., and Kaishev, R., Usp. Fiz. Nauk, No. 21, p. 408 (1939).
3. Hillig, W. B., Technology Press, p. 127. MIT (1959).
4. Hillig, W., and Turbull, D., Elementary Processes of Crystal Growth [Russian translation. p. 293. IL, Moscow (1959).
5. Chernov, A. A., Usp. Fiz. Nauk, No. 73, p. 277 (1961).
6. Wilson, M. A., Phil. Mag., No. 50, p. 238 (1900).
7. Frenkel, J., Sow. Phys., No. 1, p. 498 (1932).
8. Turnbull, D., Thermodynamics in Physical Metallurgy, A.S.M., Cleveland (1950).
9. Jackson, K. A., and Chalmers, B., Can. J. Phys., No. 34, p. 473 (1956).
10. Borisov, V. T., Dukhin, A. I., and Matveev, Yu. E., In: Problems of Metal Science and the Physics of Metals. p. 269. Metallurgiya, Moscow (1964).
11. Cahn, J. W., and Hilliard, J. E., J. Chem. Phys., No. 28, p. 258 (1958).
12. Cahn, J. W., Acta Met., No. 8, p. 554 (1960).
13. Cahn, J. W., Hillig, W. B., and Sears, G. W., Acta Met., No. 12, p. 1421 (1964).
14. Borisov, V. T., In: Growth of Crystals, Vol. 3, p. 187. Izd. Akad. Nauk SSSR (1961). [English translation: Consultants Bureau, New York (1962).]
15. Borisov, V. T., and Dukhin, A. I., Fiz. Met. i Metalloved., No. 11, p. 893 (1961).
16. Borisov, V. T., and Dukhin, A. I., In: Problems of Metal Science and the Physics of Metals. p. 363. Metallurgizdat (1962).
17. Kramer, J. J., and Tiller, W. A., J. Chem. Phys., No. 42, p. 257 (1965).
18. Herl, D. T., Growth Processes and the Growing of Single Crystals [Russian translation]. p. 301. IL, Moscow (1963).
19. Weinberg, F., Trans. Met. Soc. AIME, No. 224, p. 628 (1962).
20. Barthel, J., and Scharfenberg, R., Can. J. Phys., No. 42, p. 1411 (1964).
21. Davies, V. de L., J. Inst. Metals, No. 92, p. 127 (1963-1964).
22. Davies, V. de L., J. Inst. Metals, No. 93, p. 10 (1964).
23. Jackson, K. A., Growth and Perfection of Metals, p. 319. John Wiley (1958).

24. Wagner, R. S., and Brown, H., Trans. Met. Soc. AIME, No. 224, p. 1185 (1962).
25. Turnbull, D., J. Appl. Phys., No. 21, p. 1022 (1950).
26. Jäntsch, O., Z. F. Kristallogr., No. 108, p. 185 (1956).
27. Sosnina, E. I., and Ovsienko, D. E., Fiz. Met. i Metalloved., No. 3, p. 527 (1956).
28. Zasimchuk, I. K., and Ovsienko, D. E., Ukr. Fiz. Zh., No. 9, p. 1092 (1964).

KINETIC GROWTH OF DENDRITES
IN CERTAIN BINARY MELTS

V. V. Nikonova and D. E. Temkin

*Institute of Metal Science and the Physics of Metals of the Central
Scientific-Research Institute of Ferrous Metallurgy*

The kinetics of the growth of dendrites are of considerable interest. This is because a dendrite is one of the principal forms of growth when metals solidify. The kinetics of dendrite growth are affected by both thermal and diffusion conditions and also by kinetic processes occurring at the surface of the growing crystal. Hence a study of dendrite growth may provide information on processes taking place at the boundary between crystal and melt.

The literature contains only a very few papers of this kind for metallic systems. Thus, the authors of [1] presented data on the kinetics of tin dendrite growth, while nickel was treated in [2]. In this article we consider the relation between the growth rate of dendrites and the supercooling in pure tin melts, as well as tin containing lead, bismuth, and antimony.

The experiments were carried out in the following way. Two thermocouples were placed in the melt at a known distance from each other. A seed crystal was introduced close to one of these at a fixed supercooling. As the crystal passed the thermocouple the latter showed a rise in temperature. The time between the temperature rises at the two thermocouples was measured. This time was identified with the time required for the crystal to pass between the two thermocouples.

Samples were prepared from high-purity metals (Sn of the OVCh type, zone-refined Pb, 99.998% Sb, and 99.999% Bi) and were placed in a molybdenum−glass test tube (internal diameter 13-14 mm). The metals were prevented from oxidizing by a layer of silicone oil. We used Chromel−Alumel thermocouples (wire diameter 0.2 mm). The thermocouple junctions, covered with a thin glass film, were at 80 mm from each other. The emf of the thermocouples were recorded on the compensation principle with an ÉPP-09 one-second potentiometer. The accuracy of the relative temperature measurements was ±0.2°C.

The test tube containing the previously melted metal was placed in a special thermostat. The temperature of the thermostat was regulated by means of an ÉPP-120 potentiometer. The temperature drop in the sample was no more than 0.5°C. The method of connecting the thermocouples was such that for one position of the switch the sample temperature was measured (by means of the upper thermocouple) and for the other the thermocouples were connected differentially. The supercooling was reckoned from the temperature at the end of melting, which was taken as the temperature of the liquidus for the alloy.

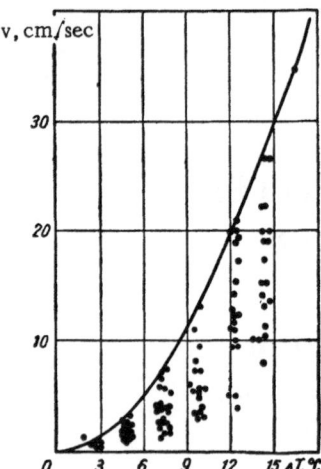

Fig. 1. Growth rate of tin dendrites as a function of the supercooling of the melt.

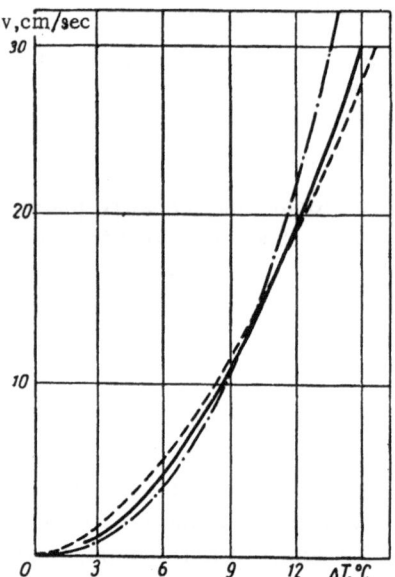

Fig. 2. Growth rate of tin dendrites as a function of the supercooling of the melt. Solid line − experimental curve; broken line − calculated for $v = 26 \, \Delta T_K$, cm/sec; dot-dash line − calculated for $v = 50(\Delta T_K)^2$, cm/sec.

The time required was recorded by means of the ÉPP-09 potentiometer (for low growth rates) or an N-100 loop oscillograph (for growth rates greater than 3 cm/sec). The error in the first case was 0.2 sec and in the second 0.01 sec.

In view of the considerable scatter in the results, probably due to an unfavorable orientation of the growing dendrites relative to the thermocouples, a series of 15-25 measurements was made for each supercooling.

Figure 1 shows the measured growth rates of pure tin dendrites as a function of supercooling. The curve drawn through the growth-rate maxima may be described by the relation

$$v = 0.14 \cdot \Delta T^2 \text{ cm/sec,} \tag{1}$$

where v is the velocity of growth, ΔT being the supercooling of the melt. Figure 2 shows a comparison between the experimental curve for pure lead and the calculated growth rate of a needle having the shape of a paraboloid of revolution. In the case of a linear relationship between the growth rate and the supercooling at the crystallization front ΔT_K ($v = K \Delta T_K$), the thermal-balance equation has the form [3]:

$$-\frac{v \varrho}{2a_2} e^{\frac{v \varrho}{2a_2}} E_i\left(-\frac{v \varrho}{2a_2}\right) = \frac{(T_{K,\infty} - T_0)\, c_2/Q}{1 + L_1/K\varrho + L_2 \sigma/v\varrho^2}, \tag{2}$$

where ρ is the radius of curvature of the tip of the needle; σ is the surface tension; T_0 is the temperature of the melt; $L_1 = \dfrac{\lambda_2}{\gamma_1 Q}\left[1.33 + 0.60\dfrac{\lambda_1}{\lambda_2}\right]$, and $L_2 = \dfrac{\lambda_2 T_{K,\infty}}{(\gamma_1 Q)^2}\left[3.86 + 2.08\dfrac{\lambda_1}{\lambda_2}\right]$, a_2, c_2, and λ_2 are the thermal diffusivity, specific heat, and thermal conductivity of the melt; γ_1 and λ_1 are the density and thermal conductivity of the crystal; Q is the heat of crystallization; and $T_{K,\infty}$ is the equilibrium temperature at the plane crystal−melt interface.

In order to describe the growth kinetics of the fastest-growing needle we obtain the following system of equations from (2):*

$$z = \frac{(x + \delta)\, x\, [1 - f(x)] + \delta f(x)}{x\, [(1 + x)\, f(x) - x]}, \tag{3}$$

$$\Theta = f(x) + \frac{(x + \delta)\, f(x)}{xz},$$

where

$$x = \frac{v_m \varrho}{2a_2}, \quad z = \frac{K\varrho}{L_1}, \quad \delta = \frac{L_2 \sigma K}{2a_2 L_1}, \quad \Theta = (T_{K,\infty} - T_0)\frac{c_2}{Q} \quad \text{and} \quad f(x) = -xe^x E_i(-x).$$

---

*In [3], system (3) is replaced by relations corresponding to a power approximation for Eq. (2).

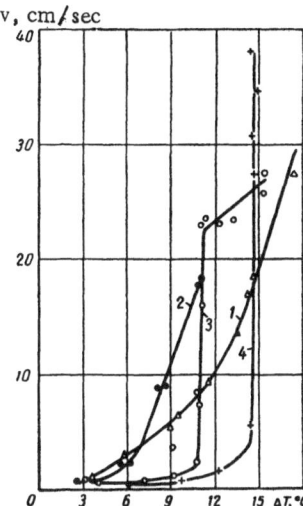

Fig. 3. Growth rate of tin dendrites as a function of the supercooling of Sn−Bi melts: 1) Sn−0.5 wt.% Bi; 2) Sn−2 wt.% Bi; 3) Sn−4 wt.% Bi; 4) Sn−8 wt.% Bi.

The broken curve in Fig. 2 represents a calculation from formula (3) with the values K = 26 cm/sec · deg, σ = 55 ergs/cm² (this value of σ is taken from [5]).

Analogous calculations may be carried out on the assumption of a quadratic relationship between the growth rate and the supercooling at the crystallization front

$$v = B(\Delta T_\kappa)^2. \tag{4}$$

A relationship of this kind was observed for the growth of tin in [6]. For the velocity range considered in [6] (below 10^{-3} cm/sec) B ≃ 50 cm/sec · deg². If we accept relation (4), then in (2) K must be replaced by $\sqrt{vB}$, and instead of Eq. (3) we obtain

$$y = \beta P + \sqrt{(\beta P)^2 + C},$$
$$\Theta = f(x)\left[1 + \frac{\beta}{y} + \frac{1}{y^2}\right], \tag{5}$$

where we have taken the following notation: $y^2 = 2a_2\rho x/L_2\sigma$,

$$\beta = \frac{L_1}{\sqrt{L_2 B\sigma}}, \qquad P = \frac{x[1 - f(x)]}{2[(1 + x)f(x) - x]}, \qquad C = \frac{(1 - x)f(x) + x}{(1 + x)f(x) - x}.$$

The dotted and dashed curve of Fig. 2 corresponds to a calculation based on equations (5) with σ = 55 ergs/cm² and B = 50 cm/sec · deg². In the supercooling range considered, the calculated curves may be approximated by a power relationship $v = A(\Delta T)^n$. For the chosen values of the parameter σ, K, and B, in the case of linear kinetics n ≈ 1.7, and in the quadratic case n ≈ 2.5. The first form is rather closer to the experimental relationship (1).

Theoretical analysis of the growth kinetics of a crystalline needle in a supercooled binary alloy [4] shows that the growth rate should fall as the concentration of the second component rises. However, the results of an experimental study of the growth rate of tin dendrites in alloys containing Bi, Pb, and Sb shown in Figs. 3-5 reflect a much more complicated relationship. These curves are drawn with respect to the maximum velocity values. The scatter in the data was analogous to that found in the case of pure tin (see Fig. 1).

For small supercoolings, the growth rate falls as the impurity concentration rises. We see from Figs. 3-5 that, at a certain critical supercooling, depending on the concentration, there is a sharp rise in dendrite growth rate for all alloys studied. The critical supercooling rises with increasing concentration of the impurity component.

The sharpest change in growth rate occurs for the tin−lead system. Thus, the growth rate in Sn−5 wt.% Pb at $\Delta T_0 \approx 24°C$ rises from 4 to 130 cm/sec. The change is not so sharp in the Sn−Sb system. If we plot a graph in coordinates of ΔT_0 and $\sqrt{(T_l - T_s)}$ (T_l and T_s are, respectively, the temperatures of the liquidus and solidus of the alloy; for Sn−5% Pb, the value of T_s is determined from the continuation of the solidus line), then, to the accuracy of determining ($T_l - T_s$) from the phase diagrams, all the experimental values of the critical supercoolings lie on a straight line passing through the origin of coordinates (Fig. 6).

The reasons underlying the growth-rate−supercooling relationships observed in these alloys are still not very clear and demand further study.

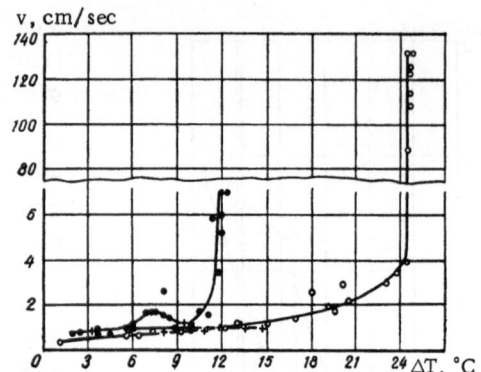

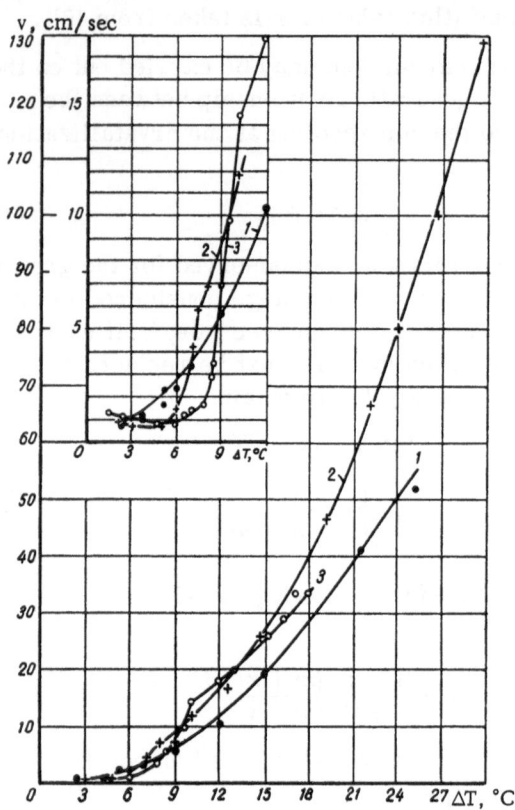

Fig. 5. Growth rate of dendrites as a function of the supercooling of Sn−Pb melts. ● − Sn + 1 wt.% Pb; + − Sn + +3 wt.% Pb; ○ − Sn + 5 wt.% Pb.

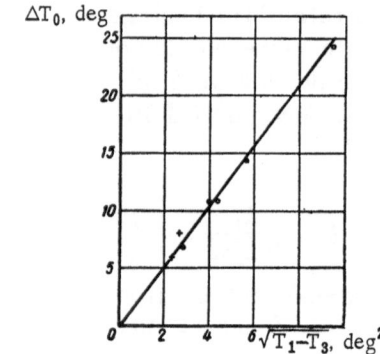

Fig. 4. Growth rate of dendrites as a function of the supercooling of Sn−Sb melts: 1) Sn−1 wt.% Sb; 2) Sn−3 wt.% Sb; 3) Sn−6 wt.% Sb.

Fig. 6. Comparison between the critical supercooling ΔT_0 and the temperature range $(T_l - T_s)$ for the alloys studied. ● − Sn + Bi; ○ − Sn + Pb; + − Sn + Sb.

Literature Cited

1. Rosenberg, A., and Winegard, W. C., Acta Met., 2(2):342-343 (1954).
2. Colligan, G. A., and Bayles, B. J., Acta Met., 10(9):895-897 (1962).
3. Temkin, D. E., Dokl. Akad. Nauk SSSR, 132(6):1307-1310 (1960).
4. Temkin, D. E., Kristallografiya, 7(3):446-450 (1962).
5. Turnbull, D., and Cech, R. E., J. Appl. Phys., 21:804 (1950).
6. Kramer, J. J., and Tiller, W. A., J. Chem. Phys., 42(1):257 (1965).

EXPERIMENTAL STUDY OF THE KINETICS OF THE CRYSTALLIZATION OF INDIUM-BASE BINARY METAL ALLOYS

V. T. Borisov and Yu. E. Matveev

Institute of Metal Science and the Physics of Metals of the I. P. Bardin Central Scientific-Research Institute of Ferrous Metallurgy

The variations observed in the structures of solid alloys with varying crystallization conditions (the rise or fall in the proportion of a particular phase, the appearance of a new or the vanishing of an old phase, the formation of a supersaturated solid solution, etc.) by no means correspond to the equilibrium phase diagrams of the systems. It is therefore important to study the kinetic diagrams relating the temperature and concentration at the phase boundary with the crystallization rate. Investigations of this kind are also required in order to understand the atomic mechanism of crystallization of the alloys.

The construction of kinetic diagrams for alloys assumes the simultaneous measurement of temperature and concentration at the crystallization front. In general, the observed deviation of the temperature at the phase boundary from the temperature of the liquidus for the given composition consists of kinetic and diffusion components: $\Delta T = \Delta T_K + \Delta T_D$; only ΔT_K represents the true supercooling at the crystallization front, while ΔT_D, called the diffusion supercooling, is the supercooling associated with the change in composition near the crystallization front. If the composition of the liquid remains constant during crystallization (i.e., $\Delta T_D = 0$), we may confine ourselves to measuring the temperature displacements at the crystal−melt boundary as a function of the growth rate. Such cases include alloys lying near minimum points and congruent chemical compounds.

The crystallization of such alloys in the indium−bismuth and indium−lead systems constituted our subject for study. We used metal of the following degrees of purity: indium, 99.9999%; bismuth, 99.99% and 99.9999%; and lead, 99.99%. In order to prevent oxidation the preparation of the samples and the conduct of the experiments were carried out under a layer of silicone oil.

1. Kinetics of the Crystallization of Indium − Bismuth Alloys. The arrangement of the apparatus for carrying out directional crystallization and measuring the temperature at the crystal−melt interface in the case of indium−bismuth alloys is shown in Fig. 1. The alloy was poured into an iron tube 1 with a wall 0.1 mm thick, diameter 7 mm, and length 120 mm. The tube was placed along the axis of a cylindrical metal vessel 2 and fixed to the bottom of this via a thermal insulator 3. The height of the column of melt in the tube

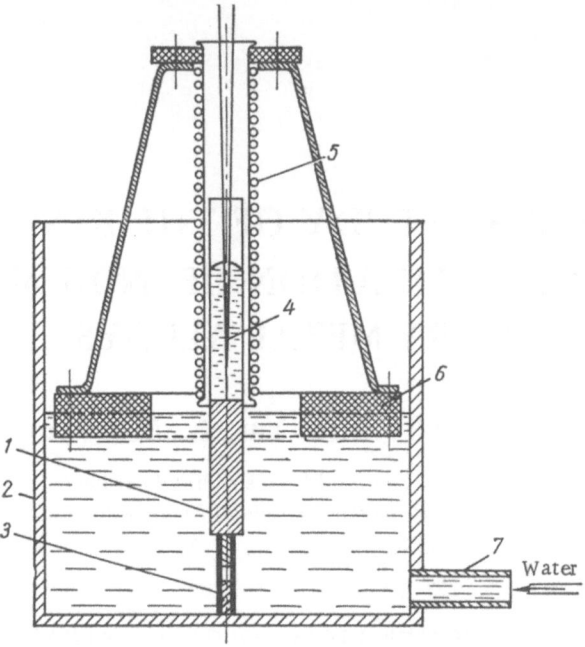

Fig. 1. Arrangement of apparatus for studying the crystallization of indium−bismuth alloys.

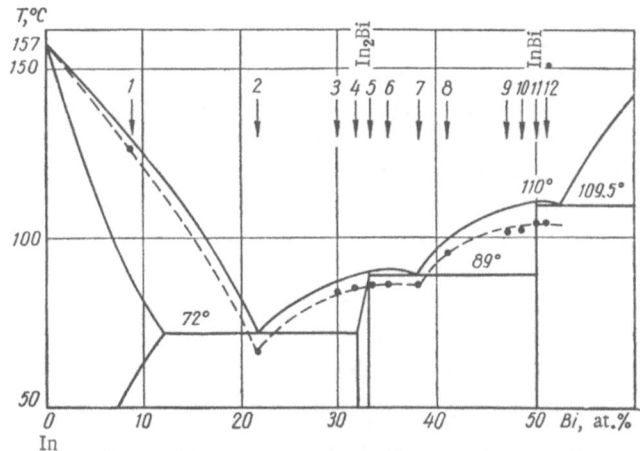

Fig. 2. Phase diagram of the indium−bismuth system.

was about 90 mm. At a depth of 30 mm from the level of the melt, the junction of a Chromel−
Alumel thermocouple 4 was placed on the axis of the tube. The diameter of the wires was 0.2
mm. Apart from the junction, the wires were insulated from the melt by a glass capillary 0.5
mm in diameter. In order to avoid dissolution of the thermocouple in the melt, the junction was
oxidized by heating in air. The heater 5 was mounted on a floating foam-plastic ring 6 and
could be moved in accordance with the level of the water in the cylindrical vessel. Water was
supplied to the vessel and reduced to the original level by means of the pipe 7. The rate of
crystallization was controlled by the rate of raising the water level in the vessel. The temperature variation was recorded with an electronic potentiometer of the BP-102 type (accuracy
±0.1°C). This arrangement made it possible to obtain temperature−time curves up to growth
rates of the order of 1 mm/sec.

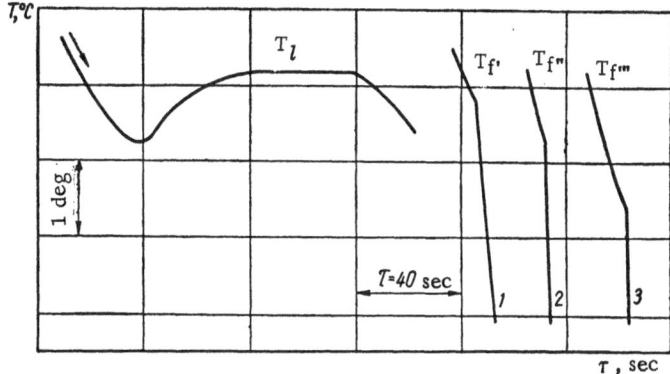

Fig. 3. Experimental cooling curves obtained with the chemical compound In_2Bi for various crystallization conditions: 1) $v' = 0.3$ mm/sec; 2) $v'' = 0.6$ mm/sec; 3) $v''' = 0.9$ mm/sec.

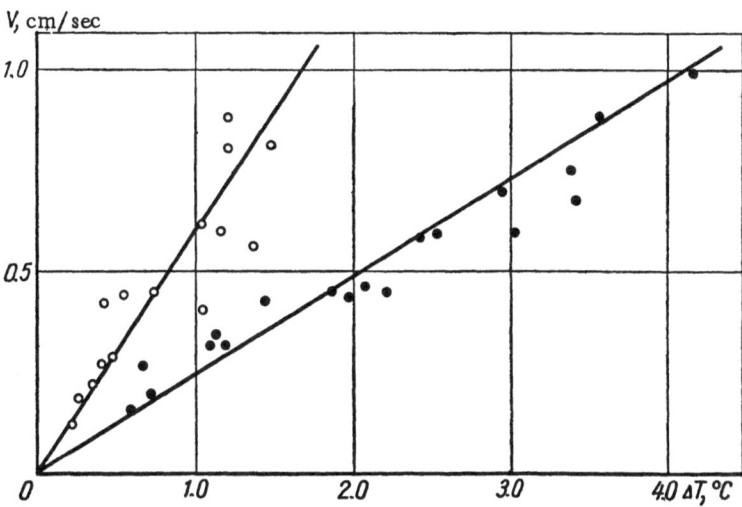

Fig. 4. Growth rate as a function of supercooling at the crystallization front for the chemical compounds In_2Bi (○) and InBi (●).

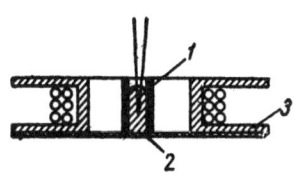

Fig. 5. Apparatus for studying the crystallization of indium in indium−lead alloys.

The phase diagram of indium−bismuth alloys is given in Fig. 2. The chemical compounds In_2Bi and InBi are congruent, melting without decomposition. The literature contains some contradictory data regarding the character of the compound In_2Bi [1, 2]. We therefore determined the course of the liquidus line in the neighborhood of compounds In_2Bi and InBi (in the range of compositions 30-50 at.% Bi) by thermal analysis. In the figure the arrows indicate the compositions of the alloys studied, and the broken line gives the kinetic diagram for a growth rate of 2 mm/sec. The characteristic experimental temperature vs. time curves

Table 1. Values of the Kinetic Coefficient
$K = V/\Delta T$ for Indium−Bismuth Alloys

No. of alloy	Bi, at. %	$K \cdot 10^2, \dfrac{cm}{sec \cdot deg}$
1	9.0	17.0
2	22.0	3.0
3	30.0	6.6
4	32.0	6.0
5	33.4 (In$_2$Bi)	6.0
6	35.0	5.5
7	38.0	8.6
8	41.0	8.1
9	47.0	3.3
10	48.5	2.3
11	50.0 (InBi)	2.4
12	51.5	3.4

are shown in Fig. 3 for various conditions of the crystallization of In$_2$Bi. The level of temperature at the crystallization front T_f was determined from the break on the curve corresponding to the instant at which the front passed through the thermocouple. Typical relationships between the growth rate V and the supercooling at the crystallization front ΔT are shown in Fig. 4 for the compounds In$_2$Bi and InBi. Table 1 gives the effective kinetic coefficient K for the alloys studied; this is the ratio of the growth rate V to the supercooling at the crystallization front ΔT, reckoned from the liquidus line T_l.

For a congruent chemical compound, the temperature displacement is determined by the kinetic supercooling at the front, and diffusion supercooling should be absent. The fact that the temperature displacements in the alloys in the neighborhood of the chemical compounds have approximately the same value as for the compound itself indicates the smallness of the diffusion supercooling in these alloys.

Thus we may suppose that the observed temperature displacements in the indium−bismuth alloys in the range 30-50 at.% Bi consist largely of the supercooling associated with the growth kinetics of the crystals. The kinetic coefficients for these alloys are 2-2.5 orders smaller than that corresponding to pure indium.

It should be noted that in the range of compositions up to about 20 at.% Bi the part played by diffusion supercooling is probably more substantial.

2. Kinetics of the Crystallization of Indium and Indium − Lead Alloys. For indium and indium−lead alloys we used the apparatus depicted in Fig. 5. The alloy was placed either in a porcelain tube 1 of internal diameter 4 mm, with its end firmly pressed against an iron plate 2 of thickness 0.5 mm, or simply placed on the plate in the form of a small drop 4 mm in diameter and 2 mm high. The plate was soldered with tin to the iron heater 3, made in the form of a coil. The height of the column of metal in the porcelain tube was 9 mm. Directional crystallization of the melt was effected by water or ice cooling via the plate 2. For good heat elimination, the point of contact of the drop of alloy with the plate was "tinned" (indium is similar to tin in this respect). In these experiments with a small volume of alloy the rate of crystal growth could be raised to 10 mm/sec. In order to reduce the inertia of the measuring circuit, Chromel−Alumel thermocouples without any junction (wire diameter 0.03-0.1 mm) were used; the electrical circuit was completed through the melt, and the temperature variation was recorded by means of an MPO−2 loop oscillograph. The signal from the thermocouple was amplified by means of a high-stability dc amplifier. The rate of growth was determined from the formula

$$V^2 = \frac{\varkappa c}{q}\left(\frac{\partial T_2}{\partial t} - \frac{\partial T_1}{\partial t}\right),$$

where $\varkappa$ is the coefficient of thermal diffusivity, q is the heat of fusion, c is the specific heat, and $\partial T_1/\partial t$, $\partial T_2/\partial t$ are the rates of temperature change in the liquid and solid phases determined from the oscillograms.

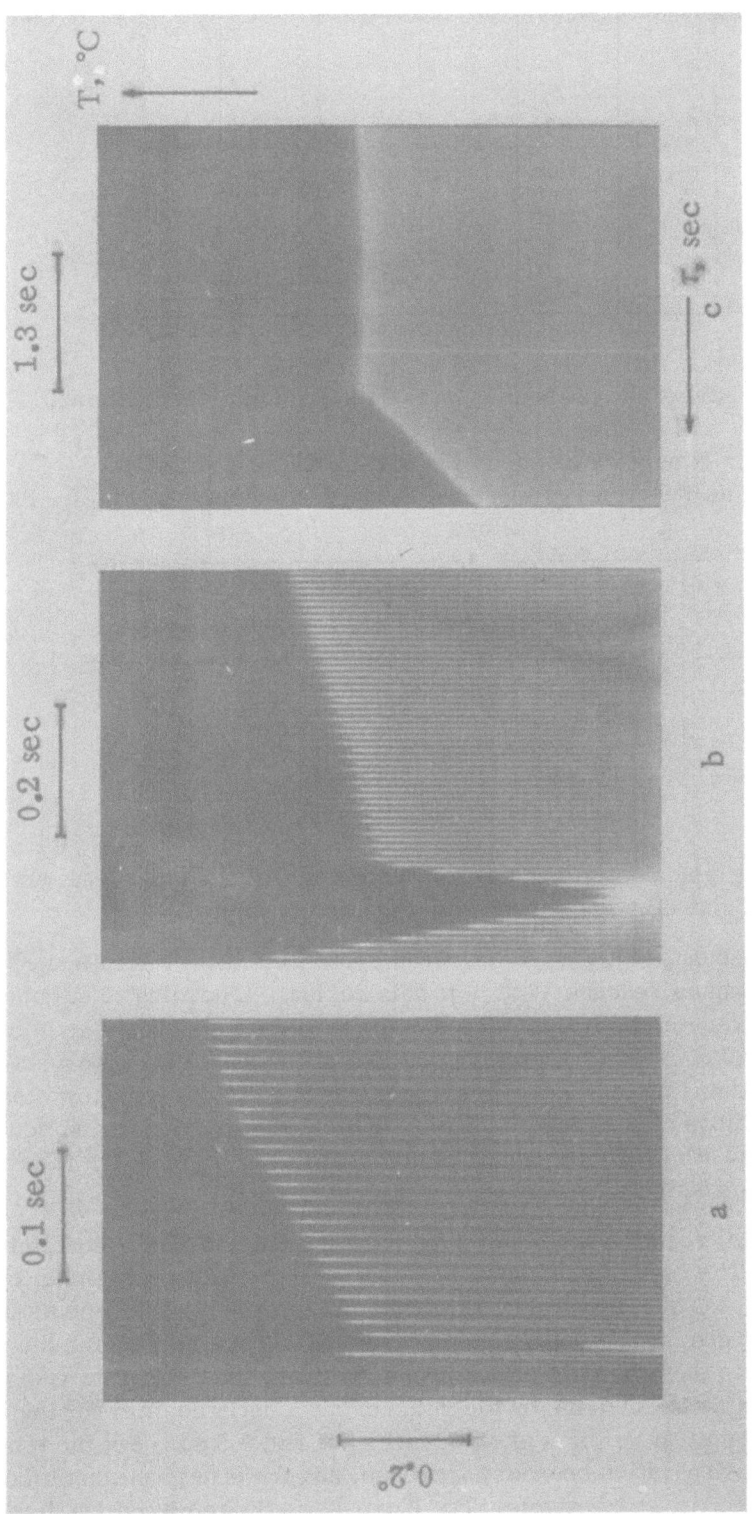

Fig. 6. Typical oscillograms obtained for indium−lead alloys at various growth rates: a) v = 3 mm/sec; b) v = 1 mm/sec; c) v = 0.1 mm/sec.

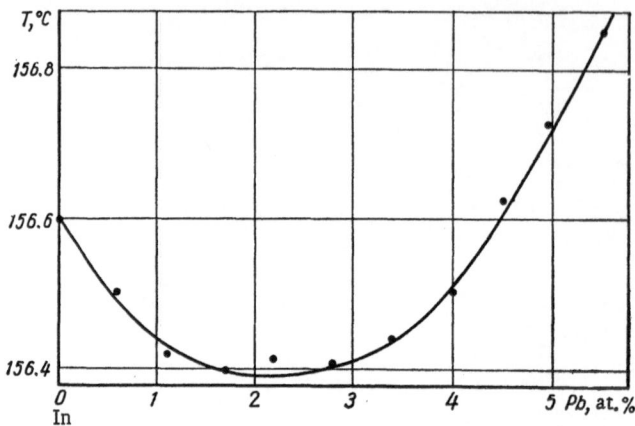

Fig. 7. Part of the phase diagram of the indium−lead system.

Table 2. Supercooling at the Front ΔT and Kinetic Coefficient K as Functions of the Lead Content in Indium−Lead Alloys

Pb, at. %	V, mm/sec	ΔT, °C	$K, \dfrac{cm}{sec \cdot deg}$
Pure In	12.0±2.5	0.11±0.05	10.9
2.3	11.0±2.0	0.09±0.04	9.9
4.6	10.0±0.5	0.10±0.04	10.0
7.1	9.0±1.0	0.34±0.03	2.6
9.5	10.0±1.0	0.80±0.03	1.3

Figure 6 shows some typical oscillograms representing a sinusoidal signal with a period of 0.01 sec, amplitude-modulated by the signal from the thermocouple.

According to published data [3], the indium−lead system contains a minimum point; the solubility of lead in solid indium reaches 10 at.% in this region. According to our measurements, the part of the diagram on the indium side has the form of Fig. 7. The minimum point corresponds to a fall of 0.20°C in the temperature and lies at 2.3 at.% Pb. On crystallization and melting of alloys containing up to 5.5 at.% lead, the temperature curves show clear extended rests analogous to those of pure indium, indicating that the liquidus and solidus lines are close together for these alloys. In the range of compositions 7-10 at.% Pb, the temperature interval between liquidus and solidus increases.

The kinetics of crystal growth were studied for pure indium and alloys containing 2.3, 4.6, 7.1, and 9.5 at.% Pb. The results are shown in Table 2. For indium and the alloys containing 2.3 and 4.6 at.% Pb, the supercooling at the front and the kinetic coefficients have practically the same values for growth rates up to 10 mm/sec. For pure indium and the alloy with 2.3 at.% Pb corresponding to the minimum on the phase diagram, the temperature displacements are in their nature kinetic. Judging by the value of the kinetic coefficient, the same is true for the alloy containing 4.6 at.% Pb. For alloys with 7.1 and 9.5 at.% Pb, the temperature displacements at the phase-separation boundary increase, and the effective kinetic coefficients fall by 2.6 and 1.3 cm/sec · deg, respectively. The increase in the temperature displacements is evidently associated with diffusion supercooling, since it is improbable that the kinetics of crystallization would change very much on changing the composition from 4.6 to 7 or 10 at.% Pb.

Results analogous to our own were also obtained in earlier work carried out on rubidium—potassium and potassium—mercury alloys [4, 5]. Bearing this in mind, we may draw the following conclusions regarding the kinetics of the crystallization of alloys.

The growth of crystals in alloys corresponding to congruent chemical compounds takes place with a considerable supercooling at the phase boundary. The supercooling is kinetic and its large value in comparison with pure metals is due to special features in the nature of the chemical-bond forces. The crystallization of alloys lying in the neighborhood of congruent chemical compounds also takes place with considerable supercooling, and the experimentally observed temperature displacements are largely associated with the kinetics of the growth process.

In alloys corresponding to minimum points on the phase diagram, growth takes place with very slight supercooling; the extent of the supercooling is of the same order as in pure metals. The same holds for alloys with liquidus and solidus lines close together. On passing away from the minimum point or the pure metal, when the temperature gap between the liquidus and solidus lines becomes greater, the value of the temperature displacement also rises; this effect evidently has a diffusion origin in which the concentration at the phase-separation boundary changes and hence so does the reduction in equilibrium temperature at the face of the growing crystal.

Literature Cited

1. Henry, O. H., and Badwick, E. L., Trans. AIME, 171:389-393 (1947).
2. Peretti, E. A., and Carapella, S. C., Trans. ASM, 41:947-960 (1949).
3. Hansen, M., and Anderko, K., Constitution of Binary Alloys [Russian translation]. Metallurgizdat, Moscow (1962).
4. Borisov, V. T., Dukhin, A. I., Matveev, Yu. E., and Rakhmanova, É. P., In: Mechanism and Kinetics of Crystallization. p. 164. Minsk (1964).
5. Borisov, V. T., Dukhin, A. I., Matveev, Yu. E., and Rakhmanova, É. P., In: Growth of Crystals. Vol. 5 (1965). [English translation: Consultants Bureau, New York (1968).]

TEMPERATURE DEPENDENCE OF THE GROWTH RATE
OF ALPHA CENTERS IN HIGH-PURITY TIN

A. I. Vykhovskii, L. N. Larikov, and V. M. Fal'chenko

*Institute of Metallophysics of the Academy of Sciences
of the Ukrainian SSR*

Presentation of the Problem

A study of the mechanism underlying the nucleation and growth of crystals in the solid phase is of great interest in developing the physical theory of the heat treatment of metals and alloys. One of the most important ways of studying this mechanism in various cases is that based on examining the kinetic laws under conditions in which the effects of subsidiary factors are eliminated. From this point of view the kinetics of polymorphic transformations deserve special attention. One of the best studied in this respect is the transformation of white tin (β-Sn) into the grey form (α-Sn).

The greatest amount of experimental data for the transformation of tin relates to the rate of growth of α-Sn(v) at various temperatures [1-7]. All these data are shown in Fig. 1. The tin used by different authors had different degrees of purity, sometimes not determined accurately enough [1-4, 7]. We may say, for example, that, although the original tin used by the authors of [3, 4, 7] was the same as in [1, 2], the resultant experimental data regarding v related to tin of a higher purity, since [3, 4, 7] employed crystals of β-tin grown electrolytically, and the original tin was greatly purified by the electrolysis. Data relating to the purity of the original tin used by various authors are presented in Table 1.

It follows from a comparison of Fig. 1 and Table 1 that the growth rate of the α-Sn centers rose systematically with increasing purity of the original tin. The temperature dependence of v for still purer tin therefore took on special interest.

On analyzing all available experimental data relating to the temperature dependence of v [8, 9] it was found that the relation between the linear transformation rate (of white tin into grey) and temperature derived from the results of [1-7] fully agreed with the view that the α-Sn centers grew by the two-dimension nucleus mechanism over a wide temperature range. It was noted in [8, 9], however, that, for small supercoolings, the growth rate of the α-Sn centers was sometimes much larger than would be indicated by the theory of two-dimensional growth. It was therefore suggested in [9] that some other growth mechanism (e.g., dislocation) might be involved in these cases. The solution of this question demands a large number of experimental values of v obtained for slight supercoolings.

54

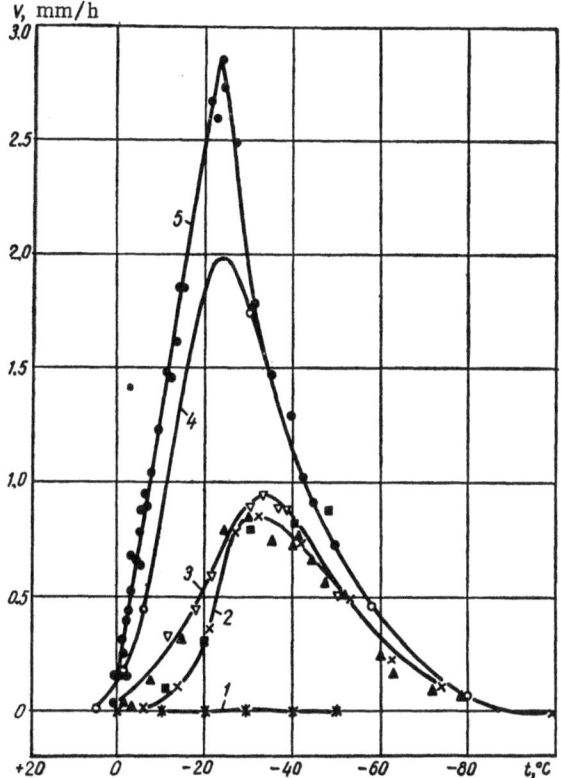

Fig. 1. Temperature dependence of the growth rate of α-Sn crystals according to various authors: $*$ — [1]; $\blacksquare$ — [5]; $\times$ — [6], C; $\bigcirc$ — [6], B; ∇ — [3]; $\blacktriangle$ — [4]; $\bullet$ — our own data.

In view of what has been said, we decided to make an experimental study of the temperature dependence of the growth rate of α-Sn centers in high-purity tin over a wide temperature range embracing both small and large supercoolings, and to analyze the results so as to establish the mechanism underlying the growth of α-Sn crystals.

Some Data Regarding the Nucleation of Grey Tin. Transformation Point of $\beta \rightleftharpoons \alpha$-Sn

All the experiments were carried out with high-purity tin of the OVCh-000 types (99.9995% Sn); the main impurities were Al < $3 \cdot 10^{-4}$%, As < $1 \cdot 10^{-4}$%, Fe < $1 \cdot 10^{-4}$%. In this tin no spontaneous transformation occurred on holding between −60 and −30°C for 10 to 12 days.

In order to facilitate the nucleation of α centers we may introduce seeds of grey tin or the semiconducting compounds InSb or CdTe [10] into the surface layer of the β-Sn, irradiate the white tin samples with neutrons at low temperatures, subsequently heating them to between −60 and −20°C [11], or subject the original white tin samples to low-temperature working [12, 13].

In order to obtain centers of grey tin, in the initial series of experiments, samples of white tin were subjected to preliminary 5% deformation (compression) at liquid-nitrogen temperature (−196°C). For this purpose a steel ring 2.5 mm high was prepared; a plate of tin 8 × 9 mm in size was placed in this, projecting by 5% from the ring. The ring and plate were flooded with liquid nitro-

Table 1

Reference	Information regarding purity of the Sn
[1, 2]	About 99.95% "Kahlbaum" and "Banka"
[3, 4, 7]	"Kahlbaum" with additional electrolytic refinement
[5]	99.9978% Vulcan tin
[6]-C	99.995% Vulcan tin
[6]-B	99.999%

gen and 5% impact compression took place in this. If after this the sample was heated to room temperature, the residual stresses were removed and there was no appreciable acceleration of the α-Sn transformation. After deformation in liquid nitrogen the samples were therefore transferred to Dewar vessels held at various low temperatures. It is interesting to note that, in a number of cases, not long before the appearance of the α-Sn center, the whole surface of the original sample became darkened. It was found that the nucleation of α-Sn centers at temperatures between −45 and −15°C was facilitated (the incubation period was reduced) if the samples were previously held at −60 or −78°C. We allowed these preliminary holding periods to last 1 h. Some of the resultant values of incubation period are shown in Table 2. If a center of grey tin was noted after it had grown to 2 or 3 mm, appropriate corrections were introduced into Table 2 so as to allow for the time required for the crystal to grow from zero size to its

Table 2

Series No.	Preliminary holding temp. (1 h), °C	Subsequent holding temp., °C	Incubation period, h	Notes
I	−60	from −41 to −30	24.5	
	−30		36	
II	−67	from −38 to −19	6.5	
	−30		25	
III	−78	from −31 to −28	3.5	Two samples were used in each experiment
			6.5	
	−60	from −31 to −28	38	
			41.5	
	−78	from −48 to −41	27	
			31	
	−60	from −48 to −41	36	
			44	
IV	−78	from −32 to −27.5	86	
	−60	from −32 to −27.5	108	

size at the instant of observation, using the subsequently obtained growth rates of α-Sn centers.

Since the temperature varied considerably during the long holding periods (Table 2), it is only reasonable to compare the accelerating influence of preliminary holdings for individual series, since all the samples of a given series were subject to identical temperature conditions on subsequent holding periods. We see from Table 2 (series I and II) that preliminary low-temperature holding for 1 h reduced the incubation period at a higher temperature, and it follows from series III and IV that holding at −78°C was more effective than holding at −60°C. The accelerating effect of preliminary low-temperature holding is probably associated with the fact that at such low temperatures the rate of nucleation of the α-Sn centers increases (see [9]), their "revelation" taking place at higher temperatures.

In order to analyze the experimental data relating to the $\beta \rightarrow \alpha$-Sn transformation we must know the transformation point T_0. The value of T_0 has been determined a number of times (see [14]). The latest determination of T_0 was carried out dilatometrically in 1958 [15] for a highly active powder consisting of 99.997% pure tin. It was found that, if the principal impurity in the tin were lead (0.003%), then $T_0 = 13.2°C$; if it were iron (0.003%), then $T_0 \approx 10.4°C$. It was also shown in [15] that the transformation point might vary over 1°, depending on the activity of the tin sample. For example, highly active tin powder might transform at 10.8°C and poorly active tin at 9.9°C. The position of the free-energy curves of white (F_β) and grey (F_α) tin was indicated schematically in [15] as a function of temperature. In view of the brittleness of grey tin, it was considered that the position of the F_α curve was independent of transformation velocity, the F_β curve lay higher, since the stresses arising in the white tin during the transformation could not be relieved. Thus, the results of [15] establish the influence of the kinetics of the transformation on the position of the transformation point. Since our own tin contained far less impurities (0.0005%) than that used in [15], we had to determine our own value of T_0. In view of the large volume effect associated with the $\beta \rightarrow \alpha$-Sn transformation (26%), we used the method of hydrostatic weighing in order to determine T_0. The principle of the experiments is described in [16]; here we shall only give a few details. The working liquid was carbon tetrachloride poured into a vessel with double walls between which circulated the thermostat liquid (acetone). Plates of β-tin, half-transformed into the α phase, 7-8 g in weight, and of the same dimensions as those on which the growth rate of the α-Sn crystals had been determined, were placed in a small glass cup. Weighing was carried out on a microbalance to an accuracy of 0.01 mg. It was found that at temperatures of T > 1.5°C the weight of the tin samples in CCl_4 increased monotonically (i.e., the transformation $\alpha \rightarrow \beta$-Sn took place) while at temperatures T < 0°C it fell monotonically (transformation $\beta \rightarrow \alpha$-Sn). The change of weight in 1 to 2.5 h was 6 to 20 mg, which greatly exceeded the accuracy of the experiments. Since on measuring the growth rate of α-Sn centers growth was reliably observed at a temperature of +1.0°C, the value of +1.5°C was taken for T_0.*

Thus our own value of T_0 differs considerably from those obtained earlier. This is no doubt due to the fact that our tin was of a much higher purity. We note that the sharp variation in the position of the transformation point from one state of aggregation to another with purity has been described earlier for various substances. Thus, the authors of [17-19] noted a sharp rise (tens of degrees) in the boiling point of very pure mercury, benzene, bromine, hydrogen sulfide, and other substances as compared with the ordinary values, and also several degrees rise in the melting point.

Experimental Determination of the Temperature Dependence of the Growth Rate of α-Sn Centers and Analysis of the Results

In addition to obtaining α-Sn centers by the method described above, in many experiments we used rolled plates of grey tin $1 \times 10 \times 15$ mm in size as seeds. For this it was sufficient to hold these plates at -20°C in test tubes in which the transformation had taken place earlier. The experiments showed that the growth rate of the α-Sn centers was independent of the means of producing these. After the formation of an α-Sn center on a tin sample the latter was placed in a glass vessel with double walls between which passed the thermostating liquid (acetone) at a given temperature (between +1 and -50°C). The method of producing and maintaining these temperatures is described in more detail in [20]. The motion of the grey tin boundary was observed through a low-power microscope ($\times 20$) furnished with an ocular micrometer. The radius of the grey tin centers r was plotted as a function of time t. Examples of these graphs for three different temperatures appear in Fig. 2. We see from this figure that the r(t) relationships are straight lines, the slopes of which give the growth rate of v of the α-Sn centers for each of the temperatures studied. As mentioned earlier in [5], in agreement with our observations, the transfer of the $\beta \rightarrow \alpha$-Sn boundary from one white tin grain to the next produced no change in the α-Sn growth rate. In other words, the relative orientation of the vanishing β grain and the developing α center had no influence on the measured growth rate v. This may be associated with the existence of large distortions at the boundary of the β and α forms.

As mentioned earlier, the formation of an α-Sn center was preceded by an incubation period not reflected in the graphs of Fig. 2. At each of the temperatures several (3-5) measurements were made. Apart from measurements at low supercoolings, this usually gave good reproducibility of the v values (scatter ± 10-15%). For measurements at temperatures between +1 and -5°C the reproducibility of the v data was much worse. Thus for different determinations of v at -2.5°C we obtained the following values: 0.60, 0.50, 0.46, 0.27, 0.20, and 0.17 mm/h.

It was found that if the grey tin center was formed at a relatively low temperature (from -30 to -20°C) and its growth was observed between +1 and -5°C, the resultant values of v were

---

*This value corresponds to the transformation of "highly active" white tin into grey, since it was obtained for samples previously cooled to -20°C. If after cooling to -20°C the samples were held at room temperature, the transformation point fell to 0°C.

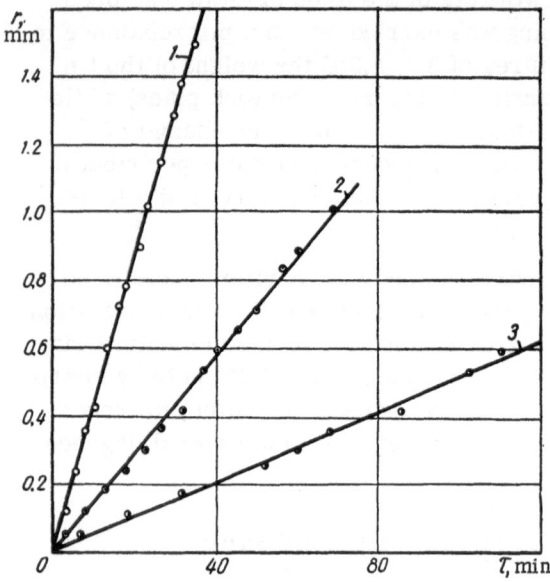

Fig. 2. Radius of the grey tin centers as a function of time. 1) t = −22°C, v = 2.65 mm/h; 2) t = −7.5°C, v = 0.91 mm/h; 3) t = −1°C, v = 0.35 mm/h.

the greatest. If, however, a β-Sn sample with an α-Sn center was kept for several hours at room temperature, the values of growth rate were much lower. The reason for this change in growth rate is apparently the fact that, after the formation of an α-Sn center at low temperature, the white tin surrounding the grey tin center develops considerable stresses, and hence there is a rise in the difference between the free energies of the white and grey tin, which constitutes the motive force of the transformation. While the sample is held at room temperature, the difference in the free energies of the β and α-Sn diminishes. For the same reason the transformation point falls, as described above. The experimental data on the growth rate of α-Sn centers resulting from our measurements are shown in Fig. 1 (curve 5).

In the region of small supercoolings we have taken the maximum velocities corresponding to the growth of α-Sn centers arising at −20°C. Correspondingly, in analyzing the experimental data, we have taken the value of T_0 = +1.5°C (corresponding to the transformation of "highly active" tin) for the transformation point.

Our values of v at all temperatures studied (except the region of large supercoolings) are much higher than published data relating to less pure tin [1-7] (see also Fig. 1). The maximum of the v = v(T) curve corresponds to a temperature of −23°C; the supercooling $\Delta T'_{max} = 24°$ (earlier it was considered that v = v_{max} at −30°C while the supercoolings $\Delta T''_{max} = 43°$). At this temperature the highest values of v $\approx$ 3 mm/h.

The collection of a large amount of data for small supercoolings also enables us to draw certain conclusions regarding the growth mechanism of the α-Sn centers. When the crystals grow from the melt or in the solid phase [21], the temperature dependence of the growth rate varies according to the particular points of the growing crystals to which atoms or molecules from the surrounding medium are attached. If such points are quite arbitrary, we have the condition known as normal growth [22-28], the rate of which is determined by the difference in the diffusion flows of atoms between the growing crystal and the mother solution

$$v = \frac{D\Delta S \cdot \Delta T}{lRT}\left(1 - \frac{r_\kappa}{r}\right),\tag{1}$$

where D is the diffusion coefficient for the transfer of material through the surface of separation, ΔS is the molar entropy of the phase transformation, l is the interatomic distance, ΔT is the supercooling, T is the absolute temperature, R is the universal gas constant, r is the radius of the center, and r_K is the critical radius of the nucleus.

If the atoms are attached at steps arising by two-dimensional nucleus formation, one of two relationships is valid. If individual atoms are attached to the two-dimensional nucleus [29], then

$$v = K_0 \exp\left(-\frac{U}{RT}\right) \exp\left(-\frac{K_1}{T\Delta T}\right), \tag{2}$$

but if atomic rows are attached to this [30], then

$$v = \frac{K_0'}{(\Delta T)^2} \exp\left(-\frac{U'}{RT}\right) \exp\left(-\frac{K_1}{T\Delta T}\right). \tag{3}$$

In formulas (2) and (3), K_0 and K_0' are factors weakly dependent on temperature, U and U' are the activation energies necessary for the passage of atoms or atomic rows, respectively, from the original phase to the crystal lattice of the nucleus, K_1 and K_1' are constants associated with the existence of an energy threshold for the formation of a two-dimensional nucleus at the face of a growing crystal, and the remaining notation is as in formula (1).

Finally, if the atoms are attached at steps associated with screw dislocations, then the temperature dependence during the growth of crystals in the solid phase [21] has the form*

$$v = \frac{AdD_b (\Delta F)^2}{4\pi RT\sigma V_m \Delta x}\left(1 - \frac{r_\kappa}{r}\right), \tag{4}$$

where A is the specific activity of the growth dislocations, d is the height of a growth step, D_b is the boundary-diffusion coefficient, ΔF is the difference between the molar free energies of the original and new phases, Δx is the width of the interphase boundary, σ is the specific free surface energy at the growth front, V_m is the volume of a gram-atom of the growing phase, and the remaining notation is the same as in formula (1).

For the transformation $\beta \rightarrow \alpha$ of tin over a wide range of supercoolings (up to $\Delta T \approx 100°$) we may put [42]

$$\Delta F = \Delta S \cdot \Delta T, \tag{5}$$

where ΔS is the molar entropy of the phase transformation at the transformation point. Substituting the quantity

$$D_b = D_0 \exp\left(-\frac{U_b}{RT}\right)$$

into formula (4), together with expression (5), we obtain

$$v = B (\Delta T)^2 \exp\left(-\frac{U_b}{RT}\right), \tag{6}$$

where U_b is the activation energy of boundary self-diffusion and the factor B depends weakly on temperature

$$B = \frac{AdD_0 (\Delta S)^2}{4\pi RT\sigma V_m \Delta x}\left(1 - \frac{r_\kappa}{r}\right). \tag{6a}$$

These temperature−crystal-growth-rate relationships [formulas (1)-(4)] differ considerably from each other. Calculations [9, 32, 33] show that in the ordinary way the experimental data for v(T) cannot simultaneously be described by several of these formulas, even with an optimum choice of parameters; only one of the three possibilities is confirmed by experiment.†

---

* Relation (4) is similar to the formula derived in [31] for the rate of the dislocation growth of crystals from the melt.

† The temperature relationships represented by (2) and (3) are similar to each other.

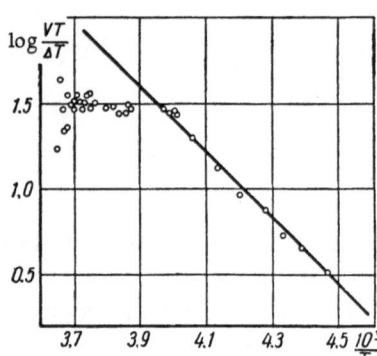

Fig. 3. Temperature dependence of the growth rate of α-Sn crystals in high-purity tin; analysis in accordance with formula (1); U = 8900 cal/g-atom.

Let us analyze our experimental data on the growth rate of α-Sn centers. In spite of the fact that in Fig. 1 (curve 5) the rising part of the v(T) curve is a straight line, the experimental data do not fit formula (1), describing normal growth. In fact, over such a wide temperature range ($\sim$20°) the proportionality factor in front of ΔT, containing the temperature-dependent factor D/T, should diminish with falling temperature, i.e., if the theory of normal growth were applicable to the case considered, the rising part of the v(T) graph should be convex toward the ordinate axis. This is supported by the graph of Fig. 3, which represents $\log(vT/\Delta T)$ as a function of $10^3/T$. If formula (1) were applicable to our case, the relationship for $\log(vT/\Delta T)$ as a function of reciprocal temperature should have been the same as the relationship $\log[D(1/T)]$, i.e., should have been a straight line, the slope of which would have given the activation energy of diffusion. We see from Fig. 3 that in the region of large supercoolings there is in fact a straight line with a reasonable value of activation energy; at low supercoolings, however, the points deviate sharply from the graph.

The experimental data also fail to fit the relationships corresponding to the mechanism of two-dimensional growth [formulas (2) and (3)]. Thus, the graph of $(\log v + U/RT)$ as a function of $1/T\Delta T$ gives a curve convex toward the axis of abscissas rather than a straight line. In order to check whether the experimental values of v(T) are described by a formula of type (6), Fig. 4 shows the values of v in coordinates of $(\log v + 0.4343U_b/RT)$ and $\log \Delta T$. For the activation energy of the boundary self-diffusion of tin U_b we take 9550 cal/g-atom, in accordance with the data of [34].* In the case of the $\beta \rightarrow \alpha$ transformation all the constants in the expression for B given in formula (6a), except A, are known. In order to estimate B we take

$$d \approx \Delta x, \qquad r_к \ll r, \qquad A = 3.3, \qquad D_0 = 6.44 \cdot 10^{-2} \text{ cm}^2/\text{sec}[34],$$

$$\Delta S = \frac{532}{274} \frac{\text{cal}}{\text{g-atom} \cdot \text{deg}} \text{ [42]}, \qquad \sigma = 10 \frac{\text{ergs}}{\text{cm}^2} \text{ [9]}, \qquad V_m = 20.4 \text{ cm}^3.$$

Substituting these values into formulas (6) and (6a), we obtain

$$v = 26 (\Delta T)^2 \exp\left(-\frac{9550}{RT}\right) \frac{\text{cm}}{\text{sec}}. \qquad (6b)$$

Formula (6b) corresponds to straight line 1 in Fig. 4, which fits the experimental points quite accurately. Hence, if we accept the principle of the transitional interphase boundary in agreement with [35], then in the case of the growth of α-Sn crystals this boundary will be quite sharp (small number of layers in the transitional zone). Thus, both the temperature dependence of the growth rate and its numerical value correspond to the dislocation growth of α-Sn crystals. As noted above, during $\beta \rightarrow \alpha$ transformation of tin, large changes of volume, and hence stresses develop. Under such conditions it is easy to picture the existence of dislocations and other defects in the crystal lattice, and also the possibility of a dislocation mechanism for the growth of α-Sn crystals.

---

*We note that a value of U = 9000 cal/g-atom, close to that obtained in [34], was obtained in [9] by analyzing the temperature dependence of v for moderately pure tin [6].

Table 3

ΔT	v_1, mm/h [1], $T_0 = 13.2°C$	v_2, mm/h [6], $T_0 = 10.4°C$	v_3, mm/h (present data) $T_0 = 1.5°C$
3.2	0.00025	—	0.38
4.4	—	0.00056	0.54
5.2	0.00030	—	0.64
8.2	0.00037	—	0.93
10.4	—	0.0036	1.20
13.2	0.00090	—	1.52
16.4	—	0.0169	1.90

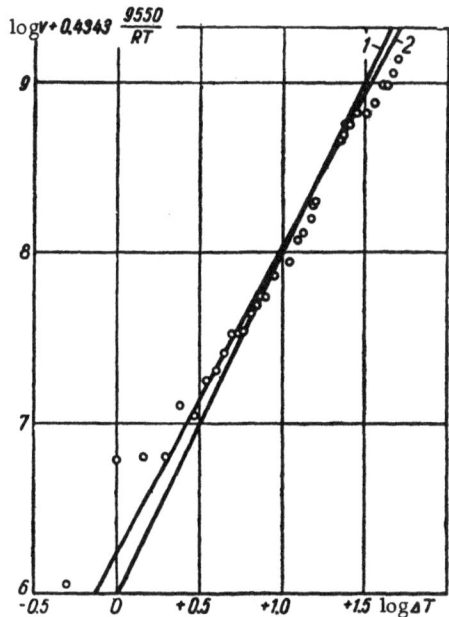

Fig. 4. Temperature dependence of the growth rate of α-Sn crystals in high-purity tin; analysis in accordance with formula (6).

It should be mentioned that the experimental points marked on Fig. 4 are rather better represented by the straight line 2 given by the empirical equation

$$v = 56.7\,(\Delta T)^{1.67} \exp\left(-\frac{9550}{RT}\right)\frac{cm}{sec}. \qquad (7)$$

In view of this we must note that in [32], in which the possibility of the dislocation growth of crystals during a polymorphic transformation in the solid phase was indicated by reference to the example of the $\beta \rightleftharpoons \alpha$ transformations of paradichlorbenzene, the power of ΔT was less than two (1.71 for the $\beta \rightarrow \alpha$ transformation and 1.62 for the $\alpha \rightarrow \beta$ transformation). For the dislocation growth of crystals from the melt the power was also smaller than two in a number of cases [31], and an attempt was made in [36] to explain this change in the character of the temperature dependence of v as being due to thermal interaction between neighboring turns of the growth spiral.

It is interesting to compare the values of v obtained for similar slight supercoolings in the present investigation and in [1, 6]. For this purpose, Table 3 shows the values for supercoolings between 0 and 16.5°C together with the values of the $\beta \rightarrow \alpha$-Sn transformation point. We see from this table that the growth rates of α-Sn centers in high-purity tin at slight supercoolings are 2-3 orders greater than in the less pure tin of [1, 6]. On the other hand, the values of v_1 and v_2 (Table 3) are larger than indicated by the theory of two-dimensional growth [9], while the temperature dependence of v_1 and v_2 is less sharp than that given by formulas (2) and (3) but sharper than that given by (4).

Thus, even from the numerical values of the growth rate of α-Sn crystals and its temperature dependence in the two cases (slight supercoolings, [1,6]) it is hard to express any definite views regarding the rival two-dimensional and dislocation growth mechanisms, and it is better to confine our attention to the views expressed in [35] regarding layer-by-layer growth. In all the remaining cases studied earlier [1-7] the temperature dependence of v corresponds to the two-dimensional growth mechanism [8, 9]. It is possible that the replacement of the dislocation mechanism of growth by the two-dimensional type is associated with the blocking of growth steps by the impurities existing in the less pure tin [36] (see also [37]). On the other hand, the existence of impurities may reduce the work of forming two-dimensional nuclei [9, 36, 38, 39] and hence increase the competitive role of this growth mechanism in the less pure tin. We note that even in the crystallization of a liquid there have been cases of a change in the growth mechanism from two-dimensional to dislocation on deformation of the growing crystal [40, 41], and sometimes this kind of change in growth mechanism has proved entirely uncontrollable [41].

Conclusions

We have determined the $\beta \rightleftharpoons \alpha$ transformation point in highly active, high-purity tin: 1.5°C. In the case of less-active tin the transformation point falls to 0°C.

In order to facilitate the nucleation of grey tin centers, an extremely effective treatment is the deformation of the tin samples at liquid-nitrogen temperature followed by 1 h at −60 or −78°C, plus a further holding period at −30 and −45°C.

We have obtained the growth rates of α-Sn crystals between +1 and −50°C. The growth rates exceed the experimental values of other authors obtained with less-pure tin. At a temperature of −23°C there is a maximum on the v/T curve: $v_{max} \approx 3$ mm/h.

These data enable the temperature dependence of growth rate and the absolute growth-rate values over a wide temperature range to be fully explained on the basis of the dislocation theory of growth. The relationship thus obtained takes the form

$$v = 56.7 \, (\Delta T)^{1.67} \exp\left(-\frac{9550}{RT}\right) \text{cm/sec.}$$

Literature Cited

1. Tamman, G., and Dreyer, K. L., Z. Anorg. Allgem. Chem., 199:97 (1931).
2. Chertok, M. M., Zh. Tekhn. Fiz., 5(4):711 (1935).
3. Komar, A. P., and Ivanov, K., Zh. Éksperim. i Teor. Fiz., 6(3):256 (1936).
4. Komar, A., and Lasarew, B., Phys. Z. Sowjetunion, 7(4):468 (1935).
5. Burgers, W. G., and Groen, L. J., Disc. Faraday Soc., 23:183 (1957).
6. Becker, J. H., J. Appl. Phys., 29(7):1110 (1958).
7. Komar, A., In: Application of X-Ray Diffraction to the Study of Materials. p. 217. ONTI (1936).
8. Bykhovskii, A. I., In: Scientific Transactions of UASKhN, 9:431 (1957).
9. Bykhovskii, A. I., Fiz. Met. i Metalloved., 12(1):64 (1961).
10. Goryunova, N. A., Dokl. Akad. Nauk SSSR, 75:51 (1950).
11. Fleeman, G., and Dienes, G. J., J. Appl. Phys., 26:652 (1955).
12. Ishikawa, H., J. Phys. Soc. Jap., 6(6):531 (1951).
13. Hedges, E. S., and Higgs, J. Y., Nature, 169(4302):621 (1952).
14. Bykhovskii, A. I., Ukr. Fiz. Zh., 8(6):609 (1953).
15. Raynor, G. V., and Smith, R. W., Proc. Phys. Soc., A244(1236):101 (1958).
16. Larikov, L. N., and Yurchenko, Yu. F., In: Questions on the Physics of Metals and Metal Science. Vol. 20, p. 191. Naukova Dumka, Kiev (1964).
17. Baker, H. B., J. Chem. Soc., 121:568 (1922).
18. Baker, H. B., J. Chem. Soc., 123:1223 (1923).
19. Baker, H. B., J. Chem. Soc., 127:1051 (1928).
20. Bykhovskii, A. I., Fiz. Met. i Metalloved., 6(3):487 (1958).

21. Larikov, L. N., In: Questions on the Physics of Metals and Metal Science., No. 10, p. 111 (1959).
22. Wilson, H. A., Phil. Mag., 50:238 (1900).
23. Frenkel, J., Phys. Z. Sowjetunion, 1:498 (1932).
24. Chalmers, B., J. Met., 6:519 (1954).
25. Larikov, L. N., Ukr. Fiz. Zh., 3(5):668 (1958).
26. Borisov, V. T., Dukhin, A. I., and Matveev, Yu. E., In: Problems of Metal Science and the Physics of Metals. p. 269. Metallurgizdat, Moscow (1964).
27. Pines, B. Ya., Zh. Tekhn. Fiz., 24:1521 (1954).
28. Lyubov, B. Ya., In: Problems of Metal Science and the Physics of Metals. p. 294. TsNIIChERMET (1958).
29. Volmer, M., Kinetik der Phasenbildung. Steinkopf, Dresden-Leipzig (1939).
30. Stranskii, I. N., and Kaishev, R., Usp. Fiz. Nauk, 21:408 (1939).
31. Hillig, W., and Turnbull, D., In: Elementary Processes of Crystal Growth [Russian translation]. p. 293. IL, Moscow (1959).
32. Bykhovskii, A. I., Larikov, L. N., and Ovsienko, D. E., Kristallografiya, 6:284 (1961).
33. Bykhovskii, A. I., In: Crystallization and Phase Transformations. p. 79. Izd. Akad. Nauk BelorussSSR, Minsk (1962).
34. Lange, W., and Bergner, D., Phys. Stat. Sol., 2(10):1410 (1962).
35. Cahn, J. W., Hillig, W. B., and Sears, G. W., Acta Met., 12(12):1421 (1964).
36. Chernov, A. A., Usp. Fiz. Nauk, 73(2):277 (1961).
37. Larikov, L. N., In: Questions on the Physics of Metals and Metal Science. Vol. 10, p. 121. Izd. Akad. Nauk UkrSSR, Kiev (1959).
38. Sears, G. W., J. Chem. Soc., 29(5):1045 (1958).
39. Hirth, J. P., and Pound, G. M., Condensation and Evaporation. pp. 102-103. Pergamon Press (1963).
40. Alfintsev, G. A., and Ovsienko, D. E., Dokl. Akad. Nauk SSSR, 156(4):792 (1964).
41. Ovsienko, D. E., and Alfintsev, G. A., In: Questions on the Physics of Metals and Metal Science. Vol. 19, p. 170. Naukova Dumka, Kiev (1964).
42. Brönsted, J. N., Z. Phys. Chem., 88:479 (1914).

GROWTH OF TIN GRAINS IN THE PRESENCE OF SURFACE-ACTIVE AND INACTIVE IMPURITIES

N. L. Pokrovskii and T. G. Smirnova

M. V. Lomonosov Moscow State University

Recently great attention has been paid to the effect of impurities on various physico-chemical and technological properties of metals and alloys. Quite recently, after the development of improved methods of producing pure metals, it has become clear how great this influence is. This is plainly seen when studying phase transformations in metal systems. As a rule, the action of impurities is especially effective at low concentrations; this is due primarily to their surface properties. It is precisely with this fact that we may associate the non-uniform distribution of impurities over the volume of an alloy, which has been repeatedly demonstrated in the course of experiments.

The most obvious representation of the surface properties of impurities may be obtained by studying their influence on the surface tension of the solvent metal. A considerable amount of experimental data has been built up regarding this question. The effect of impurities on the surface tension of mercury, tin, lead, bismuth, cadmium, copper, and many other metals has been studied in detail [1]. The effect of impurities on the primary crystallization structure, the kinetics of polymorphic transformations, and microhardness has been studied on the basis of these data. The results of these experiments prove that there is a specific link between the behavior of impurities at various phase boundaries, and in particular confirm many effects predicted by theory [1].

In recent years we have studied the action of slight impurities on the growth of tin grains in the process of selective recrystallization [2-6]. The elucidation of this question is of special interest for the theory of surface phenomena; the activity or nonactivity of impurities should be especially clearly manifested in this case, since the motive force of selective recrystallization is the free surface energy of the intergrain boundaries, which in certain model representations exhibit surface tension.

In order to illustrate the effect of impurity, let us consider the experimental data obtained for three series of samples: I) pure tin with less than $1 \cdot 10^{-3}$ wt.% impurity; II) a tin–zinc alloy containing 0.1 at.% Zn; and, III) a tin–sodium alloy containing 0.1 at.% Na, the annealing temperature in all three series being 175°C (Fig. 1). These data give a clear idea of the nature of grain growth in pure and contaminated tin, since sodium constitutes a strongly surface-active material while zinc has the greatest degree of surface inactivity relative to tin. By comparing the average grain size resulting from successive annealing in these three series of samples, we see that the largest grains are achieved in pure tin. The grains in the tin

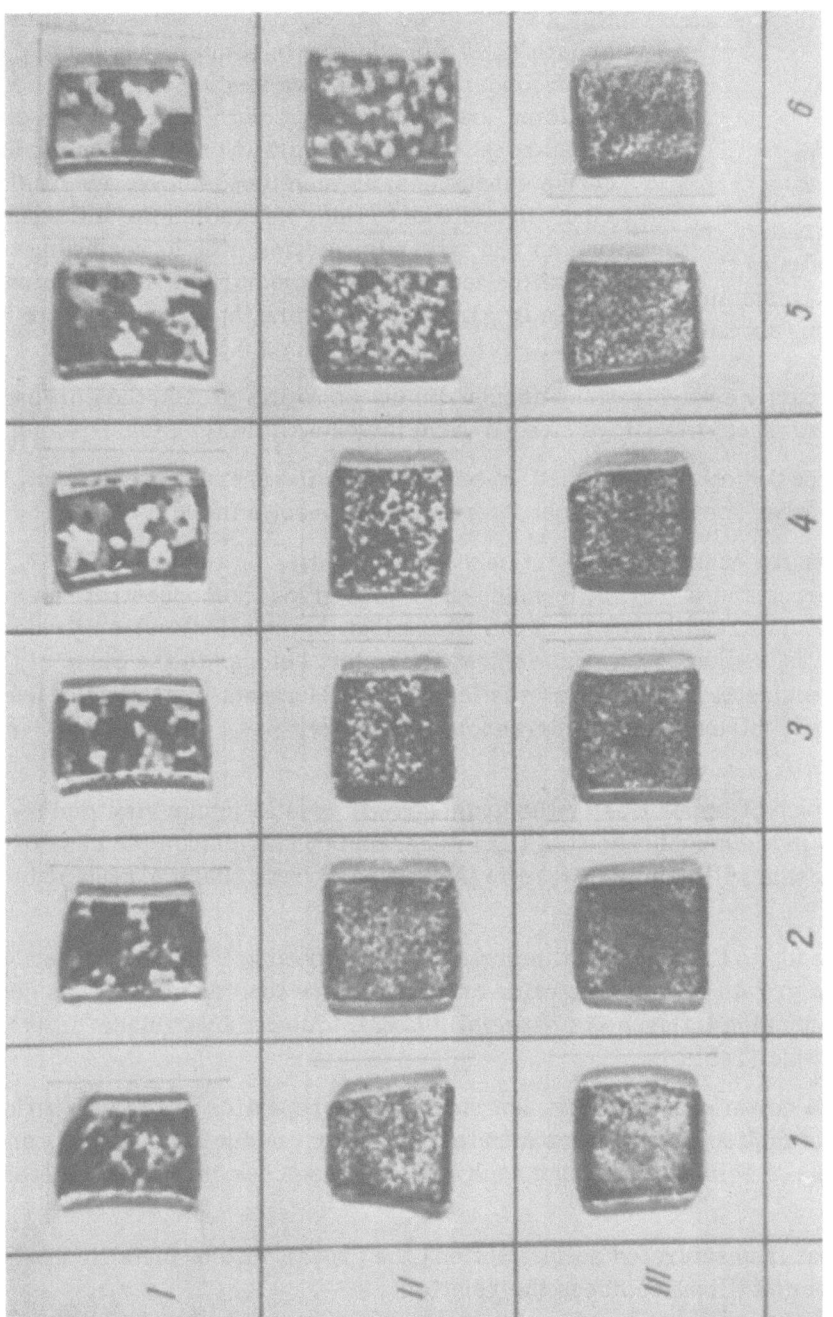

Fig. 1. Effect of impurities on the growth of tin grains in the course of selective recrystallization. Average grain diameter D for series I: 1) 0.56; 2) 0.9; 3) 1.51; 4) 1.66; 5) 2.18; 6) 1.66; for series II: 1) 0.2; 2) 0.34; 3) 0.42; 4) 0.61; 5) 0.87; 6) 1.14; for series III: 1) 0.1; 2) 0.13; 3) 0.12; 4) 0.13; 5) 0.16; 6) 0.14. Annealing time t for all series: 1) 15 sec; 2) 45 sec; 3) 2 min 15 sec; 4) 6 min 45 sec; 5) 20 min 15 sec; 6) 1 h 45 sec.

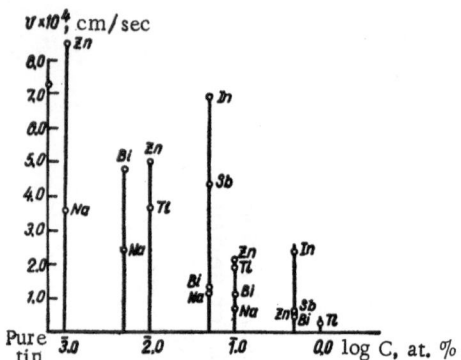

Fig. 2. Comparative diagram of the grain-growth velocities of pure tin and tin containing sodium, bismuth, thallium, antimony, indium, and zinc.

samples containing 0.1 at.% Zn also reached considerable dimensions, quite comparable with those of pure tin. At the same time there was very little grain growth in the tin samples containing 0.1 at.% sodium.

Over the past few years we have studied the influence of a great variety of impurities on grain growth [2-8]. In addition to sodium (with which we were concerned above), we studied the influence of thallium, antimony, and bismuth (also surface-active substances) on the grain growth of tin. Among inactive substances, as mentioned above, we considered zinc. In addition to this, we studied the influence of indium and silver impurities, which, according to our classification, belong to the surface-neutral class. These impurity metals were introduced into tin in quantities between 0.001 and 0.5 at.%.

We note that the impurity concentrations chosen did not as a rule exceed the corresponding solubilities in pure tin. Thus, the alloys studied were in general always single-phase.

The experimental results obtained enabled us to draw certain general conclusions. It would be desirable to consider some of the most interesting of these in more detail.

First let us consider the question of the influence of impurities on the growth rate of tin grains. These data are presented in Fig. 2, in which the x axis gives the concentrations and the y axis the velocities v. We readily see from the figure that the smallest value of v occurs for samples of tin containing sodium, which, as indicated earlier, belongs to the class of the strongest surface-active elements. Then follow the less-active elements bismuth, thallium, and antimony. The smallest influence of all corresponds to impurities of surface-inactive zinc and surface-neutral indium.

The observed influence of impurities on the grain-growth rate is in our view due to segregation of impurities in the neighborhood of the intergrain boundaries. In the case of surface-active impurities this will be higher than in that of inactive or surface-neutral impurities.

We must draw attention to the fact that the impurities are arranged in a certain order as regards their effect of the growth rate of tin grains and the surface tension of tin. The sequence starts with sodium, for which the influence is especially large, followed by surface-neutral indium and finally inactive zinc (Table 1).

Thus we may draw a certain parallel between the surface properties of the impurities in question at the boundary of molten tin with its saturated vapor (or vacuum) and at the boundaries of the growing grains of solid tin, although each of these cases has its own characteristic features.

It is shown in [9] that in unsaturated solid solutions the growth rate of metallic grains in the course of selective recrystallization obeys the relation

$$\log v = f\{\log[c(r - r_0)]\},$$

where c is the concentration of the impurity in at.%, and r and r_0 are the atomic radii of the impurity and matrix, respectively.

Table 1

Effect of impurities on surface tension of tin			Effect of impurities on growth rate of tin grains in course of selective recrystallization				
Impur-ity	Conc., at.%	$\pi=\sigma_0-\sigma$, dyn/cm	Impur-ity	Conc., at.%	$v\cdot10^4$, cm/sec		
					100°C	140°C	180°C
Na	0.06	—40	Na	0.06	0.27	0.67	1.17
Bi	0.58	—13	Bi	0.05	0.27	0.59	1.43
Tl	0.18	— 5	Tl	0.01	—	2.06	3.77
Sb	0.52	— 3	Sb	0.05	—	1.06	4.39
In	0.34	± 0	In	0.05	0.96	2.87	6.95
Zn	1.68	+1.7	Zn	0.01	2.14	3.11	5.03
			Pure tin	—	2.07	4.56	7.25

Table 2

Composition of the alloys	Activation energy, kcal/g-atom	
	according to our data	from Holmes and Wionegard [15]
Pure tin	6.4	6.0
Sn+0.005 at. % Bi	10.8	11.8
Sn+0.05 at. % Bi	—	—
Sn+0.05 at. % In	10.0	—
Sn+0.5 at. % In	10.0	—
Sn+0.01 at. % Ag	11.4	—
Sn+0.05 at. % Na	7.4	—
Sn+0.001 at. % Zn	10.1	—
Sn+0.01 at. % Zn	7.4	—

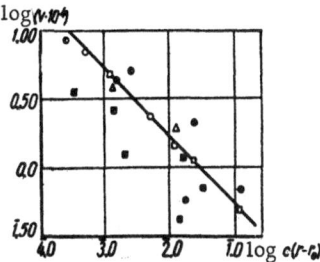

Fig. 3. The $\log v = f\{[\log c\,(r-r_0)]\}$ relationship for the alloys: tin−bismuth, □; tin−thallium, △; tin−antimony, ⊗; tin−sodium, ⊠; tin−zinc, ◐; tin−silver, ◪; tin−indium, ○.

The application of this relationship to our own results showed (Fig. 3) that the value of v for tin containing bismuth, thallium, antimony,* and indium agreed satisfactorily with this linear relationship. We note that all these impurities are characterized by a weak surface activity [10, 11], while indium even belongs to the class of surface-neutral impurities [2]. It is also an important point that the impurities are soluble in solid tin [13]. The influence of these metals on the growth rate of tin grains differs considerably (Fig. 2). This is evidently because the segregation of impurities in the regions of intergrain boundaries is due not only to their surface activity but also to upward diffusion [14], depending on the value of $r-r_0$. For this reason the effect of indium on the growth rate is negligible, since $r-r_0 = 0.01$ Å. At the same time, the effect of bismuth is much larger (Fig. 2) owing to the large value of $r-r_0 = 0.24$ Å.

*In Fig. 3 the point for the tin sample containing 0.6 at.% Sb fell out of position.

Deviations from the straight line occur for samples of tin containing sodium, silver, and zinc. The reasons for which these fail to obey the general law differ. Sodium and silver have low solubility in solid tin, so that even very small quantities of these elements may fall outside the solubility limit. The effect of zinc on v is smaller than would be expected owing to upward diffusion, which is apparently due to its inactivity. For this reason the corresponding points in Fig. 3 lie above the straight line.

The results obtained were used in order to calculate the activation energy of grain growth in pure tin and its alloys (Table 2). The calculated values of activation energy agreed satisfactorily with the data of [15].

It also follows from Table 2 that all the impurities studied increase the activation energy of tin grain growth. This is quite as it should be, since the presence of foreign atoms impedes growth processes. With increasing impurity concentration the activation energy falls, confirming certain recently developed theoretical considerations [16].

The results obtained lead to the conclusion that there is a certain connection between the behavior of impurities at various interphase boundaries. However, in order to secure a full explanation of impurity segregation at intergrain boundaries in polycrystalline material, we must consider not only the surface properties of the impurities, but also their solubility in the principal substance and the effects of upward diffusion.

Literature Cited

1. Semenchenko, V. K., Surface Phenomena in Metals and Alloys. Gostekhizdat, Moscow (1957).
2. Pokrovskii, N. L., and Smirnova, T. G., In: Growth of Crystals. Vol. 3, p. 200. Izd. Akad. Nauk SSSR, Moscow (1961).
3. Pokrovskii, N. L., and Smirnova, T. G., Fiz. Met. i Metalloved., 12(5):708 (1961).
4. Pokrovskii, N. L., and Smirnova, T. G., Fiz. Met. i Metalloved., 14(6):890 (1962).
5. Pokrovskii, N. L., Smirnova, T. G., and Chirkova, V. V., In: Surface Phenomena in Melts and Processes of Powder Metallurgy. p. 17. Izd. Akad. Nauk UkrSSR, Kiev (1963).
6. Pokrovskii, N. L., and Tissen, D. S., In: Growth of Crystals, Vol. 3, p. 207. Izd. Akad. Nauk SSSR, Moscow (1961). [English translation: Consultants Bureau, New York (1962).]
7. Pokrovskii, N. L., and Smirnova, T. G., Fiz. Met. i Metalloved., 19(3):401 (1965).
8. Pokrovskii, N. L., and Saidov, M. S., Zh. Fiz. Khim., 29(9):1601 (1955).
9. Holmes, E. L., and Wionegard, W. C., Trans. Met. Soc. AIME, 224(5):945 (1962).
10. Semenchenko, V. K., Pokrovskii, N. L., and Lazarev, V. B., Dokl. Akad. Nauk SSSR, 89(6):1021 (1953).
11. Pokrovskii, N. L., and Tissen, D. S., Dokl. Akad. Nauk SSSR, 128(6):1228 (1959).
12. Pokrovskii, N. L., and Tissen, D. S., Zh. Fiz. Khim., 34(6):1238 (1960).
13. Hansen, M., and Anderko, K., Constitution of Binary Alloys [Russian translation]. Metallurgizdat, Moscow (1962).
14. Konobeevskii, S. T., Zh. Tekhn. Fiz., 13(6):200 (1934).
15. Holmes, E. L., and Wionegard, W. C., Can. J. Phys., 29:1223 (1961).
16. Lücke, K., and Detert, K., Acta Met., 5:628 (1957).

SOME QUESTIONS RELATING TO THE GROWTH AND FORMATION OF CRYSTALS

I. V. Salli, E. V. Finagina, and É. S. Fal'kevich

Dnepropetrovsk State University

This paper constitutes a further development and refinement of the theory presented in [1, 2] and is intended to bring the conclusions of the latter closer to practical problems. Data relating to the formation of structure in steel and to the growth of silicon crystals in the reduction of silicon tetrachloride by hydrogen are also presented. According to the calculation of [1], the growth rate of a crystallization center at the diffusion stage of the process is determined by the equation

$$v = \frac{D}{\varrho}\left(\Delta c - \frac{a}{r}\right)\left(\frac{1}{x} + \frac{1}{r}\right). \tag{1}$$

Here, D is the diffusion coefficient, ρ is the density of the crystallizing substance, $\Delta c = c - c_{\sim}$, where c is the concentration of the supersaturated mother phase and $c_{\sim}$ is the equilibrium concentration of the plane surface of separation at the given temperature (T)

$$a = \frac{2\sigma M c_{\sim}}{\varrho R T},$$

where σ is the surface tension at the phase-separation boundary, M is the molecular weight of the crystallizing substance, R is the gas constant, r is the radius of curvature of the boundary, and x is the thickness of the impoverished zone surrounding the growing crystallization center. Sometimes this zone is called the "realm of crystallization."

Analysis of Eq. (1) shows that, if x is large or increases together with r, then a circular form of crystal growth with an increasing radius of curvature of the boundary is favorable. The maximum growth rate is reached at $r = 2r^*$, where

$$r^* = \frac{a}{\Delta c}.$$

This kind of growth is characteristic for the decomposition of supersaturated solid solution at a slight supersaturation.

If x remains constant during growth, the maximum rate of growth moves in the direction of large r (Fig. 1) and skeletal forms of growth (cylinders or plates with a constant curvature of the surface of separation in the growth direction) are the most favorable. If $x \leq 2r^*$, the growing crystal tends to adopt a flat-sided form.

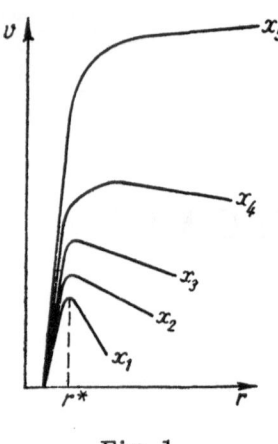

Fig. 1

One of the most important features is the fact that the greatest velocity of growth corresponds to the directions in which the crystal presents the surfaces of lowest surface tension to the mother phase.

It follows from the presentation of the problem that the outer boundary of the zone should coincide with a special type of point (inflection or break) on the $cf(r)$ curve.

Naturally x will only increase with increasing r in a diffusion process. It is only possible to keep x constant during growth by the assistance of other variables not entering into relation (1). It is known from experiments that a change in the shape of the crystals may follow a change in supercooling.†

According to [2], which was concerned with the influence of relative supercooling on x,

$$x = r\left(\frac{2\Delta TQ}{RT_0^2} - 1\right). \tag{2}$$

Here, $\Delta T = T_s - T_0$, where T_s is the equilibrium temperature and T_0 is the temperature of the medium, while Q is the activation energy required for the motion of the atoms in the medium.

The principle underlying Eq. (2) may be schematically considered as follows. The distribution of concentration at the front of the growing crystal may be obtained for spherical symmetry, without considering the temperature field, by solving the equation

$$D\frac{\partial c}{\partial r} = \frac{\text{const}}{r_1^2}.$$

The temperature field may be taken into account by using the relation between the diffusion coefficient and temperature.

If we suppose that the temperature around the nucleus is distributed in accordance with the law

$$T_1 = T_0 + \frac{r_1(T_s - T_0)}{r},$$

where T_1 is the temperature directly at the surface of separation and T_0 is that at a large distance r from the crystal, then

$$D = D_0 e^{-\frac{Q}{RT_1}} = D_0 e^{\frac{Q}{RT_0 + \frac{r_1(T_s - T_0)}{r}}}.$$

The power factor in this equation may be transformed more accurately than in [2] if we multiply the numerator and denominator by $[RT_s - (Rr\Delta T/r_1)]$ and rearrange the terms in the denominator, thus obtaining

$$\frac{Q\left(RT_s - \frac{Rr_1\Delta T}{r}\right)}{R^2 T_s T_0 + \frac{R^2 r_1}{r}\left(T_s^2 - 2T_0 T_s + T_0^2 - T_0 T_s - \frac{r_1}{r}T_0^2 + 2\frac{r_1}{r}T_s T_0 - \frac{r_1}{r}T_s^2\right)}. \tag{3}$$

---

† Yet another way of keeping x constant is the continual rotation of the growing crystal.

The second term in the denominator may be neglected (especially if r is close to r_1, the case which will always interest us most). Then the solution will appear as follows:

$$c = \text{const}_1 \cdot e^{-\frac{r_1 Q \Delta T}{RT_s T_0 r}} + \text{const}_2, \qquad (4)$$

while the distance to the inflection point on the c(r) curve, or, in other words the thickness x of the realm of crystallization, is determined, if we equate the second derivative of (4) to zero by

$$x = r_1 \left(\frac{\Delta T Q}{2 R T_s T_0} - 1 \right). \qquad (5)$$

This expression differs from (2) in that $T_s T_0$ replaces T_0^2. It follows from (5) that the concept of the realm of crystallization in the sense in which it was introduced for deriving Eq. (1), is valid when $(\Delta T Q / 2 R T_s T_0 > 1)$. If this is not so, the outer boundary of the zone is not determined and the problem is solved by means of boundary conditions for which $x \to \infty$. It also follows from this expression that by varying $\Delta T / T$ we can achieve constancy of x for varying r.

Let us consider one of the types of growth in the following form.

First, the nucleus grows with increasing radius and increasing growth rate (left-hand branch of the curve in Fig. 1). If we artificially vary $(\Delta T / T_0)$ so that x remains constant, the increase in growth velocity ceases after passing through a maximum (Fig. 1).† From this moment the crystal begins branching and growing with constant radius of curvature (r_m) in directions of low surface tension. Setting dv/dr equal to zero and putting x from (5) in the resultant expression, we obtain the radius of a crystal growing at constant velocity

$$r_m = \frac{2a}{\Delta c} + \frac{a}{\Delta c \left(\frac{\Delta T Q}{2 R T_0 T_s} - 1 \right)}. \qquad (6)$$

By using expression (6) we may consciously control the process of shape development by selecting the condition

$$\frac{\Delta T}{T_s} \geqslant \frac{2RT}{Q}. \qquad (7)$$

Condition (7) may be put in another form by using the approximate relation

$$\frac{L \Delta T}{T_s} = RT \ln \frac{c}{c_\sim}, \qquad (8)$$

where L is the heat of crystallization. Then condition (7) transforms into

$$\frac{\Delta T}{T \ln \frac{c}{c_\sim}} \geqslant \frac{RT_s}{L}, \qquad (9)$$

---

† From a certain instant x may remain constant automatically, owing to special features characterizing the separation and the distribution of the heat of crystallization.

here,

$$Q = \frac{2L^{\dagger}}{\ln \dfrac{c}{c_{\sim}}} .$$

The physical meaning of these relationships reduces to the fact that, the greater the deviation of the conditions of growth from the equilibrium condition, the more will the curvature of the surface of separation differ from plane. The equality sign in these relations corresponds to the cessation of growth and the formation of an ideally plane interface.

Let us now turn to experiment. In the production of silicon by deposition in the course of the hydrogen reduction of chlorosilanes, rods of fine-crystalline structure are usually formed. The crystallites of these rods consist of a large number of radially arranged dendrites (Fig. 2a). This structure has serious disadvantages; one of these is the fact that a large quantity of gas and impurity is captured between the dendrites and their axes. When the material is subsequently remelted and used for the pulling of single crystals, this has a deleterious effect on the properties of the latter.

Our problem was to discover the optimum conditions for the crystallization of silicon during hydrogen reduction, such that the resultant rods should be compact and of monolithic form. Naturally the ideal conditions would be those in which the seed crystal would be converted into a perfect single crystal of arbitrarily large size.

If we use the degree of drusiness (cavity content) as a characteristic of the structure

$$\varepsilon = \frac{d_a - d_c}{d_c} ,$$

where d_a is the actual diameter of the resultant rod and d_c is that calculated on a weight basis, then the experimental data show that increasing the ratio of the tetrachloride to hydrogen from $\frac{1}{60}$ to $\frac{1}{35}$ at 1120°C reduces the dendrite factor from 30 to 4%. Further increasing the concentration does not reduce the factor any further. Increasing the temperature from 980 to 1120°C for a constant concentration of $\frac{1}{40}$, however, roughly halves the dendrite factor.

Thus, optimum conditions are determined by due choice of both factors.

We see from condition (6) that we can only avoid the formation of dendrites if in the course of growth the crystallization center increases the radius of curvature of the surface of separation (growth in the form of a sphere is excluded, since under this condition the rate of growth falls rapidly with increasing crystal radius).

Such growth is possible if we approximate the conditions of growth to the equilibrium conditions determined by relation (9), in which an optimum temperature of the growing crystal surface corresponds to each given supersaturation.

In fact, the structure of silicon (Fig. 2b), which has a low degree of drusiness, is characterized by the fact that the thickness of the principal axes of the dendrites is in this case far

---

† This relation is also valid for single-component systems, since

$$\ln \frac{c}{c_{\sim}} = \ln \frac{p}{p_{\sim}} ,$$

where p is the vapor pressure. For a single-component system we replace Δc in (1) by $\Delta \rho$, the density difference, while D is the coefficient of self-diffusion.

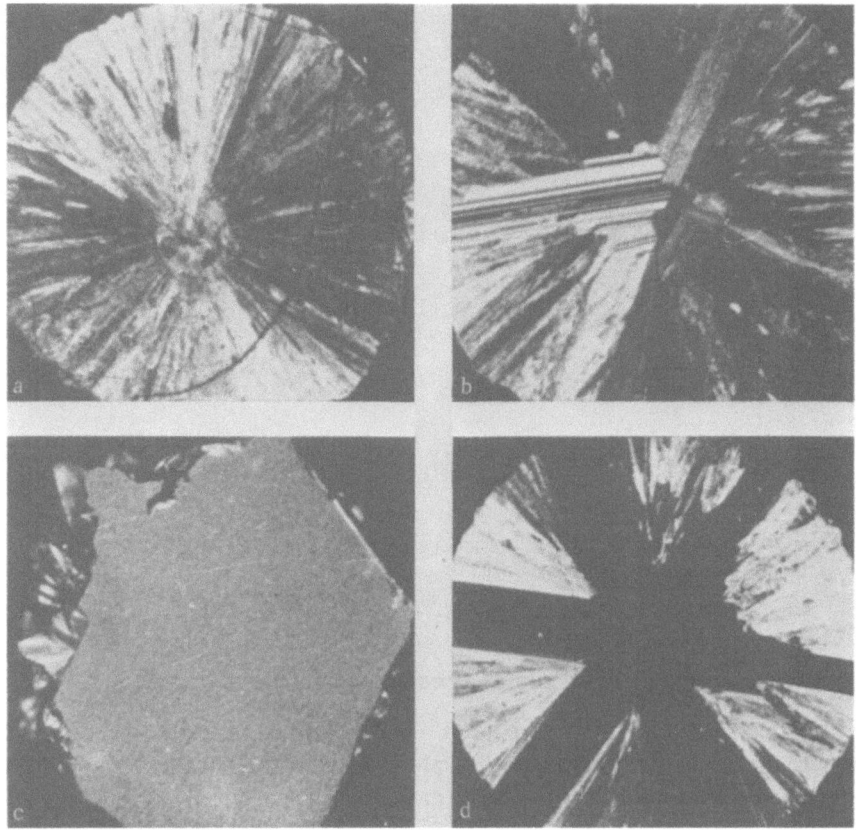

Fig. 2

greater. Second-order axes only begin to develop at the periphery of the rod.

Since the most promising procedure appeared to be that of producing a single-crystal bar, it was natural to use a single-crystal seed. We shall not describe the results of the numerous experiments associated with peculiarities in the generation of the centers, the effect of substrate structure, the formation of defects, and the special crystallographic aspects of the nucleation of silicon, since this would fall outside the scope of the present article.

We shall simply consider the data directly relating to the principal consequence of the theory in question. Single-crystal growth occurred under optimum conditions very similar to those calculated. In view of the fact that a large number of rods were grown simultaneously by reduction in the apparatus used for the experiments, strict control could naturally only be exerted over the temperature of the rods. The silicon concentration deviated from the mean value not only in different rods but also in different parts of the surface of one and the same rod.

Hence the more or less perfect sections of rod (Fig. 2c) were accompanied by other less perfect regions. These regions consisted of monolithic plates of different thicknesses growing from the seed and bordering the gas phase predominantly with faces having a low surface tension, i.e., the (111). The gaps between the strictly directional plates were occupied by plates oriented at random, which were formed at later stages of growth.

If we consider that single-crystal growth (radius of curvature nearly infinite) corresponds to a temperature of 1150°C and that the ratio of chloride to hydrogen is $\frac{1}{30}$, we obtain from Eq.(8) [here $(\Delta TQ/2RT_ST) - 1 = 0$]

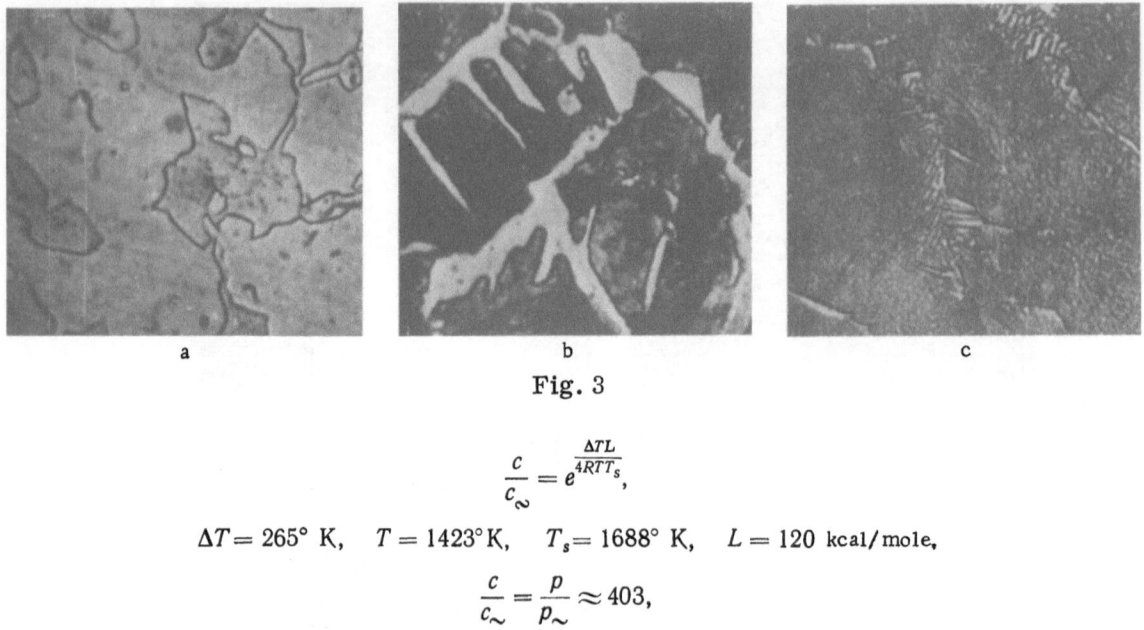

a   b   c

Fig. 3

$$\frac{c}{c_\infty} = e^{\frac{\Delta T L}{4RTT_s}},$$

$\Delta T = 265°$ K, $T = 1423°$ K, $T_s = 1688°$ K, $L = 120$ kcal/mole,

$$\frac{c}{c_\sim} = \frac{p}{p_\sim} \approx 403,$$

where p and $p_\sim$ are the corresponding partial pressures of silicon vapor.

If we consider that for $T_s = 1688°$K, $p_\sim \simeq 5 \cdot 10^{-5}$ mm Hg, then the partial pressure of silicon vapor over the sample for growth with a plane boundary should be $p \simeq 0.02$ mm Hg, which is very near the conditions of the experiment.

A change in concentration may produce a deviation from the equilibrium condition and correspondingly change the size of the realm of crystallization, then r_m will fall, leading to the formation of plates (Fig. 2d).

Unfortunately, establishing the true values of vapor pressure over various parts of the sample under conditions of a production experiment involves almost insuperable difficulties; hence, the formulas simply act as a qualitative aid for developing a correct technology for growing silicon of an assigned quality.

It is interesting to verify the theoretical data presented by reference to experimental results on the $\gamma - \alpha$ transformation of hypoeutectoid steel.

Let us determine the supercooling for which the ferrite crystals separating from the austenite will have a plane surface of separation from the austenite.

For this purpose we determine ΔT from condition (7) and substitute the corresponding values for steel containing 0.6% C

$$\frac{\Delta T}{T} = \frac{2RT_s}{Q} = \frac{2 \cdot 2 \cdot 823}{32000} \simeq 0.1,$$

i.e., $(T_s - T)/T = 0.1$. Therefore, $T = T_s/1.1 = 746°$K, and, hence, $\Delta T = 77°$K. Thus, starting from supercoolings greater than 70°, the ferrite crystals should assume ramified form. Naturally it is difficult to determine the plane-boundary shape experimentally, since this corresponds to a narrow temperature range. This is all the more difficult to do in the case of thick samples, for which the supercooling is not established at the same time over the whole volume of the samples at such a high temperature.

We studied the structure of hypoeutectoid steels of the U-4 and U-6 types at different supercoolings. We prepared samples $3 \times 3 \times 6$ mm in size, held these for half an hour in a furnace at a temperature of 950°C, and immediately transferred them to a salt bath in which the required isothermal-holding temperature was maintained.

Each of the samples was held in this bath at a strictly maintained temperature for 15 min and then water-quenched.

Figure 3 shows the resultant characteristic structures of U-6 steel obtained for different supercoolings. The structure of the steel corresponding to a supercooling of 50° is shown in Fig. 3a. We see from the figure that on growing from the boundary into the depths of the austenite grain the ferrite grains have a predominantly circular form. For a supercooling of 100° (Fig. 3b), however, the ferrite grains start branching, and almost everywhere their growth into the austenite grain takes place in the form of thickish needles (Fig. 3b). For a still greater supercooling (150°) the thickness of the needles becomes smaller (Fig. 3c).

Different structures appear for different treatments of the matrix. On analyzing the experimental data we may conclude that the transformation from the rounded to acicular (lamellar) form occurs in the region of 70°.

The growth rate of the ferrite grain in hypoeutectoid steel was determined in [3]. By comparing the experimental data with those corresponding to diffusion theories of crystal growth, in which no allowance was made for the realm of crystallization, we find a discrepancy which is especially substantial for large supercoolings and in the latter stages of growth. This discrepancy could no doubt be eliminated if the size of the realm of crystallization could be taken into account.

In conclusion we note a particular aspect of the martensite transformation. In [2] we determined the surface tension at the martensite−austenite interface by means of a growth-rate formula similar to formula (1) for the case of the diffusionless transformation. Here we assumed that the growth rate of the martensite crystals was entirely determined by the rate at which the heat of crystallization was carried away from the front of the growing crystallite. The activation energy may be determined from formula (7):

$$Q = \frac{2RTT_s}{\Delta T}.$$

If we consider that the equilibrium temperature for the austenite−martensite system is close to 500°K, then for a supercooling equal to 300°K the activation energy is approximately equal to 1200 cal, which is extremely close to the value of activation energy determined by Kurdyumov.

Literature Cited

1. Salli, I. V., Zh. Éksperim. i Teor. Fiz., 25 : 2 (1953).
2. Salli, I. V., Physical Basis of Structure Formation in Alloys. Metallurgizdat, Moscow (1963).
3. Salli, I. V., and Finagina, E. V., In: Scientific Reports of Dnepropetrovsk State University. Vol. 45. Dnepropetrovsk (1959).

STRUCTURE AND CRYSTALLIZATION MECHANISM
OF SELENIUM SPHERULITES

I. E. Bolotov and E. A. Murav'ev
Ural Scientific-Research Institute of Ferrous Metallurgy

In the majority of cases crystallization consists of the generation and growth of crystals having a definite lattice orientation, i.e., single crystals. However, another very frequent case is that in which textured polycrystalline aggregates grow out from a single center; these aggregates have a rounded form and are called spherulites. The elucidation of the mechanism underlying the development of such aggregates is one of the most important and difficult problems in the theory of crystal growth.

Selenium [1] is one of the substances capable of crystallizing in the form of spherulites. In the crystal lattice of selenium the atoms are arranged in the form of long spiral chains similar to large molecules. Hence selenium may serve as a model for the study of spherulite formation over a wide class of high-molecular substances, in which spherulite formation usually takes place on crystallization [2]. Our present problem was to study the formation and structure of selenium spherulites.

Method

Revelation of the initial stages of selenium spherulite formation is facilitated by the fact that the spherulites may be obtained on crystallization of thin films (about 1000 Å thick) of amorphous selenium [3]. Films of such thickness may be subjected to direct electron-microscope and microdiffraction study. The crystallization of amorphous selenium takes place at a temperature of about 100°C. The initial stages in the growth of the spherulites may be observed by continuous heating.

The low temperature required for the growth of the spherulites enabled us to observe this process directly under the optical microscope. For this purpose we used selenium (99.7% Se) films obtained by vacuum evaporation onto freshly cleaved mica. The kinetics of spherulite growth were observed by means of a motion-picture camera mounted in the MIM-7 microscope in place of the ocular.

For electron-microscope observation the selenium film was strengthened with a vacuum-evaporated carbon film. Then the mica was placed on the surface of water in such a way that it floated with the selenium and carbon films on top. The selenium film was separated from the mica and remained on the surface of the water. Then the two films were caught up and placed on the copper-grid specimen holders of the electron microscope.

Fig. 1. Optical micrographs of the same spherulite at different
stages of growth. ×340. Polarized light, crossed Nicols.

The electron-microscope and microdiffraction studies were carried out with an electron microscope of the UÉMV-100 type. There were some difficulties in view of the fact that evaporation of the selenium occurred under ordinary beam intensities. The samples were therefore studied with a reduced beam intensity. It was found that the heating of the film due to the electron beam was sufficient for the development and growth of the crystals. By carefully raising the beam intensity, the growth of the crystals could be observed directly in the electron microscope.

Experimental Data

The optical micrographs of selenium spherulites in polarized light (Fig. 1) show radial structure of a kind typifying the spherulites of a number of substances.

By means of electron microdiffraction we determined the orientation of the lattice in individual large spherulites. We found that the C axis of the hexagonal selenium lattice was approximately perpendicular to the radius of the spherulite, so that the directions of the spiral chains of the selenium "molecules" were concentric. This agrees with data obtained in [1] for three-dimensional spherulites crystallizing from molten selenium, and also with the general laws relating to the spherulites of high-molecular substances, viz., that the direction of the molecule is perpendicular to the radius of the spherulite.

Study of crystal growth under the optical microscope shows that in the initial stages the spherulites are not rounded crystals. Figure 1 shows the same crystal at different stages of growth. We see that the regular circular form is only reached in the later stages of growth. Initially the crystal has an elliptic or oval form and shows no regular radial structure in polarized light.

Detailed study of this kind of nucleus (and still smaller ones) was carried out in the electron microscope. The finest nuclei (up to about 0.5 μ) have an irregular oblong form and (as indicated by electron-diffraction pictures) constitute perfect single crystals (Fig. 2). The C axis (direction of the spiral chains of "molecules') coincides with the plane of the crystals. The crystals are drawn out in a direction perpendicular to the C axis. In the majority of cases the electron-diffraction zone axis is the [010], i.e., one of the [1210] prismatic planes of the second kind coincides with the plane of the crystal. As size increases (from one to several microns) the crystals take on the form of distorted (flattened along the small axis) ellipses (Fig. 3).

Electron microdiffraction shows that the structure of the crystals deviates from the single-crystal form more and more as size increases. This may be judged from the azimuthal spreading of the point reflections, which are transformed into arcs. This corresponds to

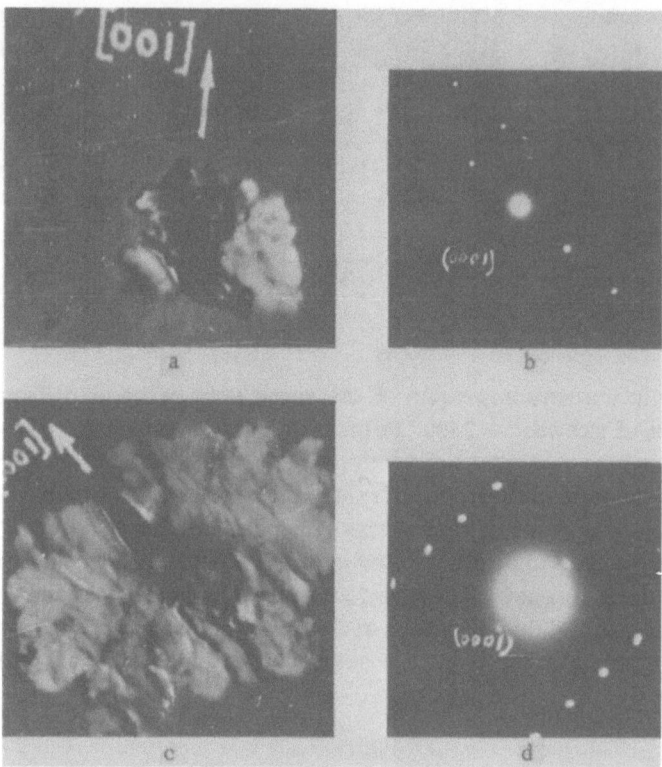

Fig. 2. Single-crystal nuclei of selenium spherulites: a,c) electron micrographs, × 30,000 (direction of the C axis of the hexagonal selenium cell indicated by the arrow); b, d) electron-diffraction
photographs of the corresponding crystals.

bending of the crystal lattice such that the C axis (direction of the chains of "molecules") ro-
tates slightly around an average position, remaining in the plane of the crystal. The rotation
of the lattice relative to the original position (C axis perpendicular to the major axis of the ellipse) increases on moving away from the major axis, and the greatest spreading of the reflections occurs near the poles of the minor axis of the ellipse.

The electron beam often has the effect of etching the crystal, the direction of the etch
figures (lines) coinciding with that of the C axis (as confirmed by electron microdiffraction examination of different parts of the same crystal, using a 1-μ diaphragm). This enables us to
follow the variation in orientation within the crystal very clearly (Fig. 4).

The direction of the C axis in a larger crystal may be seen in Fig. 5. Here the circles
indicate the points at which the orientation was determined and the lines show the direction of
the C axis. This crystal already has a texture similar to that of a spherulite. Although we can
still see the position of the minor axis of the ellipse, the point of emergence of which on the
surface corresponds to the maximum deviation of the orientation from the correct position (the
C axis makes a maximum angle with the outer surface of the crystal), this deviation is nevertheless smaller than in the case of the smaller crystals. The variation of orientation in crystals of different sizes is shown schematically in Fig. 6.

We see from the electron micrographs that the boundary of the crystal is not smooth but
has many projections reminiscent of dendrites or teeth (see Figs. 2 and 3). Such teeth are
especially clearly visible in crystals grown in the electron microscope as a result of heating

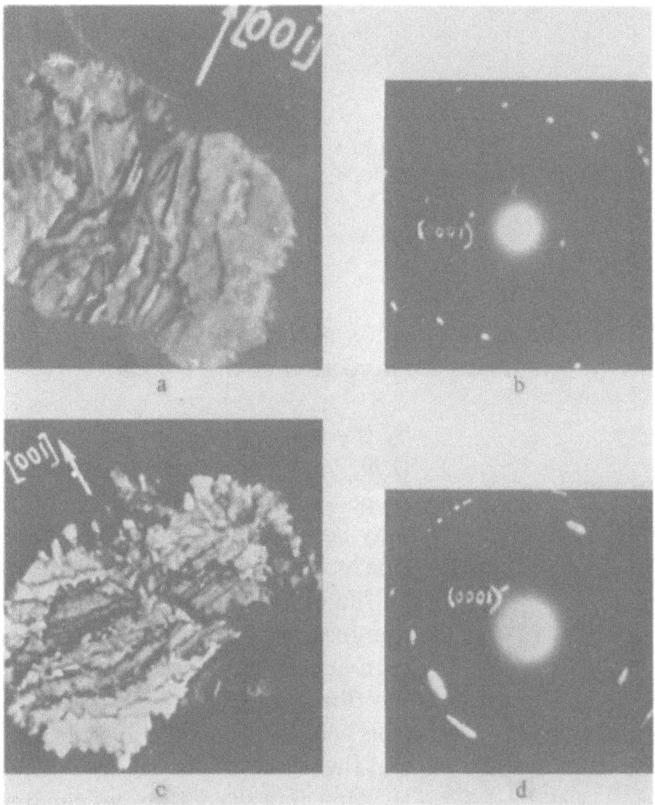

Fig. 3. Further stages in the growth of nuclei. Onset of departure from single-crystal structure. a,c) Electron micrographs, ×15,000 and 10,000; b, d) electron-diffraction photographs.

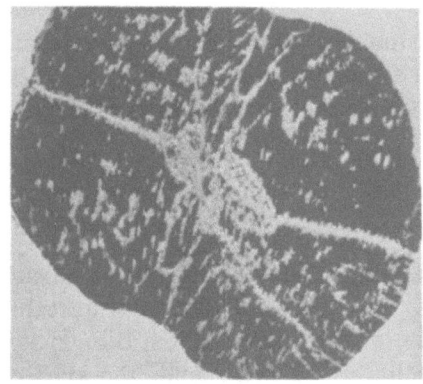

Fig. 4. Etch figures in an Se crystal (lines) formed under the electron microscope and showing the direction of the C axis. ×8000.

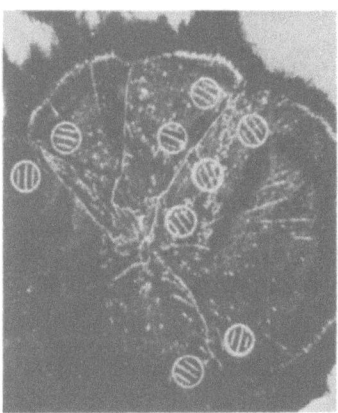

Fig. 5. A nearly formed spherulite, × 6000. Circles indicate the parts in which the lattice orientation was determined (to scale) and the lines show the direction of the C axis.

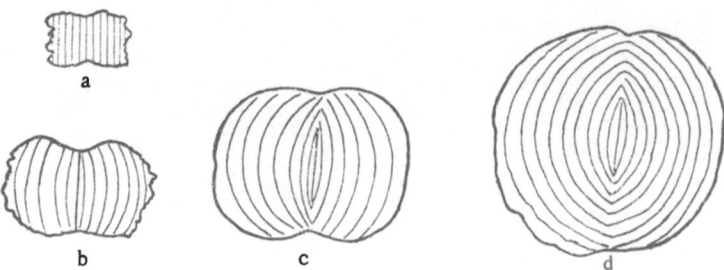

Fig. 6. Structure of Se crystals at various stages of growth. Lines
indicate the direction of the C axis.

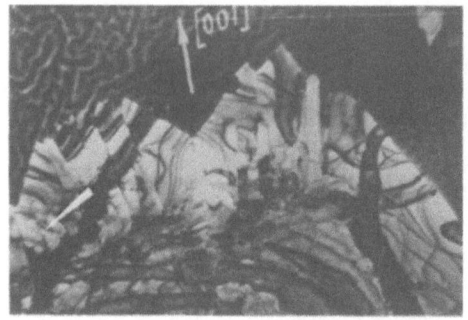

Fig. 7. Teeth on the crystallization
front and fibrous structure in an Se
crystal grown in the electron microscope. × 12,000.

by the electron beam, i.e., without a mica substrate
(Fig. 7). The existence of teeth on the crystallization
front leads to the formation of a fibrous structure, as
may be seen in Fig. 7. The boundaries between the
fibers constitute the trace left by the concave angle
as the crystallization front moves. Experiments with
the growth of crystals under the influence of the electron beam enabled us to follow the development of
the fibrous structure in detail. We noticed a connection between the velocity of the crystallization front
and the width of the fibers. The velocity of crystallization in a direction perpendicular to the C axis is
usually several times greater than in the direction of
the C axis. The width of the fibrils in the first direction is smaller than in the second. The higher rate
of crystal growth in the direction perpendicular to
the C axis explains the formation of nuclei drawn out
in this direction.

Discussion of Results

The electron-microscope and electron microdiffraction data clearly indicate that each
spherulite develops from one single crystal. One therefore wonders why it is that the lattice
in the single-crystal nucleus gradually bends during growth, so that the single crystal finally
turns into a textured spherulite.

Plane single-crystal nuclei are drawn out in a direction perpendicular to the C axis, this
axis being the most closely packed.

Hence the most developed faces in the nucleus are the side faces having a comparatively
low reticular density and high surface energy. Such crystals will be energetically unstable.
Owing to the fact that different surface-tension forces act along different faces, the crystal
will be deformed under the influence of these forces in such a way as to reduce the area of the
face having the maximum surface tension. The internal energy of the crystal is increased as
a result of the consequent distortions in the crystal lattice, and is compensated by a reduction
in the total surface energy. Thus, the development of bending in the crystal lattice may be regarded as a result of a fall in the surface energy of the crystal.

The literature contains attempts at explaining the formation of spherulites in other ways.
The authors of [1,4] in particular suggest that in polymers the formation of a spherulite is due
to ordering (owing to volume change in the course of crystallization) in the arrangement of

noncrystallized parts of the molecules projecting from the crystal. If crystallization involves an increase in volume, then the projecting parts of the molecule in fact take up such a position that the surrounding amorphous medium is "prepared" for the texture of a spherulite. In the case of selenium, however, crystallization is accompanied by a reduction in volume. Hence, if the "self-orientation" of the molecules had any significance, we should obtain a texture in which the molecules were arranged radially, contrary to fact. It is also hard to explain the formation of a spherulite as a result of the splitting of a single-crystal nucleus under the influence of stresses arising on crystallization [6]. These stresses would have to be mainly directed in such a way as to stretch the crystal along its major axis (since as a result of the close packing of the molecules it is in this direction that the maximum reduction in dimensions occurs), whereas, in order to obtain the observed texture the forces would have to be directed perpendicular to this axis.

At the same time it is easy to see that, as a result of the contraction of the face perpendicular to the C axis under the influence of the large surface-tension forces existing on this face, lattice curvature of exactly the right kind to give the observed texture will take place.

It is shown in a number of papers [2, 5] that spherulites are formed when the dimensions of the crystal are equal to or greater than the quantity $\delta = D/g$ (where D is the diffusion coefficient of the atoms of the crystallizing material in the medium and g is the crystal growth rate), the width of the fibers formed on crystallization being δ. The authors consider that in this case "noncrystallographic splitting" takes place and results in the formation of a spherulite; they do not, however, disclose the nature of this splitting.

We may suppose that the physical nature of the "noncrystallographic splitting" is in fact splitting under the influence of surface-tension forces, different on different faces. It then becomes clear why splitting starts appearing when the dimensions of the crystal are greater than δ, since, in this case, the crystal consists of fibers and contains weakened, imperfect sections of lattice (boundaries between the fibers), along which splitting may take place.

It should be noted that the fibers occurring in selenium crystals are the finer the greater the crystallization velocity, i.e., as in the case of polymers, their formation is due to the existence of an impurity-enriched layer (impurity in the wide sense of the word) in front of the crystallization front. Such "impurities" include, for example, "poorly crystallized" selenium molecules.

Conclusions

In selenium spherulites the spiral chains of "molecules" parallel to the C axis of the hexagonal cell have approximately concentric directions.

A spherulite develops from a single-crystal nucleus drawn out in a direction perpendicular to the C axis lying in the plane of the crystal. As a result of this, the nucleus in question has a high surface energy and becomes unstable.

During the growth of the single-crystal nucleus the lattice becomes curved, reducing the area of the face perpendicular to the C axis and producing spherulite texture.

The curving of the lattice is apparently a result of surface-tension forces acting on different faces.

Literature Cited

1.   Eninger, H., Z. Kryst., 115, 3/4 (1961).
2.   Keith, H., and Padden, F., J. Appl. Phys., 34(8): 2409 (1963).

3.   Fourie, D., and Van der Walt, C., Z. Phys., 169(2) : 326 (1962).
4.   Gruenwald, G., J. Polymer Sci., 61(172) : 381 (1962).
5.   Keith, H., J. Polymer Sci., 2(P.A. 10) : 4339 (1964).
6.   Palatnik, L. S., and Kosevich, M. V., In: Growth of Crystals, Vol. 3. Izd. Akad. Nauk
      SSSR, Moscow (1961). [English translation: Consultants Bureau, New York (1962).]

GROWTH OF CRYSTALS DURING THE
SOLIDIFICATION OF EUTECTIC ALLOYS

Yu. N. Taran

Dneptropetrovsk Institute of Ferrous Metallurgy

Advances recently achieved in the technology of growing metallic single crystals from the melt, together with recent investigations into the mechanism underlying the growth of single crystals, have created a foundation for detailed study of the mechanism and kinetics of complex forms of multiphase crystallization such as the eutectic transformation in alloys. As in the study of single-phase crystallization, experiments on the directional solidification of eutectic alloys and the microanalysis of decanted samples have proved extremely fruitful. A tendency toward the formation of lamellar eutectic structures on increasing the purity of alloys and reducing the solidification rate has been established [1, 2, 3], and the development of a cellular front of eutectic crystallization has been studied [4, 5]. Preliminary attempts at a quantitative description of the process have been made; in particular, the interrelation between the differentiation of columns and the linear velocity of crystallization has been determined [6]. The theoretically established laws have been experimentally confirmed by studying low melting-point alloys and white [7] and grey [8] cast irons.

In theory and experiments on directional crystallization, however, one usually analyzes the stationary growth of the eutectic pile of phases rather than studying the generation of the eutectic columns and their idiomorphic forms. The micromorphological characteristics of eutectic structures are also often unreliable, since the structures are classified according to the microscopic form of single random sections or the shape of the surface of decanted samples. The results of this kind of analysis make up the main bulk of accumulated data on the morphology of eutectics. A common feature of eutectics in different alloys is the existence of a finely dispersed two-phase mixture in which some parts often appear as inclusions in a monolithic base. A widely accepted model of the eutectic column is accordingly a system of many fine crystals of one phase distributed in a monolithic matrix of another phase. In order to justify this kind of model, the authors of [9, 10] propose a mechanism based on the multiple generation of one of the phases on the growing surface of a single crystal of the other phase; this concept is even retained in recent papers [6, 11]. In accordance with the model it is a generally accepted practice to divide the eutectic phases into continuous and discontinuous classes.

Stereometric microanalysis of metal alloys does not support the model in question, but rather indicates the continuity of both phases in the eutectic column. It is shown in [12, 13] that the "point" cross sections of the discontinuous phase in low melting-point eutectics of

Sn−Pb, Cd−Zn, Sn−Cd, and Pb−Cd constitute sections of elongated fibers. A study of eutectics in Al−Mn and Al−Fe alloys [14] suggests that the phase described as the "dispersed" one is in fact a skeletal single crystal. The concept of the continuity of the phases in eutectic columns had also been developed in earlier papers [15, 16]. Let us consider the structure of the eutectics in the widely used and technologically important alloys of the iron−carbon system.

The austenite−graphite eutectic is the main structural constituent of grey cast irons. Microanalysis of samples of grey cast iron shows isolated sections of graphite plates or lamellas arranged in the solid solution. This view of the structure is due to the still widely held concept of the discontinuity of the graphite phase. Thus Chalmers considers the iron−carbon eutectic to be a good example of discontinuous eutectic structure, the formation of which requires the dense generation of one of the phases [17]. Tiller regards the austenite−graphite eutectic as anomalous, considering that the graphite plates are not only isolated but formed in the liquid in front of the crystallization front [6]. Meanwhile it was shown in [18] that the column of the austenite−graphite eutectic in ordinary grey cast iron consisted of a single graphite inclusion in an austenite matrix.

Under certain conditions (superheating of the melt, vacuum treatment, acceleration of cooling) a fine austenite−graphite eutectic containing so-called supercooled graphite is formed in grey cast iron. The fine, whorling sections of the supercooled graphite are an extremely convincing example of the discontinuity of the graphite phase. However, fine grinding [19] and etching of the samples [20] indicates that within the limits of a column the supercooled graphite really constitutes a single highly branched formation.

In grey cast irons treated with rare-earth metal additives, a finely differentiated austenite−graphite eutectic also develops [21, 22]; many authors describe the graphite in this as "point," "finely dispersed," etc. A spherulitic eutectic column in cerium cast iron quenched in the course of solidification is shown in Fig. 1a. Thanks to the dark troostite matrix, the column is easily visible on the background of fine ledeburite into which the liquid phase transforms on quenching. Point and fine-lamellar sections of graphite may be seen in unetched microsections of such columns (Fig. 1b). However, layer-by-layer analysis convinces one that, within the limits of the column, the graphite constitutes a single inclusion formed by the systematic branching of a central graphite nucleus. Figure 2 shows several successive sections of one column. In the section passing through the center (a), we clearly see the connection between the compact central nucleus and the radial branches. After removing a layer 0.5μ thick, the dimensions of the nucleus cross section become smaller (b). In addition to this, the extent of the cross sections of the radial plates becomes smaller (Fig. 1c, layer of 1μ removed). Later, point graphite appears in the place of the central nucleus (Fig. 1d, $1.5-\mu$ layer), and this partly transforms into the fine, whorled form (Fig. 1e, f; layers respectively 2.5 and 3.5μ). Comparison of numerous sections and empirical modeling of the column clearly confirm the continuity of the graphite phase.

The second most frequently occurring eutectic component of iron−carbon alloys is ledeburite, the austenite−cementite eutectic of white cast irons. Ledeburite represents a typical variety of eutectic of honeycomb structure. Under the microscope, the ledeburite columns most frequently appear to consist of ellipse-like sections of austenite distributed in a cementite matrix.

According to earlier investigations [23, 18], the crystallization of ledeburite starts with the development of cementite plates. The column grows along a cementite plate (longitudinal growth) and along the normal to the latter (transverse growth). It was supposed that the formation of the honeycomb structure took place during transverse growth by the thickening of the basal cementite plate, on the crystallization front of which new austenite inclusions were

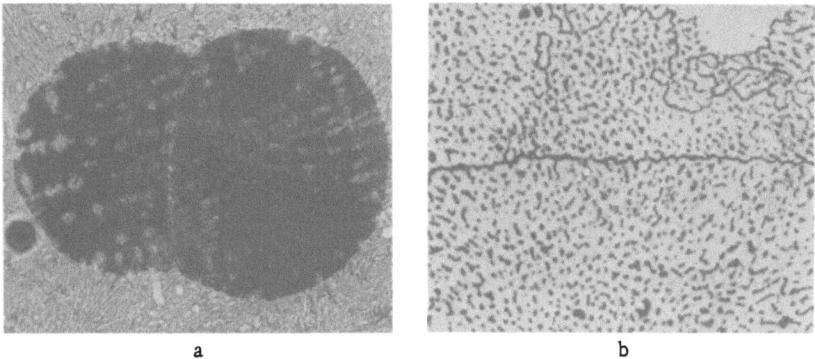

Fig. 1. Form of a eutectic column (a, ×50, etched with Nital) and section (b, ×600, not etched) of the graphite in a fine graphite eutectic.

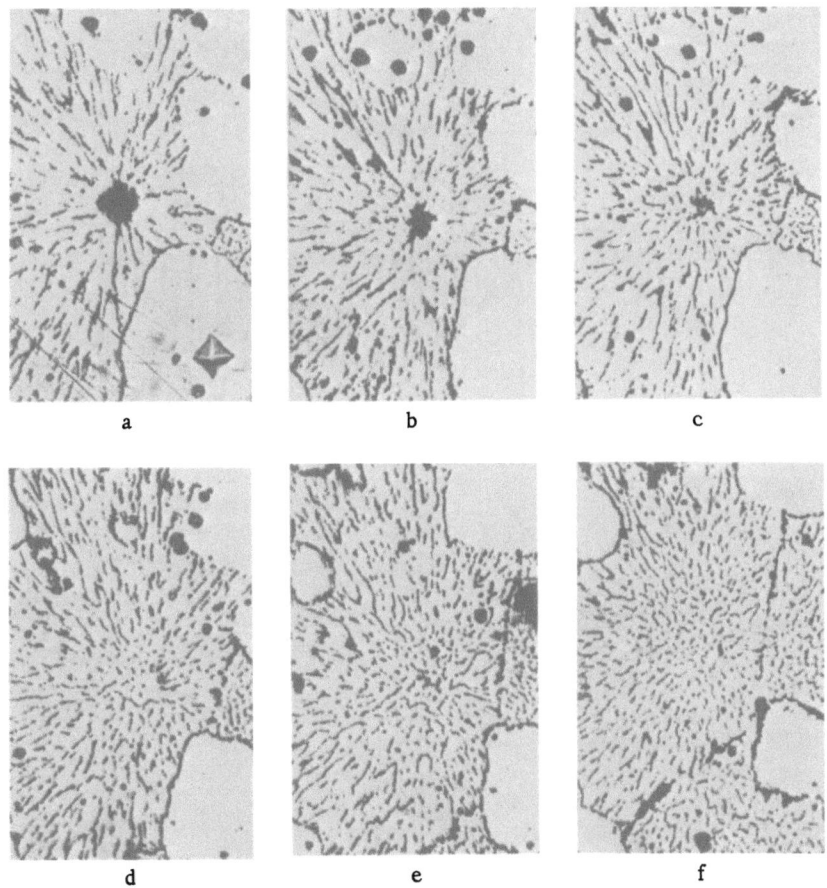

Fig. 2. Successive sections of a column of a fine graphite eutectic, × 600; not etched.

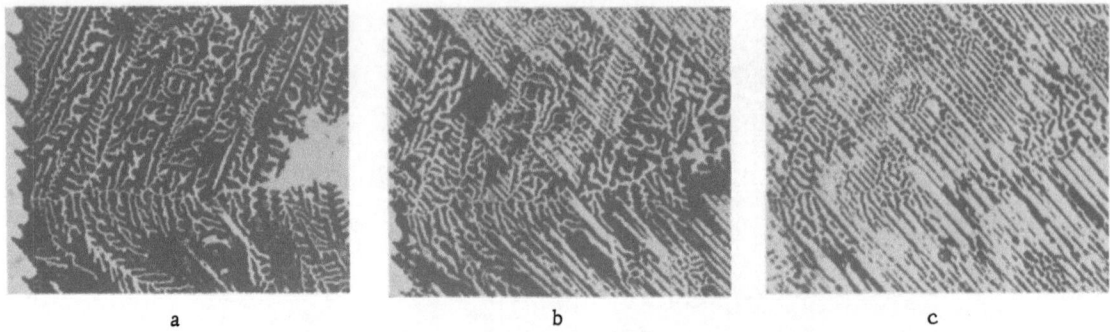

Fig. 3. Successive sections of the base of a ledeburite column, × 100; etched with Nital.

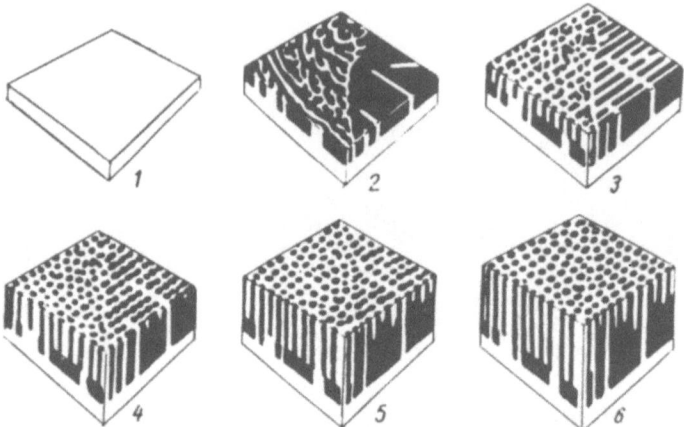

Fig. 4. Scheme representing the formation of a ledeburite (cementite light, austenite dark).

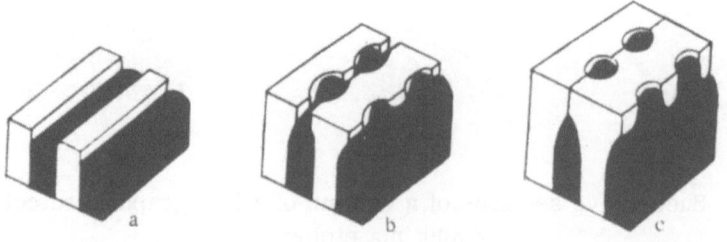

Fig. 5. Scheme representing the degeneration of the lamellar structure of ledeburite into rod structure.

generated, the earlier-grown ones dying out. In [24] the transverse growth of ledeburite was described as the formation of rod-like branches of a flat austenite dendrite, playing the part of a leading phase, cementite crystallizing in the gaps between this.

Detailed microanalysis showed that the leading phase in the ledeburite transformation is cementite. The characteristics of the growth of the cementite, due to its crystallographic nature, determine not only the shape of the ledeburite columns but also their most important micromorphological characteristics.

The base of the ledeburite column is a cementite plate having a block growth texture [25]. The second phase of the eutectic, austenite, is generated on the surface of the plate; this grows in the form of plane dendrites along the plate. As the dendrites grow, the cementite base plate grows in the spaces between them and their branches, forming a system of branched veins. The simultaneous growth of the austenite dendrites and the cementite veins constitutes the onset of the essentially eutectic crystallization. The microscopic picture of the initial layer of the column may be observed by finely repolishing a sample of a hypoeutectic alloy cleaved along the cementite base plate (Fig. 3a). In the left half of the picture we see the toothed edge of one branch of the base plate; this is thicker and still not fully polished.

Subsequent repolishing of the sample of Fig. 3a shows successive stages in the growth of the ledeburite column. From the cementite veins grow arrow-like piles of cementite (Fig. 3b). These are strictly oriented in accordance with the texture of the base plate. This is shown by x-ray diffraction and supported by the agreement between the directions of the toothed contours and piles. Between the cementite arrows grow plates of austenite, taking their origin from the branches of the plane dendrites. The rectilinearity and regular orientation of the cementite arrows indicates the leading role of the cementite in eutectic decomposition. Completion of the growth of the arrow-like piles leads to the formation of the finely differentiated eutectic structure.

Stereometric microanalysis of a large number of columns shows that at the initial stages of growth the column has basically a lamellar structure. Only at those parts in which the arrow-like piles of cementite are superimposed on the branched system of interdendritic veins are conditions created for the direct formation of rod structure. However, as the column develops, even the parts with lamellar structure are transformed into a rod-like eutectic, so that very soon after nucleation the whole ledeburite column acquires its characteristic honeycomb structure.

Figure 4 shows the main stages in the formation of the column: the formation of the basic cementite plate (1), the crystallization of plane austenite dendrites and cementite veins (2), the formation of arrow-like piles of cementite separated by austenite (3), and the subsequent development of rod structure (4-6). The left halves of the pictures represent the development of this structure on a base with thickly distributed veins and the right halves correspond to an austenite base. In the latter case the rod structure is formed as a result of the degeneration of the plate. The mechanism of this process may be represented (Fig. 5) by using data relating to the development of a cellular crystallization front [26, 27]. We see from Fig. 5a, in agreement with microanalysis data, that in the transverse growth of the column the cementite proceeds slightly in advance of the austenite plates. This promotes the enrichment of the liquid in front of the austenite with impurities and thus increases the concentration supercooling. Projections appear at the edges of the austenite plates (Fig. 5b), and this leads to the enrichment of the liquid separating them with carbon and to the growth of cementite at such points. The cementite arrows join up, establishing the separation of the austenite plates (Fig. 5c).

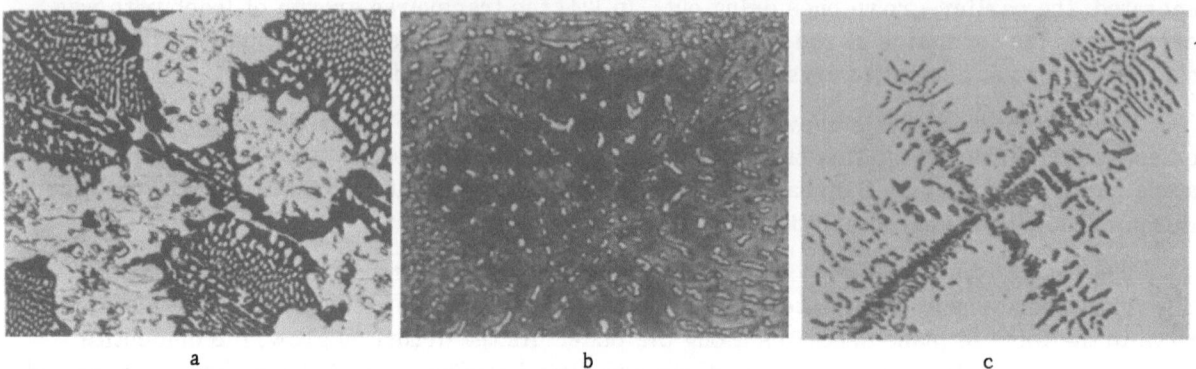

Fig. 6. Vanadium carbide eutectic in white cast iron (a, × 100, etched with sodium picrate, ledeburite dark; b, × 500, thermal etching; c, × 100, not etched).

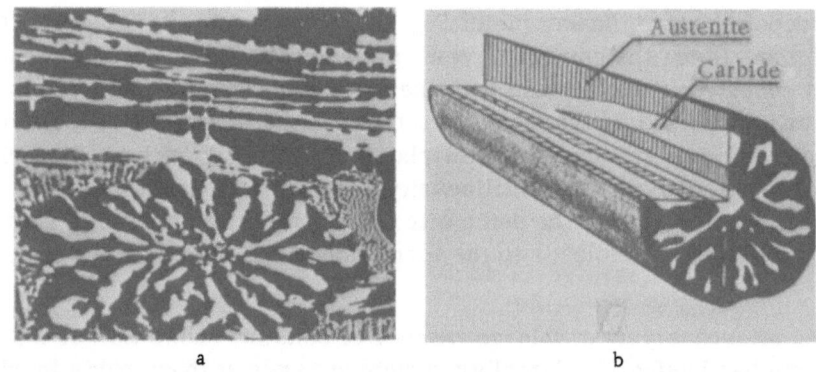

Fig. 7. Chromium carbide eutectic in white cast iron (× 500) after thermal etching.

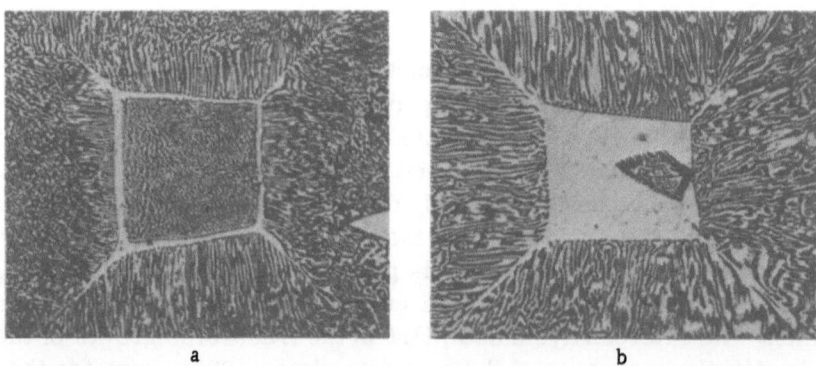

Fig. 8. Successive sections of a column in Bi−Sn alloy (× 200) etched with sodium picrate.

Thus the ledeburite column (or at any rate the substructural unit of the column, the cell based on one plane austenite dendrite) constitutes two crystals of eutectic phases growing in one another. We come to the same conclusion by analyzing the structure of other eutectics in iron−carbon alloys.

On alloying cast irons with vanadium, an austenite−vanadium carbide eutectic $\gamma + V_4C_3$ is formed. Increasing the vanadium content in cast iron containing 4% C from 2-10% leads to the gradual replacement of ledeburite by this eutectic. Figure 6a shows micrographs of an Fe−C−V alloy containing 6.22% V in which the ledeburite and the austenite−vanadium carbide eutectic occupy equal volumes. Just as the ledeburite, in microsection, appears to consist of a cementite base containing austenite inclusions, so also the austenite−vanadium carbide eutectic appears in the form of an austenite matrix containing fine vanadium carbide inclusions (Fig. 6b). However, layer-by-layer microanalysis of an individual column shows that the latter consists of a single branched V_4C_3 crystal included in the austenite. If the plane of the microsection passes through the center of the column, we can see that the eutectic vanadium carbide forms the dendritic skeleton of the column (Fig. 6c). The dendritic branches lead the solidification front. On these grows the eutectic austenite, also developing from a single nucleus. The single-crystal nature of the austenite is indicated by the similar orientation of the thin plates of secondary cementite at the boundaries of the column.

The structure of the austenite−chromium carbide eutectic formed in Fe−C−Cr alloys is analogous. Figure 7a shows a micrograph of cast iron containing 3.6% C and 13% Cr quenched in the course of solidification. The photograph represents longitudinal (top) and transverse (bottom) sections of two $\gamma + (Cr, Fe)_7C_3$ columns growing normal to one another. By considering a great deal of stereometric-microanalysis data and comparing the cross sections, we see that the austenite−chromium carbide eutectic develops on a basal rod of the carbide $(Cr, Fe)_7C_3$, which constitutes the axis of the column. The branches of the rod carbide grow mainly along the axis and develop less in a radial direction; in the gaps between these, crystallization of the eutectic austenite takes place. On the whole, the $\gamma + (Cr, Fe)_7C_3$ column has the form of a cylinder (or cone) and constitutes a crystal of trigonal carbide branching out in a single-crystal austenite matrix (Fig. 7b).

Summarizing our study of eutectics in cast irons, we may conclude that the eutectic column is a bicrystallite formation in which the two phases are both continuous and grow in one another. The same conclusion follows from the stereometric analysis of low melting-point alloys. For example, the eutectics in Bi−Sn and Bi−Pb alloys are analogs of the ledeburite in white cast irons. Bismuth growing in the form of rhombohedra with almost square faces [28] shows a tendency toward layer-like growth in the same way as cementite [29]. As in the case of austenite, the characteristic form of growth for the tin and lead in the melt is a three-dimensional dendrite. Figure 8a shows a column of Bi−Sn eutectic of honeycomb structure formed on a crystal of excess bismuth. Fine repolishing of the sample reveals the same successive stages in the development of the bicrystallite eutectic structure as occurred in ledeburite. Figure 8b, in particular, catches the instant at which a plane tin dendrite starts growing on the face of a basic bismuth crystal. An analog to the austenite−graphite eutectic of grey cast irons is the aluminum silicide eutectic in silumins, since there are good grounds for supposing [30] that the silicon constitutes a continuous branched phase within the limits of the column.

The results of the morphological study of eutectics in metal alloys tend to favor the model of eutectic-column formation proposed by Bochvar [31]. According to Bochvar, at the instant at which two separately generated crystals of eutectic phases join together, combined growth sets in; in the course of this, one of the phases leads the crystallization and creates the skeleton of the eutectic grain, while the second phase lays itself out in the interstices of the skeleton. It does indeed follow from the data presented that the crystallization of binary eutectics in metal alloys constitutes the cooperative growth of two crystals branching or splitting in accordance with their crystallographic nature. This process is characterized not by the specific individual growth of two crystals [31], but by the generation of a crystal of the "led" phase on a base crystal of the "leading" phase. In any particular system, only one phase stands out as a generator of the eutectic column, quite independently of the degree of eutectic composition of the melt. This is confirmed by the results of [32].

Common characteristics of leading (generating) phases include a higher melting point and anisotropy of growth resulting from the presence of directional bond forces in the crystal lattice. This latter circumstance explains the tendency of leading phases to splitting or "noncrystallographic" branching, which plays an important part in the formation of finely differentiated eutectic structures. As a result of the fine differentiation of the eutectics, short-range diffusion of the components in front of the crystallization front rapidly removes supersaturation of the liquid and thus eliminates the possibility of the multiple generation of phases at the crystallization front. Acceleration of cooling retards the diffusion processes but favors the further splitting of the leading phase, and the conditions for cooperative growth of the two crystals remain favorable.

In view of the model of the eutectic column developed in this paper (viz., a bicrystallite formation), it is reasonable to replace the existing definitions of eutectic phases as "continuous" and "dispersed" by the concepts of "matrix" and "branched" phases. These concepts are suitable for describing the ultimate structure of the column. We must not identify them with the concepts of the leading and led phases which characterize the dynamics of column growth. For example, in the crystallization of ledeburite, the leading phase is the cementite, which also constitutes the matrix phase of the eutectic, within which the branched austenite crystal is embedded. In the austenite—graphite eutectic the leading phase is the graphite, but the matrix phase of the column is formed by the led phase, the solid solution.

Literature Cited

1. Chadwick, G. A., J. Inst. Met., 91:298 (1963).
2. Yue, A. S., J. Inst. Met., 92:248 (1964).
3. Hunt, J. D., and Chilton, J. P., J. Inst. Met., 91:260 (1963).
4. Weart, H. W., and Mack, D. J., Trans. Met. Soc. AIME, 212:664 (1958).
5. Kraft, R. W., and Albright, D. L., Trans. Met. Soc. AIME, 221:95 (1961).
6. Tiller, W. A., In: Liquid Metals and Their Solidification [Russian translation]. Metallurgizdat, Moscow (1962).
7. Wilkinson, M. P., and Hellawel, A., BCIRA J., 11(4):439 (1963).
8. Lakeland, K. D., BCIRA J., 12(5):634 (1964).
9. Grechnyi, Ya. V., Zh. Fiz. Khim., 30(1):184 (1956).
10. Grechnyi, Ya. V., Zh. Fiz. Khim., 30(2):392 (1956).
11. Cooksey, D. J., et al., Phil. Mag., 10(107):745 (1964).
12. Nikonova, V. V., and Bartenev, G. M., Izv. Akad. Nauk SSSR, Met. i Toplivo, Moscow, 1:100 (1962).
13. Nikonova, V. V., and Bartenev, G. M., In: Crystallization and Phase Transformations. Izd. Akad. Nauk BelorussSSR, Minsk (1962).
14. Parkhutik, P. A., Fiz. Met. i Metalloved., 18(2):307 (1964).
15. Avakyan, S. V., and Lashko, N. F., Dokl. Akad. Nauk SSSR, 6:847 (1949).
16. Avakyan, S. V., Kislyakova, E. N., and Lashko, N. F., Zh. Fiz. Khim., 9:1057 (1950).
17. Chalmers, B., Physical Metallurgy [Russian translation]. Metallurgizdat, Moscow (1963).
18. Bunin, K. P., Ivantsov, G. I., and Malinochka, Ya. N., Structure of Cast Iron. Mashgiz, Moscow (1952).
19. Bunin, K. P., Malinochka, Ya. N., and Fedorova, S. A., Liteinoe Proizv., 9:21 (1953).
20. Morrogh, H., Brit. Foundryman, 5:53 (1960).
21. Gorev, K. V., Proskurina, Z. N., and Shevchuk, L. A., In: Mechanical Properties of Cast Metal. Izd. Akad. Nauk SSSR, Moscow (1963).

22. Taran, Yu. N., et al., Izv. Akad. Nauk SSSR, Metally, 3 : 131 (1965).
23. Bunin, K. P., Iron−Carbon Alloys. Mashgiz, Moscow (1949).
24. Hillert, M., and Steinhauser, H., Jernkontorets Ann., 144(7) : 520 (1960).
25. Novik, V. I., and Taran, Yu. N., This collection, p. 202.
26. Tiller, W. A., et al., Acta Met., 1 : 428 (1953).
27. Chadwick, G. A., Progr. Mat. Sci., 12(2) : 100 (1963).
28. Kuznetsov, V. D., Crystals and Crystallization. GITTL, Moscow (1954).
29. Tiller, W. A., J. Appl. Phys., 29 : 611 (1959).
30. Bell, J. A. E., and Winegard, W. C., J. Inst. Met., 130 :318 (1965).
31. Bochvar, A. A., Mechanism and Kinetics of the Crystallization of Eutectic Alloys. ONTI, Moscow (1936).
32. Sundquist, B. E., Bruscato, R., and Mondolfo, L. F., J. Inst. Met., 91 : 204 (1963).

THEORETICAL DETERMINATION OF TEMPERATURES IN GROWING CRYSTALS OF AXIALLY SYMMETRIC FORM

B. I. Birman

All-Union Scientific-Research Institute of Single Crystals

When growing single crystals from the melt, the thermal conditions have a considerable effect on the growth processes and quality of the crystals. These conditions are primarily determined by the shape and size of the crystals, the character and intensity of heat elimination from the outer surfaces of the crystal, the character and intensity of heat transfer in the melt, the velocity of drawing, and the thermophysical properties of the liquid and solid phases. In crystallization from many-component systems, the thermal conditions are also affected by diffusion processes in the melt and the variation in the temperature at the phase boundary with the concentrations of the components.

The thermal conditions in turn determine the shape of the crystallization front, the growth rate and the thermal gradient at the phase boundary, and also the thermal stresses in the crystals during growth. A knowledge of these parameters is necessary in order to obtain perfect single crystals both on the micro- and macroscopic scales.

At the present time various methods of directional crystallization are widely used for producing single crystals from the melt [1, 2]. Of these the most widespread are those of Kyropulos [3] and Czochralski [4], in which crystallization takes place from top to bottom, and those of Stockbarger [5] and Stöber [6], in which it takes place from bottom to top. In all these methods the heat is carried away predominantly through the solidified crystal, and the temperature field in the crystal may be considered to have axial symmetry. The crystallization front is not always plane, but on a macroscopic scale constitutes a surface of revolution with a curvature varying over the radius and also with time. Usually we distinguish plane, convex, and concave forms of the crystallization front, although in the practical growth of single crystals more complex cases may occur.

It is shown in [7, 8] that the most favorable shapes of crystallization front for the growth of single crystals are the plane and convex types; with a concave front, it is more probable that parasitic nuclei will be generated on the side of the crystal, and there is also no possibility of the tapering out of boundaries between twins or blocks with large angular disorientations, so that polycrystalline growth is favored.

On the other hand, the existence of a plane crystallization front at all stages of growth is associated with uniform heat elimination from the crystal over its end surfaces, uniform heat transfer in the melt, and the absence of side heat leakage, since the existence of the latter to any appreciable extent tends to make the crystallization front concave. Hence, an analysis of the effect of thermal conditions on the shape of the crystallization front is certainly of great interest.

The problem of determining the laws governing the motion of the crystallization front in the associated temperature fields has so far only been solved for bodies of the simplest shapes [9]: half-plane [10, 11], sphere [12], cylinder [13, 14], paraboloid of revolution [15], and ellipsoid of revolution [16].

In this paper we shall solve a similar problem for the general case of the crystallization of axially symmetric bodies of revolution.

The following assumptions will be made:

1. The temperature in the solid phase $T(r, t)$, depending only on the axial coordinate z, the radial coordinate r, and the time t, is determined from the solution of the heat-conduction equation

$$\frac{\partial^2 T}{\partial r^2} + \frac{1}{r}\frac{\partial T}{\partial r} + \frac{\partial^2 T}{\partial z^2} = \frac{1}{\chi}\frac{\partial T}{\partial t},$$

(1)

where χ is the thermal diffusivity.

2. All the thermophysical parameters of the substance (thermal conductivity, density, specific gravity) are considered independent of temperature. Radiant heat transfer inside the solid phase and melt is assumed negligibly small.

3. The crystallization front $z = y(r, t)$ is supposed to be a certain function of time which may be represented in the form of a series of even powers of r:

$$y(r, t) = a(t) + b(t)r^2 + c(t)r^4 + \ldots .$$

(2)

The function $a(t)$ characterizes the displacement of the crystallization front along the growth axis (for $r = 0$). The derivative of a with respect to time $v_0(t) = da/dt$ represents the velocity of crystallization along the growth axis.

The functions $b(t)$, $c(t)$, etc., characterize the curvature of the crystallization front and its variation with time. The derivatives of these functions: $v_1(t) = db/dt$, $v_2(t) = dc/dt$, etc., characterize the rate of change in the shape of the crystallization front.

4. The intensity of heat transfer in the melt is supposed axially symmetric, and the thermal flux in the melt at the crystallization front is considered to be a known function of time expressible as a series in even powers of r:

$$q_l(r, t) = q_0(t) + q_1(t)r^2 + q_2(t)r^4 + \ldots .$$

(3)

On the basis of the thermal balance at the crystallization front, the thermal flux along the normal to the crystallization front on the side of the solid phase is

$$q_f(t) = Q\gamma \frac{dy/dt}{\sqrt{1 + (dy/dr)^2}} + q_l(r, t).$$

(4)

This is also a known function, since $y(r, t)$ and $q_l(r, t)$ are known.

5. The temperature at the crystallization front is supposed to be a known function of time and radial coordinate:

$$T_f(r, t) = T_{f0}(t) + T_{f1}(t)r^2 + T_{f2}(t)r^4 + \ldots .$$

(5)

6. The heat outflow over the cooled surfaces of the crystal (the end at z = 0 and the side at r = R, where R is the outer radius of the crystal, are considered unknown, since the temperatures at these surfaces are also unknown.

On the basis of assumptions 1-6, the boundary conditions for Eq. (1) may be written in the form

$$T(r, z, t)|_{z=y(r,t)} = T_f(r, t),\tag{6}$$

$$\lambda \frac{\partial T}{\partial n}\bigg|_{z=y(r,t)} = q_f(r, t).\tag{7}$$

Condition (7) may be written in another form by using the well-known geometrical relations for the derivatives along the normal in terms of the derivatives with respect to the coordinates:

$$\left[\frac{\partial T}{\partial z} - \frac{dy}{dr}\frac{\partial T}{\partial r}\right]_{z=y(r,t)} = f(r, t),\tag{8}$$

where

$$f(r, t) = \frac{q_f(r, t)}{\lambda}\sqrt{1 + (dy/dr)^2} = \frac{Q\gamma}{\lambda}\frac{dy}{dt} + \frac{q_1(r, t)}{\lambda}\sqrt{1 + (dy/dr)^2}.\tag{9}$$

The solution of Eq. (1) with boundary conditions (6) and (8) will be sought by the method of successive approximations in the following form [17]:

$$T(r, z, t) = \sum_{n=0}^{\infty} T_n(r, z, t).\tag{10}$$

As zero approximation we take the solution of the stationary problem of heat conduction

$$\frac{\partial^2 T_0}{\partial r^2} + \frac{1}{r}\frac{\partial T_0}{\partial r} + \frac{\partial^2 T_0}{\partial z^2} = 0,\tag{11}$$

and in order to obtain the successive approximations we solve the recurrent differential equations

$$\frac{\partial^2 T_n}{\partial r^2} + \frac{1}{r}\frac{\partial T_n}{\partial r} + \frac{\partial^2 T_n}{\partial z^2} = \frac{1}{\chi}\frac{\partial T_{n-1}}{\partial t}.\tag{12}$$

For the zero approximation the boundary conditions are given in the form

$$T_0(r, z, t)|_{z=y(r,t)} = T_f(r, t) = \sum_{s=0}^{\infty} T_s r^{2s},\tag{13}$$

$$\left[\frac{\partial T_0}{\partial z} - \frac{dy}{dr}\frac{\partial T_0}{\partial r}\right]_{z=y(r,t)} = f(r, t) = \sum_{s=0}^{\infty} f_s r^{2s},\tag{14}$$

and for subsequent approximations in the form

$$T_n(r, z, t)|_{z=y(r,t)} = 0,\tag{15}$$

$$\left[\frac{\partial T_n}{\partial z} - \frac{dy}{dr}\frac{\partial T_n}{\partial r}\right]_{z=y(r,t)} = 0.\tag{16}$$

We shall seek the functions $T_n(r, z, t)$ satisfying equations (11) for $n = 0$ and (12) for $n > 0$ in the form of expansions in double series of even powers of r and even and odd powers of z

$$T_n(r, z, t) = \sum_{j=0}^{\infty} \sum_{l=0}^{\infty} u_{n,j,l} r^{2j} z^{2l} + v_{n,j,l} r^{2j} z^{2l+1}. \tag{17}$$

Substituting these series in (11) and (12) and equating terms with equal powers of r and z, we obtain recurrence relations between the coefficients $u_{n,j,l}$ and $v_{n,j,l}$

$$u_{n,j,l} = \begin{cases} \dfrac{(-4)(j+1)^2}{2l(2l-1)} u_{0,j+1,l-1} & (\text{for } n = 0) \\[3mm] \dfrac{(-4)(j+1)^2}{2l(2l-1)} u_{n,j+1,l-1} + \dfrac{1}{\chi} \dfrac{1}{2l(2l-1)} \dfrac{du_{n-1,j,l-1}}{dt} & (\text{for } n \geqslant 1) \end{cases} \tag{18}$$

$$v_{n,j,l} = \begin{cases} \dfrac{(-4)(j+1)^2}{(2l+1)2l} v_{0,j+1,l-1} & (\text{for } n = 0) \\[3mm] \dfrac{(-4)(j+1)^2}{(2l+1)2l} v_{n,j+1,l-1} + \dfrac{1}{\chi} \dfrac{1}{(2l+1)2l} \dfrac{dv_{n-1,j,l-1}}{dt} & (\text{for } n \geqslant 1) \end{cases} \tag{19}$$

By successively applying recurrence relations (18) and (19) and making a number of transformations, we obtain a general expression for the temperature in the n-th approximation

$$T_n(r, z, t) = \sum_{i=0}^{\infty} \sum_{k=0}^{i} \sum_{m=0}^{n} \frac{(i!)^2 (-4)^k (k+m)!}{[(i-k)!]^2 \, k! m!} r^{2i-2k} \frac{1}{\chi^m} \left\{ \frac{z^{2k+2m}}{(2k+2m)!} \frac{d^m A_{i,n-m}}{dt^m} + \frac{z^{2k+2m+1}}{(2k+2m+1)!} \frac{d^m B_{i,n-m}}{dt^m} \right\}, \tag{20}$$

where $A_{i,k} = u_{0,i,k}$, $B_{i,k} = v_{0,i,k}$ are two-index coefficients depending on the time.

On the basis of Eq. (20) we shall write down a few of the first terms for the zeroth, first, and second approximations:

$$T_0(r, z) = A_{00} + B_{00}z + A_{10}(r^2 - rz^2) + B_{10}\left(r^2 z - \frac{2}{3} z^3\right)$$

$$+ A_{20}\left(r^4 - 8r^2 z^2 + \frac{8}{3} z^4\right) + B_{20}\left(r^4 z - \frac{8}{3} r^2 z^3 + \frac{8}{15} z^5\right)$$

$$+ A_{30}\left(r^6 - 18r^4 z^2 + 24r^2 z^4 - \frac{16}{5} z^6\right) + B_{30}\left(r^6 z - 6r^4 z^3 + \frac{24}{5} r^2 z^5 - \frac{16}{35} z^7\right) + \cdots, \tag{21}$$

$$T_1(r, z) = A_{01} + B_{01}z + \dot{A}_{00}\frac{z^2}{2} + A_{11}(r^2 - rz^2) + \dot{B}_{00}\frac{z^3}{6}$$

$$+ B_{11}\left(r^2 z - \frac{2}{3} z^3\right) + \dot{A}_{10}\left(\frac{r^2 z^2}{2} - \frac{z^4}{3}\right) + A_{21}\left(r^4 - 8r^2 z^2 + \frac{8}{3} z^4\right)$$

$$+ \dot{B}_{10}\left(\frac{r^2 z^3}{6} - \frac{z^5}{15}\right) + B_{21}\left(r^4 z - \frac{8}{3} r^2 z^3 + \frac{8}{15} z^5\right) + \dot{A}_{20}\left(\frac{r^4 z^2}{2}\right.$$

$$\left. - \frac{4}{3} r^2 z^4 + \frac{4}{15} z^6\right) + A_{31}\left(r^6 - 18r^4 z^2 + 24r^2 z^4 - \frac{16}{5} z^6\right)$$

$$+ \dot{B}_{20}\left(\frac{r^4 z^3}{6} - \frac{4}{15} r^2 z^5 + \frac{4}{105} z^7\right) + B_{31}\left(r^6 z - 6r^4 z^3 + \frac{24}{5} r^2 z^5 - \frac{16}{35} z^7\right) + \cdots, \tag{22}$$

$$T_2(r, z) = A_{02} + B_{02}z + \dot{A}_{01}\frac{z^2}{2} + A_{12}(r^2 - 2z^2) + \dot{B}_{01}\frac{z^3}{6} + B_{12}\left(r^2z - \frac{2}{3}z^3\right) + \ddot{A}_{00}\frac{z^4}{24} + \dot{A}_{11}\left(\frac{r^2z^2}{2} - \frac{z^4}{3}\right)$$

$$+ A_{22}\left(r^4 - 8r^2z^2 + \frac{8}{3}z^4\right) + \ddot{B}_{00}\frac{z^5}{120} + \dot{B}_{11}\left(\frac{r^2z^3}{6} - \frac{z^5}{15}\right) + B_{22}\left(r^4z - \frac{8}{3}r^2z^3 + \frac{8}{15}z^5\right) + \ddot{A}_{10}\left(\frac{r^2z^4}{24} - \frac{z^6}{60}\right)$$

$$+ \dot{A}_{21}\left(\frac{r^4z^2}{2} - \frac{4}{3}r^2z^4 + \frac{4}{15}z^6\right) + A_{32}\left(r^6 - 18r^4z^2 + 24r^2z^4 - \frac{16}{5}z^6\right)$$

$$+ \ddot{B}_{10}\left(\frac{r^2z^5}{120} - \frac{z^7}{420}\right) + B_{21}\left(\frac{r^4z^3}{6} - \frac{4}{15}r^2z^5 + \frac{4}{105}z^7\right) + B_{32}\left(r^6z - 6r^4z^3 + \frac{24}{5}r^2z^5 - \frac{16}{35}z^7\right) + \ldots .$$

$$(23)$$

We see from formula (21) that the expression giving the zeroth approximation for the thermal flux contains 2i coefficients of integration A_{i0} and B_{i0}, which depend on time but not on the coordinates. These coefficients may be determined from the 2i boundary conditions obtained from (13) and (14) by equating terms with equal even powers of r, after putting the equation of the crystallization front into $T_0(r, z, t)$ and $\frac{\partial T_0}{\partial z} - \frac{dy}{dr}\frac{\partial T_0}{\partial r}$.

Then we see from formulas (22) and (23), and in general form from (20), that, on integrating each successive n-th approximation, in addition to the earlier-determined coefficients of the previous approximations (differentiated with respect to time), there appear 2i new coefficients of the current approximation. Substituting the equation of the crystallization front into $T_n(r, z, t)$ and $\frac{\partial T_n}{\partial z} - \frac{dy}{dr}\frac{\partial T_n}{\partial r}$ and, in accordance with boundary conditions (15) and (16), equating terms with equal even powers of r to zero, we obtain 2i equations from which these coefficients A_{in} and B_{in} may be determined.

Thus the boundary conditions given are sufficient for determining the integration coefficients, and in principle the problem may be taken to completion in any approximation.

However, the solution of such a problem in general form is extremely cumbersome owing to difficulties of an algebraic character associated with the solution of a system of equations with an infinite number of variables. Hence, for the purpose of subsequent analysis, we shall confine ourselves to the simplest case of considering the zero approximation for a crystallization front in the form of a paraboloid of revolution with time-varying parameters $y(r,t) = a(t) + b(t)r^2$.

As a result of solving this problem, we may write the formula for the distribution of temperatures in the zero approximation in the following compact form:

$$T(r, z) = T_{f0} + \sum_{s=0}^{\infty} (-1)^s \frac{r^{2s}}{4^s(s!)^2} \frac{\partial^{2s}}{\partial z^{2s}} M(a, b, z),$$

$$(24)$$

in which $M(a, b, z)$ represents the expression

$$M(a, b, z) = \sum_{m=0}^{\infty}\left[(-1)^m \frac{f_m}{4^{m+1}b^{2m+1}} F_m(u) + (-1)^m \frac{mT_m}{4^m b^{2m}} G_m(u)\right],$$

$$(25)$$

where T_m and f_m are the coefficients of the expansions, in even powers of r, of the temperature $T_f(r, t)$ at the crystallization front and its normal gradient $f(r, t)$, determined from the boundary conditions (13) and (14). Here u represents the dimensionless parameter

$$u = 1 + 4b(t)[a(t) - z], \tag{26}$$

characterizing the distance from the crystallization front and its curvature.

The functions of parameter u, $F_m(u)$ and $G_m(u)$, are given by the formulas

$$F_m(u) = \sum_{p=2m+1}^{\infty} \frac{[(p-m-1)!]^2}{p!\,(p-2m-1)!} (1-u)^p, \tag{27}$$

$$G_m(u) = \sum_{p=2m+1}^{\infty} \frac{(p-m)!\,(p-m-1)!}{p!\,(p-2m-1)!} (1-u)^p, \tag{28}$$

which are suitable for calculations at short distances from the crystallization front when the latter is only slightly bent, or by the formulas

$$F_m(u) = \sum_{n=0}^{m} \frac{(m!)^2}{[(m-n)!]^2 (n!)^2} u^n (d_{mn} - \ln u), \tag{29}$$

$$G_m(u) = \frac{1}{u} - u^m + \sum_{n=0}^{m-1} \frac{m!\,(m+1)!\, u^n (d_{mn}^{\bullet} + \ln u)}{(m-n)!\,(m-n-1)!\, n!\,(n+1)!}, \tag{30}$$

which are suitable for any positive values of parameter u, in which the constant numbers d_{mn} and d_{mn}^* equal

$$d_{mn} = 2 \left\{ \sum_{k=1}^{m-n} \frac{1}{k} - \sum_{k=1}^{n} \frac{1}{k} \right\}, \tag{31}$$

$$d_{mn}^{\bullet} = \left\{ \sum_{k=1}^{m-n} \frac{1}{k} + \sum_{k=1}^{m-n-1} \frac{1}{k} - \sum_{k=1}^{n} \frac{1}{k} - \sum_{k=1}^{n+1} \frac{1}{k} \right\}. \tag{31a}$$

Below we list a few particular values of the functions $F_m(u)$ and their derivatives:

$$\left. \begin{aligned} F_0(u) &= -\ln u, \\ F_1(u) &= -(u+1)\ln u + 2(u-1), \\ F_2(u) &= -(u^2 + 4u + 1)\ln u + 3u^2 - 3, \end{aligned} \right\} \tag{32}$$

$$\left. \begin{aligned} \frac{\partial F_0}{\partial z} &= 4b \cdot \frac{1}{u}, \\ \frac{\partial F_1}{\partial z} &= 4b \left(\frac{1}{u} + \ln u - 1 \right), \\ \frac{\partial F_2}{\partial z} &= 4b \left[\frac{1}{u} + (2u+4)\ln u - 5u + 4 \right], \end{aligned} \right\} \tag{32a}$$

$$\left. \begin{aligned} \frac{\partial^2 F_0}{\partial z^2} &= 16b^2 \cdot \frac{1}{u^2}, \\ \frac{\partial^2 F_1}{\partial z^2} &= 16b^2 \left(\frac{1}{u^2} - \frac{1}{u} \right), \\ \frac{\partial^2 F_2}{\partial z^2} &= 16b^2 \left(\frac{1}{u^2} - \frac{4}{u} - 2\ln u + 3 \right), \end{aligned} \right\} \tag{32b}$$

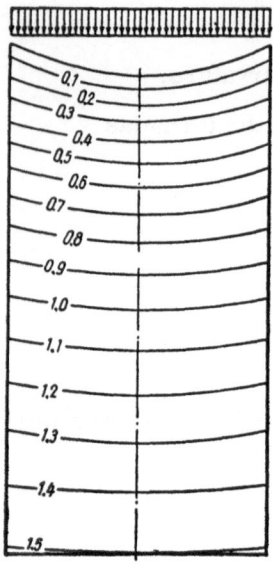

Fig. 1. Thermal field in the solid phase for a concave crystallization front ($\Delta h/R = 0.25$). Uniform distribution of thermal flux along the phase-separation boundary ($\beta = 0.5$, $f_0 = 1$, $f_1 = 0$).

$$\frac{\partial^3 F_0}{\partial z^3} = 64b^3 \cdot \frac{2}{u^3},$$

$$\frac{\partial^3 F_1}{\partial z^3} = 64b^3 \left(\frac{2}{u^3} - \frac{1}{u^2} \right),$$

$$\frac{\partial^3 F_2}{\partial z^3} = 64b^3 \left(\frac{2}{u^3} - \frac{4}{u^2} + \frac{2}{u} \right), \qquad (32c)$$

$$\frac{\partial^4 F_0}{\partial z^4} = 256b^4 \cdot \frac{6}{u^4},$$

$$\frac{\partial^4 F_1}{\partial z^4} = 256b^4 \left(\frac{6}{u^4} - \frac{2}{u^3} \right),$$

$$\frac{\partial^4 F_2}{\partial z^4} = 256b^4 \left(\frac{6}{u^4} - \frac{8}{u^3} + \frac{2}{u^2} \right). \qquad (32d)$$

Calculations have been made on the basis of these formulas for the thermal fields in crystals with plane, convex, and concave crystallization fronts.

For a plane crystallization front and a uniform distribution of temperature gradient along it ($f_0 = 1$, $f_1 = f_2 = \ldots = 0$), the thermal field in the solid phase is well known to be characterized by an equidistant set of plane isotherms. Side heat leakage is here absent and the axial thermal flux remains constant at any distance from the crystallization front.

For a concave crystallization front and a uniform distribution of thermal flux along it ($f_0 = 1$, $f_1 = f_2 = \ldots = 0$) (Fig. 1), the isotherms are curved surfaces, the distance between which increases on passing away from the crystallization front. In order to preserve a stable form of the temperature profile, there must be a side leakage of heat from the crystal, the intensity of this diminishing on moving away from the front. With increasing distance from the front the axial thermal flux also diminishes, and hence the outflow of heat from the end must also be reduced in order to keep the shape of the front stable.

The actual cooling conditions associated with the Czochralski method of crystal growth (and with other methods of directional crystallization at the final stages) promote stabilization of a concave crystallization front; hence this shape is often found in the practical growth of crystals.

On the other hand, the creation of a stable convex form of the crystallization front is a more complex technological problem. It follows from theoretical analysis (Fig. 2) that, for a convex shape of the front and a uniform distribution of thermal flux along it ($f_0 = 1$, $f_1 = f_2 = \ldots = 0$), the isotherms become denser on passing away from the crystallization front. Maintenance of a stable temperature profile of this kind requires additional heating of the crystal along its side surface, the intensity of this increasing (with time). Hence, as the crystal grows the end heat outflow must be increased and made more nonuniform (more intensive in the center). In such a case only crystals of limited height can be grown. The greater the curvature of the front, the smaller will be the possible height of the crystals.

Practical conditions of crystal cooling are not favorable toward the stabilization of a convex front with a uniform access of heat from the melt. Owing to the insufficient intensity of end cooling and the existence of side heat outflow, it becomes necessary to have a more

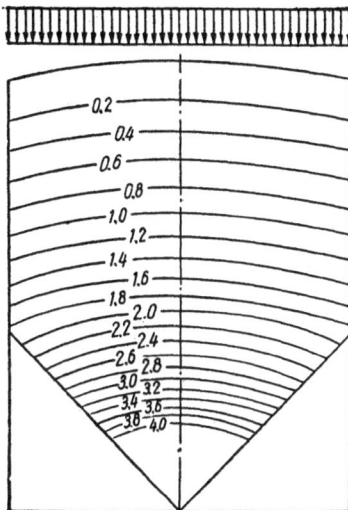

Fig. 2. Thermal field in the solid phase for a convex crystallization front ($\Delta h/R = -0.1$). Uniform distribution of thermal flux along the phase-separation boundary ($\beta = -0.2$, $f_0 = 1$, $f_1 = 0$).

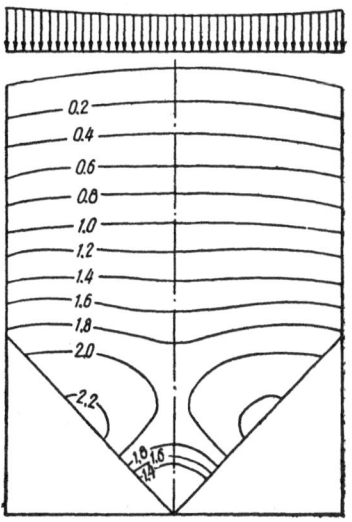

Fig. 3. Thermal field in the solid phase for a convex crystallization front ($\Delta h/R = -0.1$). Distribution of thermal flux along the phase-separation boundary obeys a parabolic law: $f(r, t) = f_0 + f_1(r^2/R^2)$ ($\beta = -0.2$, $f_0 = 1$, $f_1 = 0.1$).

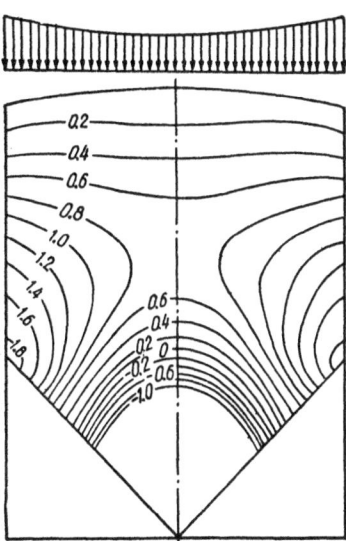

Fig. 4. Thermal flux in the solid phase for a convex crystallization front ($\Delta h/R = -0.1$). Distribution of thermal flux along the phase-separation boundary obeys a parabolic law: $f(r, t) = f_0 + f_1(r^2/R^2)$, ($\beta = -0.2$, $f_0 = 1$, $f_1 = 0.5$).

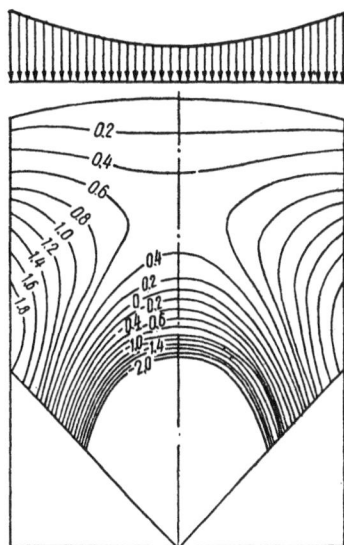

Fig. 5. Thermal flux in the solid phase for a convex crystallization front ($\Delta h/R = -0.1$). Distribution of thermal flux along the phase-separation boundary obeys a parabolic law: $f(r, t) = f_0 + f_1(r^2/R^2)$ ($\beta = -0.2$, $f_0 = 1$, $f_1 = 1$).

intensive access of heat into the peripheral parts of the crystal. Thus, a convex front has a tendency to straighten, turning first into a plane front and then into a concave one. The higher the growth velocity, the greater is the probability of passing into a concave front.

For a nonuniform distribution of thermal flux over the surface of the crystallization front, there is a considerable change in the character of the thermal field in the crystal. Figures 3, 4, and 5 show the calculated temperature distribution in a crystal with a convex crystallization front when the heat-intake function $f(r, t)$ varies parabolically

$$f(r, t) = f_0 + f_1 r^2 \tag{33}$$

with values of $f_0 = 1$, $f_1 = 0.1$, 0.5, and 1, for which the intensity of the inflow of heat at the periphery of the crystal is greater than at the center. (The temperature-gradient distributions along the normal are shown at the tops of Figs. 3-5.)

We see by comparing Fig. 2 with Figs. 3-5 that the additional supply of heat to the crystal around the periphery of the convex crystallization front leads to the appearance of a "saddle shape" in the form of the isotherms. In order to produce this kind of thermal field we require first cooling at the end of the crystal and then even additional heating. At the side of the crystal, zones of reduced temperature and local side heat loss develop. The greater the nonuniformity of the temperature gradient at the center and periphery of the crystal, the closer to the crystallization front will be the region of minimum temperature along the axis of the crystal, and the more local side heat outflow will be required.

The thermal fields shown in Figs. 3-5 differ essentially from the real thermal fields in crystals with a convex crystallization front. In practical conditions the shape of the crystallization front is not stable in time, nor is it strictly parabolic; in these respects it differs from the cases analyzed.

Our analysis of thermal fields indicates that, by creating a nonuniform flow of heat from the melt, it should be possible to grow long crystals stably, even with a convex crystallization front. One of the most important methods of perfecting the technology of growing crystals from the melt indicated by the present theoretical analysis is that of securing a control-led directional heat flow. The present analysis enables us to calculate the intensity of the heat outflow required.

It follows from formula (24) that the axial and radial temperature gradients at the cooled surfaces of the crystal are given by the expressions

$$\frac{\partial T}{\partial z}\bigg|_{z=0} = \sum_{m=0}^{\infty} \sum_{s=0}^{\infty} \frac{r^{2s}}{(s!)^2} \frac{(-1)^{m+s}}{(4b^2)^{m-s}} \left[f_m \frac{\partial^{2s+1}}{\partial z^{2s+1}} F_m(z)\bigg|_{z=0} + 4bmT_m \frac{\partial^{2s+1}}{\partial z^{2s+1}} G_m(z)\bigg|_{z=0} \right], \tag{34}$$

$$\frac{\partial T}{\partial r}\bigg|_{r=R} = \sum_{m=0}^{\infty} \sum_{s=1}^{\infty} \frac{2sR^{2s-1}}{(s!)^2} \frac{(-1)^{m+s}}{(4b^2)^{m-s}} \left[\frac{f_m}{4b} \frac{\partial^{2s}}{\partial z^{2s}} F_m(z) + mT_m \frac{\partial^{2s}}{\partial z^{2s}} G_m(z) \right]. \tag{35}$$

We find by analyzing formulas (34) and (35) that an axial temperature gradient in the furnace which decreases on moving away from the crystallization front promotes the establishment of a concave front, while a rising temperature gradient in the furnace in combination with nonuniform end heat outflow promotes a convex front.

The thermal fields in axially symmetric crystals calculated in this paper may also serve as a basis for the calculation of thermoelastic stresses in crystals during growth; this is extremely important for obtaining very large perfect crystals.

Literature Cited

1. Él'baum, K., Usp. Fiz. Nauk, 79(3) : 545 (1963).
2. Tiller, W. A., The Art and Science of the Growing Crystals (Ed. J. J. Gilman). London-New York (1963), p. 276.
3. Kyropulos, S., Z. Anorg. Allgem. Chem., 154 : 308 (1925).
4. Czochralski, J., Z. Phys. Chem., 92 : 219 (1918).
5. Stockbarger, D., Rev. Sci. Instr., 7 : 133 (1936).
6. Stöber, F., Z. Krist., 61 : 299 (1925).
7. Stepanov, I. V., and Vasil'eva, M. A., In: Growth of Crystals, Vol. 3, pp. 223-238. Izd. Akad. Nauk SSSR, Moscow (1961). [English translation: Consultants Bureau, New York (1962).]
8. Bolling, G. F., and Tiller, W. A., J. Appl. Phys., 31 : 1345 (1960).
9. Lyubov, B. Ya., Roitburd, A. L., and Temkin, D. E., In: Growth of Crystals, Vol. 3, pp. 68-74. Izd. Akad. Nauk SSSR, Moscow (1961).
10. Stefan, F., Monatsschr. Math. Phys., 1 : 1 (1890).
11. Borisov, V. T., Lyubov, B. Ya., and Temkin, D. E., Dokl. Akad. Nauk SSSR, 104(2) : 223 (1955).
12. Lyubov, B. Ya., and Temkin, D. E., In: Problems of Metal Science and the Physics of Metals. pp. 311-316. Metallurgizdat, Moscow (1958).
13. Temkin, D. E., Inzh. Fiz. Zh., No. 4, pp. 89-92 (1962).
14. Birman, B. I., In: Growth of Crystals, Vol. 5. Izd. Akad. Nauk SSSR (1965). [English translation: Consultants Bureau, New York (1968).]
15. Ivantsov, G. P., Dokl. Akad. Nauk SSSR, 58 : 567 (1947).
16. Horvay, G., and Cahn, J. W., Acta Met., 9 : 695-705 (1962).
17. Kreith, F., and Romie, F. E., Proc. Phys. Soc., 68 : 277 (1955).

EFFECT OF CERTAIN GROWTH PARAMETERS ON THE SHAPE OF THE CRYSTALLIZATION FRONT DURING THE GROWTH OF SINGLE CRYSTALS

L. A. Sysoev, É. K. Raiskin, and Yu. N. Gavrilyuk

*Khar'kov All-Union Scientific-Research Institute
of Single Crystals*

A large proportion of the single crystals of piezoelectric semiconductors, scintillation single crystals, etc., are now grown by the method of directional crystallization in vertical moving ampoules [1, 2]. When growing by this method, the shape of the crystallization front, which may be concave, plane, or convex, has a great influence on the quality of the single crystal. A concave form arises when there is a radial outflow of heat from the center to the periphery in the crystallization zone. A plane crystallization front is formed when there is only an axial outflow of heat. If there is a radial flow of heat directed from the periphery to the center, a concave crystallization front is produced.

In the course of growing single crystals, parasitic nuclei tend to grow out into the melt along the normal to the crystallization front [3]. Thus, with a concave shape of the front, the main single crystal may be tapered out by parasitic blocks; the greater the concavity, the shorter will be the distance within which this occurs (Fig. 1).

There are some indications in the literature regarding the effect of various growth parameters on the bending of the crystallization front, but these are all of a qualitative character, and no quantitative relationships are given [3, 4].

The influence of individual parameters on the extent of the bending K may be given in general form by means of the following functional relationship:

$$K = f\left(v, D \ \lambda, \frac{dt}{dl}, a, L\right),$$

where v is the velocity of the ampoule; D is the diameter of the crystal being grown, a is the thermal diffusivity of the material, λ is the thermal conductivity of the material cf the crystal, L is the heat of fusion of the material, and dt/dl is the temperature gradient of the furnace in the crystallization zone.

Fig. 1. Shapes of crystallization front.

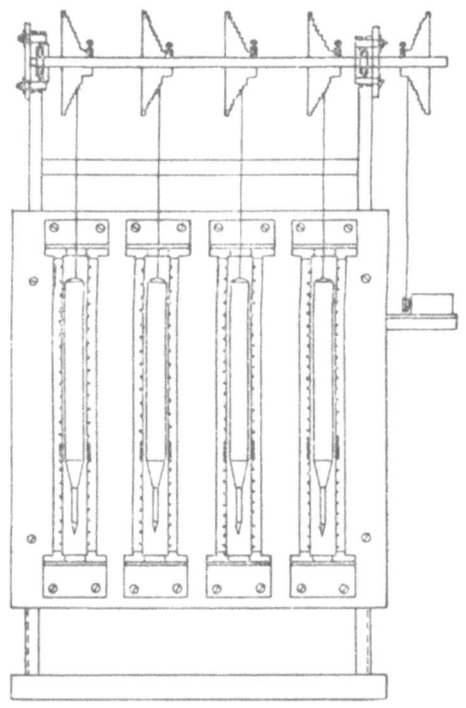

Fig. 2. Test bed of model furnaces.

In view of this it is very important to obtain quantitative relationships for the bending of the crystallization front as a function of the rate of crystallization, the diameter of the crystal being grown, the thermal conductivity and thermal diffusivity of the crystal, the heat of fusion of the material, and the temperature gradient of the furnace.

We therefore decided to study the effects of certain of the parameters mentioned on the bending of the crystallization front. As model subjects for study we took tin, bismuth, and naphthalene crystallized in ampoules 20 mm in diameter. The choice of these substances was due to their relatively low melting points, which facilitated the experiments, and also their wide range of thermal conductivities, from 0.373 (for naphthalene) to 66.2 W/m · deg (for tin), which made it possible to extend the results to a large number of crystallizing materials.

The work was carried out on a special test bed (Fig. 2). The test bed comprised a set of four tubular Pyrex furnaces each having independently controlled feeding of the upper and lower zones. The temperatures of the furnaces were regulated by means of an LATR-1 transformer. The test bed was supplied with stabilized voltage.

Motion of the Pyrex ampoules containing the crystallizing substance was effected by means of an electric drive with a system of graduated pulleys enabling the velocity to be varied between $2.78 \cdot 10^{-6}$ to $278 \cdot 10^{-6}$ m/sec. The bending of the crystallization front was measured during the experiment by means of a mechanical gauge. The accuracy of the measurements was ±0.2 mm.

The temperatures were measured with a Chromel–Copel thermocouple 0.3 mm in diameter, the readings of which were recorded with a PP-1 potentiometer.

Separate control of the supply to the upper and lower windings of the furnace made it possible to create any desired temperature distributions in the inner space of the furnace.

In order to compare the temperature fields in the crystallizing material and the furnace, respectively, we determined the temperature distribution in the substance during growth; owing to the creation of convective flows in the melt, the temperature fell off very little along the ampoule at all points above that corresponding to the maximum furnace temperature, and was indeed only related to the temperature in the upper part of the furnace in respect of the absolute value of the temperature (Fig. 3).

At a distance of 20 to 30 mm below the point of maximum furnace temperature, the convection currents captured and agitated the melt, owing to its viscosity; this explains the difference between the melt and furnace temperatures in this region. This phenomenon was easily observable through the transparent furnace. Below this region the temperatures of the melt and crystal approached that of the furnace. Hence we may subsequently consider that the axial gradient of the furnace equals the axial gradient of the growing crystal to a fairly good accuracy.

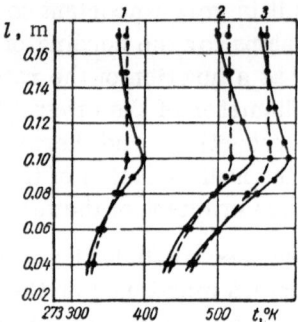

Fig. 3. Temperature distribution
inside the furnace (solid lines) and
in the crystallizing substances
(broken lines). 1) Naphthalene;
2) tin; 3) bismuth.

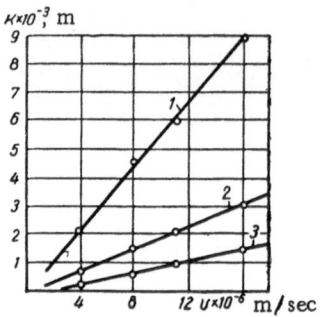

Fig. 4. Bending of the crystallization
front as a function of the velocity of
the ampoule (dT/dl = 500 deg/m).
1) Naphthalene; 2) bismuth; 3) tin.

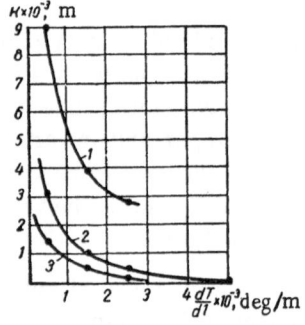

Fig. 5. Bending of the crystallization front as a function of temperature gradient near the front (v =
16 · 10^{-6} m/sec). 1) Naphthalene;
2) bismuth; 3) tin.

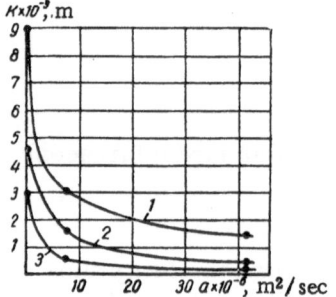

Fig. 6. Bending of the crystallization
front as a function of the thermal
diffusivity of the material. 1) v =
16 · 10^{-6} m/sec, dT/dl = 500 deg/m;
2) v = 7.8 · 10^{-6} m/sec, dT/dl = 500
deg/m; 3) v = 16 · 10^{-6} m/sec,
dT/dl = 2500 deg/m.

We studied the effect of the velocity of the ampoule on the bending of the crystallization
front for the three substances (naphthalene, tin, and bismuth) with furnace temperature gradients of 500 and 2500 deg/m and velocities between 3.7 · 10^{-6} and 16 · 10^{-6} m/sec. We see
from Fig. 4 that the bending is linearly related to the velocity of the crystallization front. The
rate of rise in the extent of the bending increases from tin to naphthalene, i.e., is inversely
proportional to the thermal conductivity of the crystallizing material.

The effect of the furnace temperature gradient on the bending of the crystallization front
may be seen from Fig. 5, from which it follows that small temperature gradients lead to large
bendings of the crystallization front. Usually when growing single crystals by the method of
directional crystallization one tries to have large temperature gradients at the crystallization
front. Until now, however, the importance of having large gradients has only been explained by
the necessity of eliminating concentration supercooling. The curves of Fig. 5 show that there
is an optimum in the temperature gradient at the crystallization front of the order of 4000 deg/m,

above which the effect of the gradient on the extent of the bending is almost absent owing to the reduction in the inclination of these curves to the x axis.

The establishment of an optimum temperature gradient is of considerable importance, since excessively high gradients during crystal growth produce large thermal stresses, leading to plastic deformation, distorting the crystal lattice, and to structural defects.

The effect of the thermal diffusivity of the material may be seen from the graph of Fig. 6, in which the bending of the crystallization front starts rising sharply on reducing the thermal diffusivity below $5 \cdot 10^{-6}$ m^2/sec; hence, for low values of thermal diffusivity crystallization should be carried out at velocities of the order of $1 \cdot 10^{-6}$ m/sec.

Analysis of the experimental results shows that analogous relationships for the bending of the crystallization front will also hold on growing single crystals by zone recrystallization.

Literature Cited

1. Ékshtein, Yu., and Iindra, I., In: Growth of Crystals, Vol. 3, p. 300. Izd. Akad. Nauk SSSR, Moscow (1961). [English translation: Consultants Bureau, New York (1962).]
2. Belyaev, L. M., et al., In: Growth of Crystals, Vol. 3, p. 338. Izd. Akad. Nauk SSSR, Moscow (1961). [English translation: Consultants Bureau, New York (1962).]
3. Growth Processes and the Growing of Single Crystals [Russian translation]. p. 370. IL, Moscow (1963).
4. Stepanov, I. V., and Vasil'eva, M. A., In: Growth of Crystals, Vol. 3, p. 223. Izd. Akad. Nauk SSSR, Moscow (1961). [English translation: Consultants Bureau, New York (1962).]

PRODUCTION OF GERMANIUM CRYSTALS BY THE METHOD OF A. V. STEPANOV

S. V. Tsivinskii

A. F. Ioffe Leningrad Physico-Technical Institute

In view of the increasing demands for single crystals now being made by their users, it is desirable to effect crystallization in such a manner as to obtain single crystals (1) of given shape, size, and orientation, (2) with a specified concentration of impurities and impurity distribution over the volume of the crystal, and (3) with a specified number of structural defects and defect distribution.

The various existing methods of producing crystals (Czochralski method, etc.) are not yet capable of satisfying these requirements. Considerable prospects have been opened in this regard by the method of Professor A. V. Stepanov, designed for growing single crystals and polycrystalline specimens.

The present paper is devoted to the use of this method for producing germanium crystals.

Essence of the Stepanov Method

On drawing a solid rod from a liquid which it wets, a liquid column of melt will be drawn behind it. The same takes place when drawing single crystals from the melt by the Czochralski method.

Stepanov proposed using phenomena associated with the existence of this column for controlling the crystallization process. By regulating the shape and parameters of the molten column we may vary the shape of the growing crystal and obtain crystals of prearranged form [1-6]. Thus, this method may be regarded as a development of that of Czochralski.

The crystallographic orientation of the growing crystal is determined, as in all other methods, either by the orientation of the seed or by that of a crystal "dislodged" by its neighbors.

Producing Germanium Crystals in the

Form of a Thin Strip

Let us briefly consider how we used the Stepanov method to obtain germanium crystals in the form of a thin strip. These crystals were obtained by drawing from a slit into which the melt was fed under a pressure of 5-6 g/cm^2. This pressure was created because the slit was "sunk" to a depth of 8 or 9 mm below the level of the melt in the crucible. Germanium passed from the crucible into the slit as drawing progressed. Provided that the size of the crucible

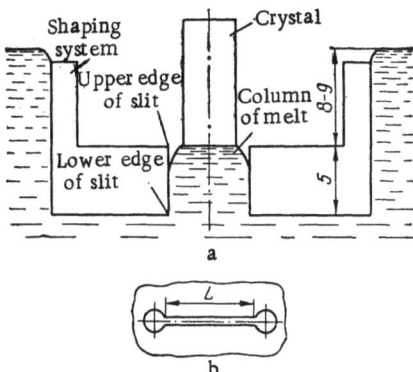

Fig. 1. Components of the shaping system for producing germanium crystals in the form of a thin strip. a) Mode of operation of the shaping system; b) shape of the slit from which the germanium strip crystals are drawn.

was adequate, strip crystals 70-150 mm long could be drawn without adding further germanium to the crucible. The arrangement of the process is illustrated in Fig. 1a.

Before drawing began, a seed in the form of a single-crystal plate of similar shape to the strip was introduced into the slit to a depth sufficient for it to touch the melt in the crucible. After this a convex column of melt was formed and this was retained throughout the whole drawing process.

Under favorable temperature conditions the crystallization front lay inside the slit after introduction of the seed (Fig. 1a).

As shown in [5, 7], the capillary conditions depend on the radius of curvature of the surface of the column of melt. A large curvature at the ends of the thin strip leads to unfavorable capillary conditions, preventing the establishment of a stable drawing process. In order to eliminate this disadvantage, we adopted a special arrangement to reduce the curvature of the column of melt at the edges of the slit by drawing a strip with thickened edges. With the same thickness of the edges, strips of varying width and thickness, including very wide, thin strips, may be obtained.

The slit used to obtain strip with thickened edges is shown in Fig. 1b. By means of this slit we grew germanium single crystals in the form of both wide and narrow strip. The cross section of the strip was similar to the shape of the slit. The narrow strips had plane sections between 7.5 and 8 mm wide, and a total width of about 13 mm; the diameter of the thickened ends was between 2.0 and 2.9 mm. The thickness of these strips varied between 0.17 and 1.1 mm, being determined by the width of the slit. The wide strips had a plane section 22 mm wide and a total width of about 27 mm; the diameter of the thickened ends was 2.0 to 2.9 mm. The thickness of the wide strips was 0.27 mm.

The plane part of the strips had constant thickness (± 5-10%). In some cases, the surface of the strip was almost specular over considerable regions.

Narrow strips were grown with four different crystallographic orientations. A large proportion of the strips were obtained by drawing along the [1$\bar{1}$0] and [11$\bar{2}$] directions. The plane of these strips coincided with the (111) plane. The orientation of the strips may be briefly represented by [1$\bar{1}$0] (111), [11$\bar{2}$] (111). Wide strips were only grown with one orientation: [1$\bar{1}$0] (111). Clearly, single-crystal strips of any orientation may be obtained. Photographs of a wide strip and its cross section are shown in Figs. 2a, b.

Stepanov's method makes it possible to control the concentration and distribution of impurities in the growing crystal. In particular, if the crystallization front is plane and if its position in the slit is kept at the same level (with respect to height), crystals with a uniform impurity distribution may be obtained. We did in fact obtain such crystals.

Figure 3 shows the variation in the specific electrical resistance R_{sp} along and across a narrow strip, and Fig. 4 shows the same characteristics for a wide strip. We see that R_{sp} is quite constant (± 3-15%).

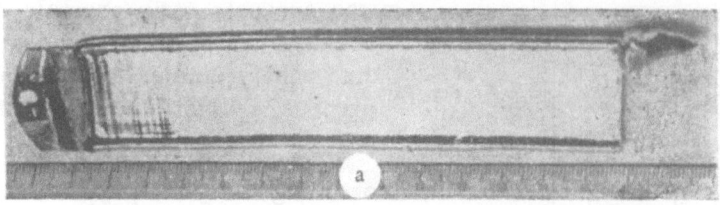

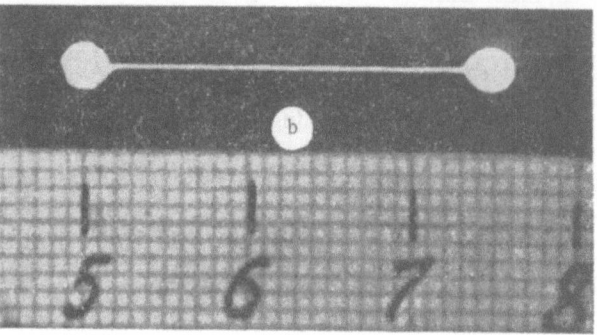

Fig. 2. External form of a wide germanium single-crystal strip-plate 0.27 mm thick obtained by the Stepanov method (a) and its cross section (b).

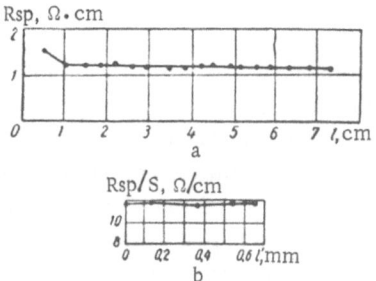

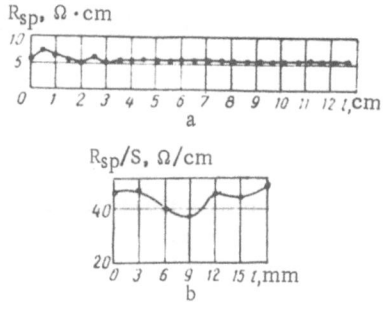

Fig. 3. Variation in the specific electrical resistance (R_{sp}) over the length (a) and width (b) of a narrow (~13 mm) germanium strip.

Fig. 4. Variation in the specific electrical resistance over the length (a) and width (b) of a wide (27 mm) germanium strip.

We also studied the dislocation distribution in the strip crystals. The dislocations were revealed by etching the plane of the strip, the (111) plane, with an alkaline solution of potassium ferricyanide [24 g KOH + 16 g $K_3Fe(CN)_6$ + 200 ml H_2O].

The results of the investigation reduce to the following. There were some cases in which the dislocations were distributed nonuniformly (Fig. 5a). Sometimes the dislocation density varied from one part of the crystal to another by several orders, $10-10^5$ cm^{-2}; however, there were samples of narrow strip (Fig. 5b) in which the dislocations were distributed uniformly with a density of about $5 \cdot 10^4$ cm^{-2}. Among the wide strips there were samples in which the dislocation density varied between 10^4 and 10^5 cm^{-2} over extensive regions.

Figure 6 shows the variation in dislocation density over the length and breadth of a wide single-crystal strip. In dislocation density such samples are similar to the crystals used in making semiconductor apparatus.

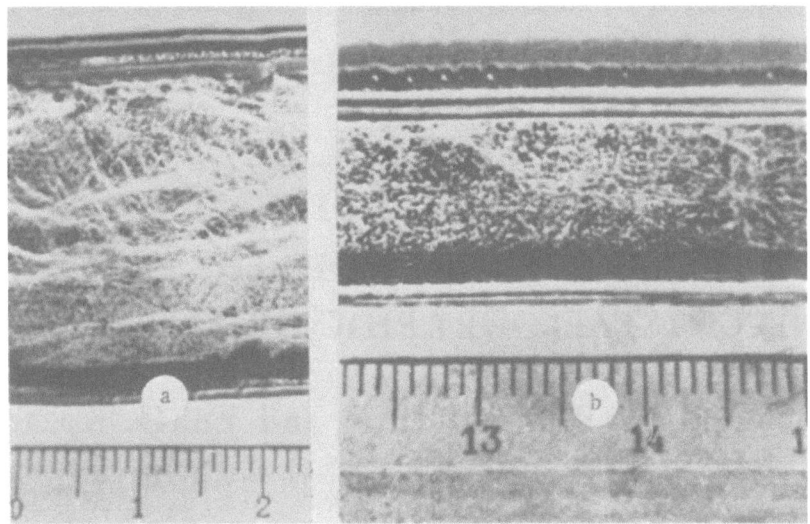

Fig. 5. Sections of wide and narrow strips with nonuniform (a) and
uniform (b) dislocation distributions.

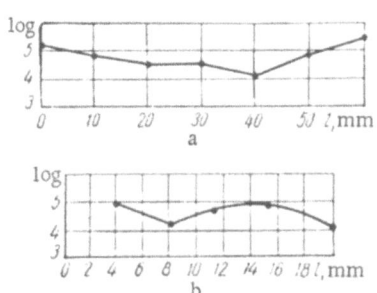

Fig. 6. Variation in dislocation
density over the length (a) and
breadth (b) of a wide germanium
strip.

Conclusions

We see from the results of the foregoing that the
Stepanov method will give germanium crystals of assigned
shape, size, and crystallographic orientation, especially
crystals in the form of thin strip plates. The strip crystals have homogeneous electrical properties and in a number of cases quite uniform dislocation distributions. These
properties may well be improved later.

We see from what has been said that the Stepanov
method constitutes one of the most versatile methods of
crystallization, satisfying the requirements listed at the
beginning of this paper as regards promising methods of
producing single crystals.

The possibility of growing long, thin crystal strips
similar in thickness and properties to those required for
semiconductor apparatus, should simplify the production of such apparatus and aid automation.

Literature Cited

1.   Stepanov, A. V., Zh. Tekhn. Fiz., 29(3): 381 (1959).
2.   Stepanov, A. V., and Shakh-Budagov, A. L., Zh. Tekhn. Fiz., 293: 394 (1959).
3.   Stepanov, A. V., Future Metal Treatment. Lenizdat, Leningrad (1963).
4.   Tsivinskii, S. V., and Stepanov, A. V., Fiz. Tverd. Tela, 7(1): 194 (1965).
5.   Gaule, G. K., and Pastore, J. R., Met. Elem. Comp.-Semicond. p. 207. Boston (1960).
6.   Brissot, J. J., and Raynod, H., Electrochem. Technol., 1(9-10): 304 (1963).
7.   Tsivinskii, S. V., Inzh. Fiz. Zh., 5(9): 59 (1962).

GROWING SINGLE CRYSTALS OF ZINC AND BISMUTH WITH PREARRANGED SHAPE AND CRYSTALLOGRAPHIC ORIENTATION

F. F. Lavrent'ev, V. P. Soldatov, and Yu. G. Kazarov

Khar'kov Physico-Technical Institute of the Academy of Sciences of the Ukrainian SSR

When studying various physical characteristics of plastic deformation, such as the starting stresses for initiating the motion of dislocations [1], the velocity of dislocations as a function of applied stress, the dimensions of an elastic twin as a function of the load [2], etc., we require single crystals of a definite geometrical form, with a specific orientation of the crystallographic elements of slip and twinning, and of perfect structure. A number of interesting methods of growing metal crystals have now been proposed [3-5]. The use of these, however, does not always ensure the production of single crystals suitable for carrying out various physical tests. It thus becomes necessary to perfect the old methods of growth and to develop new ones.

In this paper we propose a simple and convenient method of growing zinc and bismuth single crystals of assigned shape and crystallographic orientation, with a relatively low dislocation density.

Growing Method

The principal material for making the crystallization apparatus is the mineral pyrophyllite, which can easily be machined and polished in unannealed form. Pyrophyllite has good refractory properties. Previously annealed at 1000°C, it does not alter its shape on subsequent heating. A very important property of pyrophyllite is the fact that it is not wetted by molten metals and that it is nonporous (its density is 2.8 g/cm³). This property is extremely important, since it enables the grown crystals to be easily removed from the mold, eliminating the possibility of damage to the single crystal, which is extremely likely to occur if the surface of the growing crystal sticks to the mold walls. Growth was carried out in the apparatus illustrated in Fig. 1.

For clarity we show a section of the apparatus in a plane perpendicular to the crystal growth axis. The apparatus consists of a pyrophyllite base in the form of a box 1 containing a ceramic 2 carrying a Nichrome heater 3. The ceramic comprises three pyrophyllite plates 10 mm thick, 45 mm high, and 170 mm long, with grooves cut uniformly along it. Each groove has a cross section of 4 × 4 mm, and the distance between grooves is also 4 mm. The

110

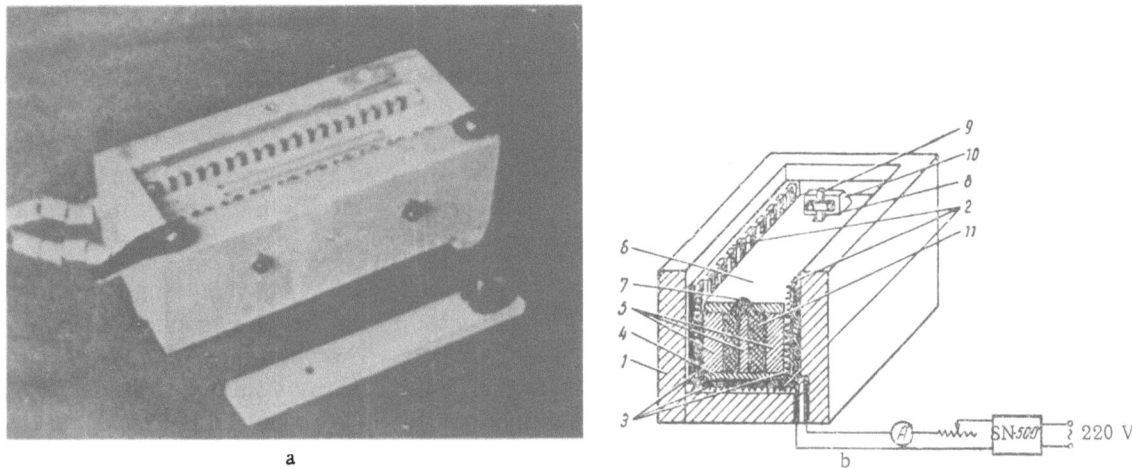

Fig. 1. General view (a) and detailed diagram (b) of the apparatus for growing Zn and Bi single crystals.

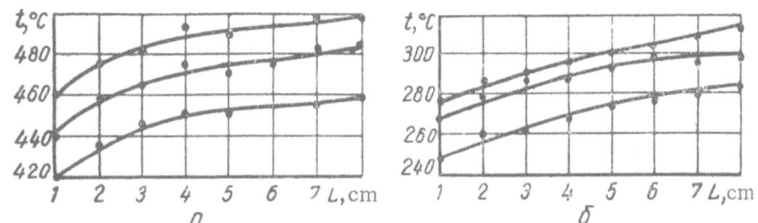

Fig. 2. Temperature distribution along the mold at various periods of growth. a) For growing zinc crystals; b) for growing bismuth crystals.

Nichrome heater 3, in the form of a spiral, is placed in the grooves in such a way as to make the number of turns in each successive groove smaller by 3-4 turns than in the preceding one. This arrangement of the spiral made it possible to obtain a definite temperature gradient along the mold. At the bottom of the base 1 near the lowest turns of the heater 3, we placed a pyrophyllite block 4; two molds determining the shape and size of the growing crystals were attached to this by means of special prismatic bars 5. The surfaces of the block 4 and bars 5 were carefully polished, since these surfaces formed the mold and the quality of the polishing determined the surface perfection of the growing crystals. A top 6 was placed over the bars 5; this contained apertures 7 and 8 for pouring the melt and regulating the filling of the mold. The orientation of the growing crystals is given by the seed 9, fixed in a special orienting block 10 made of copper (for better heat outflow). The seed 9 is introduced into the melt 11 through the aperture 8 and is partly fused. Crystal growth is effected as a result of the motion of the temperature gradient along the axis of the mold, moving away from the seed, the temperature being reduced in accordance with a specified program. The temperature at the initial point of the mold, near the seed aperture 8, was 430-440°C (when growing zinc), rising to between 480 and 490°C at the end at which crystal growth was completed. The mean value of the gradient along the mold was 0.5 deg/mm.

Figure 2 shows the variation of the mold temperature along the direction of crystal growth on reducing the temperature when growing zinc and bismuth. We see that as the temperature is reduced, the value of the gradient remains almost unaltered, but the gradient

Fig. 3. Two zinc single crystals grown from one seed.

producing crystal growth moves along the growth direction from the seed to the other end of the mold. The rate of growth is determined by the rate at which the temperature is reduced. In our apparatus the fall in temperature was achieved by changing the current in the heater (by smoothly increasing its resistance).

This was achieved by moving the slide of a rheostat connected in series with the heater circuit. The velocity of the slide was governed by the linear velocity of a pulley mounted on the shaft of an SD-1/300 motor and a cord connected to the rheostat slide. The optimum velocities for growing zinc and bismuth crystals on our apparatus lay between 5 and 10 mm/h. The heater was supplied with stabilized voltage from an SN-500 voltage stabilizer. Two single crystals were grown simultaneously from one seed in the apparatus of Fig. 1. Figure 3 shows a zinc crystal after growth and chemical polishing (to remove the oxide film). Bismuth crystals have the same form. The dimensions of the crystals obtained may be estimated from the millimeter scale in Fig. 3. We see from the figure that the artificial prismatic plane $(10\bar{1}0)$ of the single crystal thus grown has a degree of perfection similar to the (0001) cleavage plane.

It should be noted that for successful growth of the crystals severe overheating of the mold should be avoided. Overheating leads to intense oxidation of the surface of the growing crystals and (especially harmful) to the development of a thick film of oxides which prevents good adhesion between seed and melt, produces parasitic nuclei, and hence results in the formation of bi- and tricrystals. If the required conditions are met, the method described is very efficient and gives an 85-90% useful yield of single crystals.

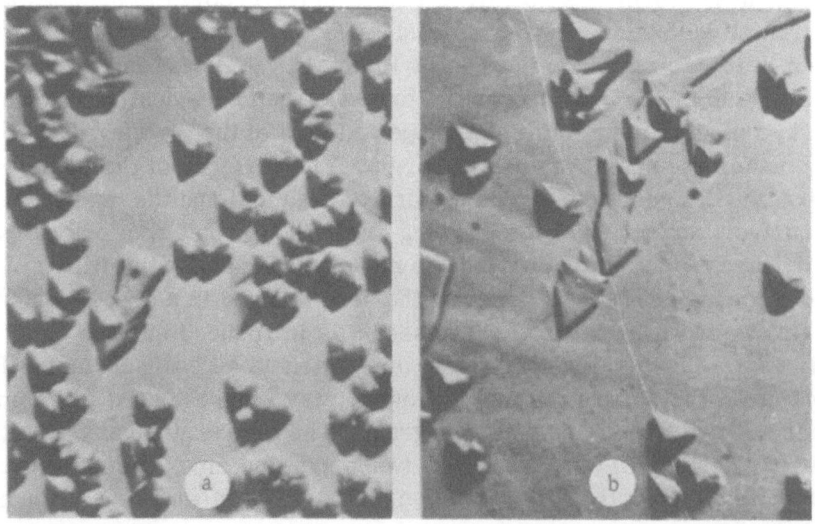

Fig. 4. Dislocation structure of bismuth single crystals. a) Grown in a rigid mold (× 450); b) grown in a "soft" mold (× 250).

Perfection of the Crystals

The degree of structural perfection of the crystals thus grown depends on a number of factors, in particular on the rigidity of the mold. The use of a rigid mold unable to change its dimensions during crystal growth prevents the formation of perfect crystals, since, under these circumstances, stresses develop owing to changes in crystal volume during growth. This is especially so for bismuth, which increases its volume by more than 3.3% on solidification. Bismuth crystals grown in a rigid mold have a dislocation density of 10^6 cm^{-2}. This may be seen from Fig. 4a, which illustrates the (111) cleavage plane of a bismuth crystal after etching.

The use of a mold capable of changing its size during crystal growth gives single crystals with a dislocation density of about 10^4 cm^{-2}. In Fig. 4b we present a picture of the etched cleavage plane of a bismuth crystal obtained in a "soft" mold. The "softness" of the mold is achieved by allowing the bars 5 forming the mold to move freely in the block 4 (see Fig. 1b), i.e., they are able to move during the crystallization process. The geometric characteristics of the mold are secured by thoroughly grinding the bars 5 to the block 4.

In conclusion it should be noted that the proposed growth method may also be used for the production of other low-melting metals.

Literature Cited

1.   Lavrent'ev, F. F., and Salita, O. P., Dokl. Akad. Nauk SSSR, 151: 1071 (1963).
2.   Soldatov, V. P., and Startsev, V. I., Fiz. Tverd. Tela, 6(6): 1671 (1964).
3.   Bazhenov, V. V., and Labzin, V. A., Fiz. Met. i Metalloved., 12(2): 289 (1961).
4.   Rybalko, F. P., Bainov, M. A., and Katanov, L. M., Fiz. Met. i Metalloved., 9(5): 796 (1960).
5.   Sharvin, Yu. V., and Gantmakher, V. F., Pribory i Tekhn. Éksperim., 6: 165 (1963).

PRODUCING SINGLE CRYSTALS OF COPPER, ALPHA BRASS, AND BETA BRASS

E. M. Savitskii and N. I. Novokhatskaya

A. A. Baikov Moscow Institute of Metallurgy

It is well known that the study of single crystals opens up new possibilities in the understanding of metals, since it eliminates the influence of grain boundaries. For copper and copper alloys, especially brasses, it is of great practical and theoretical interest to develop a theory for the falls in ductility occurring between 300 and 700°C. Although much work has been carried out on this subject, most has been concerned with polycrystalline samples [1-3]. In some papers the falls in ductility have been attributed to the effects of grain boundaries; hence, a corresponding study with single crystals appeared desirable.

Many papers have been written on methods of growing copper single crystals. Kunzler and Wernik used graphite molds for producing these [4]. Nouwelle [5] proposed a method of zone refining in vacuum or in a hydrogen atmosphere with a cooling rate of 6 mm/h. An interesting method involving a horizontal graphite boat for growing single crystals and bicrystals was proposed by Gow and Chalmers [6].

In general, copper single crystals may be grown by any of the methods: Bridgman, Czochralski, or recrystallization.

We decided on the Bridgman method, which gave single crystals of the required shape and quite large size. We constructed two systems for growing copper single crystals. In the first case we used a graphite heater-mold, the configuration of which gave the required temperature gradient. In choosing the shape of the heater we used the data of [7], the authors of which obtained aluminum crystals by this method, whereas we used it for copper single crystals. In contrast to the proposed method, however, we dispensed with the thermal-insulation layer, the use of which in vacuum apparatus is undesirable, owing to severe outgassing when heated; we replaced this by screens of sheet molybdenum.

The vacuum system consisted of backing and oil-vapor diffusion pumps, giving a rarefaction of

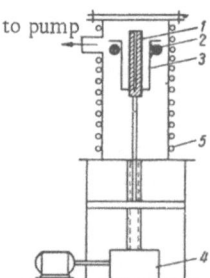

Fig. 1. Apparatus for growing single crystals by the Bridgman method.
1) Graphite mold; 2) current lead;
3) molybdenum heater; 4) reducer;
5) water cooling.

114

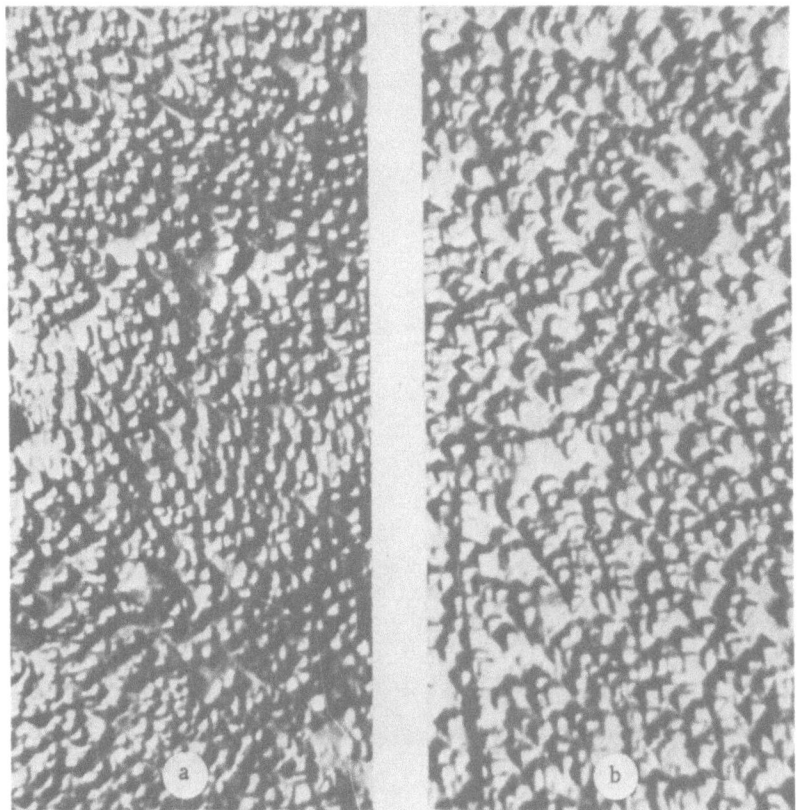

Fig. 2. Dislocation density of single crystals grown from: a) oxygen-
free copper; b) high-purity copper.

10^{-4} mm Hg. Heating was carried out by means of an OSU-40 reducing transformer, the required power being at least 1-1.2 kW. Cooling was effected at a rate of 1 mm/min by reducing
the electrical power fed to the furnace. In order to obtain single crystals of copper, the heating and cooling process had to be repeated two or three times without opening the chamber. In
this way, 25 copper single crystals were obtained; the orientation had no regular character,
despite keeping the growth conditions constant.

The advantages of the method include simplicity of construction and the absence of any
need to move the sample; the disadvantages include the conical form of the sample, the low percentage yield of useful single crystals, and the impossibility of obtaining single crystals of desired orientation.

In order to obtain samples of cylindrical form we set up an apparatus for growing single
crystals by the classical Bridgman method [8]. The arrangement of this is indicated in Fig. 1.
In the vacuum chamber, a graphite mold containing the molten metal, fixed to a water-cooled
rod, was slowly let down (by means of a mechanical drive and electric motor) through a heater
cut out of sheet molybdenum. The velocity of the mold was 4 cm/h. The construction of the
drive mechanism allowed the velocity to be varied from 2 to 120 mm/h. The required furnace
power was 1.2-1.5 kW. In this way we grew cylindrical copper single crystals from 1 to 30
mm in diameter and 120 to 150 mm long. A diameter of this order made it possible to cut out
several samples with the same orientation for mechanical tests.

On using the cylindrical mold and moving it through the heating zone at 4 cm/h, single
crystals of the diameters indicated were obtained at every attempt.

Fig. 3. General view of apparatus for the deformation-free machining of single crystals.

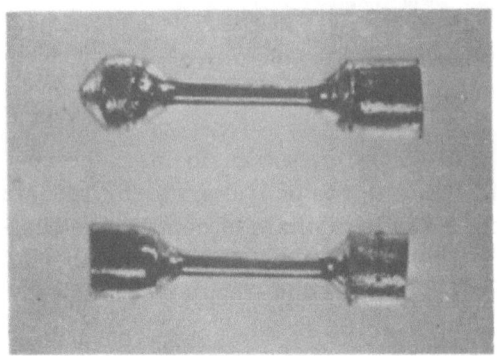

Fig. 5. Microsamples for tensile testing obtained from copper single crystals.

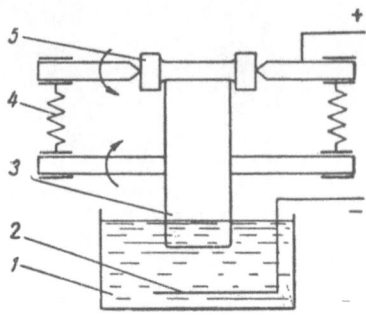

Fig. 4. Electrolytic method of preparing samples by means of a rotating disc: 1) electrolyte; 2) cathode; 3) rotating metal wheel covered with woolen cloth; 4) pressing spring; 5) sample-anode.

The original material included copper of the following types: M2 (99.7% pure), MO (99.95%), oxygen-free (99.99%), and highly purified V3 copper (99.996%). Gas analysis of the resultant single crystals showed that the oxygen content was much lower than in the original material (M2$_{orig}$, 0.10% O_2; M2$_{s.c.}$, 0.0007% O_2); in oxygen-free copper the impurity content was also reduced, but to a lesser extent (original: 0.0006% O_2, 0.0002% H_2, 0.0002% N_2; single-crystal: 0.0004% O_2, 0.0001% H_2, 0.0001% N_2).

In order to estimate the degree of perfection of single crystals of different purities we used a method based on the dislocation etching of the (100) plane. The samples were cut from single crystals with a fine abrasive wheel (thickness 0.75 mm) on a special machine designed for cutting crystals along a specified plane. After this, the crystals were etched to a depth of 0.5 mm and vacuum-annealed at 850°C for 5 h. Electropolishing and etching took place in a 60% solution of orthophosphoric acid.

Figure 2 shows dislocation etching of copper single crystals of different purities grown under the same conditions. We see from the figure that the dislocation density depends to a considerable extent on the purity of the material. Copper is a very ductile substance, so that we cannot link the dislocation density solely with the purity of the material. The dislocation density is here greatly influenced by the rate of cooling during growth of the single crystals, and by the care taken in preparing the sample for etching. In studying the mechanical properties of single crystals it is very important to preserve the single-crystal structure of the sample in the course of giving it the specified form. This is achieved by using the electrolytic method, which enables samples of the required form to be obtained from almost all materials having electrical conductivity. An advantage of this method is its simplicity and the fact that the original structure of the material is not destroyed. Figures 3 and 4 show the general form and the functional diagram of apparatus for the deformation-free machining of the single crystals. The working tool is the metal wheel constantly wetted with electrolyte, rotating at 6-8 rpm. The cathode is placed in the electrolyte and the anode is the sample itself. In order to eliminate direct electrical contact between the sample and the wheel, the latter is covered with woolen cloth resistant to the electrolyte. In order to obtain a circular profile, the sample must also rotate. The sample rotation rate may be equal to or greater than that of the wheel (8-10 rpm are suitable). If it is required to produce plane or square samples, the continuous rotation of the sample is replaced by periodic rotation of 90 or 180° around the axis. Figure 5 shows samples obtained by this method from copper single crystals. The electrolyte is a 50% solution of orthophosphoric acid, and the working voltage 10 V. Fluctuations in the diameter along the working part are no greater than 0.05 mm. No damage is done to the structure.

In order to obtain brass single crystals, a method based on that of Elam devised in 1926 was used [9]. Graphite molds were moved through a furnace at a high speed, so as to produce single crystals without losing zinc. A considerable amount of zinc nevertheless did evaporate and the single crystal obtained was inhomogeneous in composition. We proposed a method of producing single crystals of α-brass (L-95, L-90, L-85, L-75, L-65) and β-brass (L-52) in quartz ampoules sealed without evacuation. The diameter of the ampoules was made roughly equal to that of the sample. The free space in the ampoule over the sample was roughly one-third of the height of the ampoule. In order to create a temperature gradient, the porcelain heater tube of the furnace had a nonuniform winding. In the course of melting the zinc vapor had a partial pressure of about 1 atm. Under these conditions, and on moving the samples through the furnace at 13.55 cm/h, single crystals quite homogeneous in chemical composition were obtained. Thus, on taking samples from bottom to top every 15 mm, the zinc content in single crystals of the various types of brass were as follows: L-65, 34.10, 36.60, 36.02, 36.60, 37.70%; L-70, 31.78, 30.62, 31.01, 31.78, 32.17%; L-75, 24.46, 25.62, 25.04, 26.38, 32.55%; L-80, 22.92, 22.92, 23.10, 22.15, 25.62%.

In this way we obtained brass single crystals between 10 and 30 mm in diameter, from which we then prepared samples for mechanical testing. In contrast to copper, these samples may be prepared without harming the original single-crystal structure by very careful machining on a lathe with subsequently etching of the work-hardened layer and annealing at 650°C for 5 h.

Literature Cited

1. Gubkin, S. I., Deformability of Nonferrous Alloys. Metallurgizdat, Moscow (1947).
2. Savitskii, E. M., Effect of Temperature on the Mechanical Properties of Metals and Alloys. Izd. Akad. Nauk SSSR, Moscow (1957).
3. Presnyakov, A. A., Ductility of Metal Alloys. Izd. Akad. Nauk KazSSR, Alma-Ata (1959).
4. Kunzler, J. E., and Wernik, J. H., Trans. AIME, 212:856 (1958).
5. Nouwelle, H. J., Prop. phys. et chim. des metaux de trés hout pureté. CNRS, Paris, p. 221 (1960).
6. Gow, K. V., and Chalmers, B., Brit. J. Appl. Phys., 2:300 (1951).
7. Preobrazhenskii, A. Ya., and Stepanov, V. A., Pribory i Tekhn. Éksperim., 3:20 (1961).
8. Bridgman, P. W., Proc. Am. Acad. Arts Sci., 60:305 (1925).
9. Elam, C. F., Proc. Roy. Soc., 115(A):133 (1927).

PRODUCTION OF SINGLE CRYSTALS OF
THE RARE-EARTH METALS AND
A STUDY OF THEIR PROPERTIES

E. M. Savitskii, V. F. Terekhova,
and V. E. Kolesnichenko

A. A. Baikov Moscow Institute of Metallurgy

Single crystals of the rare-earth metals (rem) were first studied in 1957 [1]. Work by Soviet scientists began to appear in 1962 [2-4]. The majority of published papers relate to the magnetic and electrical properties of rem single crystals.

Rare-earth metals and their alloys are of great interest as subjects for study in connection with the physics of magnetism. In view of the fact that rem atoms have incomplete 4f-electron shells and layers lying quite deeply inside the atom, these metals exhibit extremely peculiar characteristics as regards variations of their magnetic properties.

Six of the rem exhibit ferromagnetism (gadolinium, terbium, dysprosium, holmium, erbium, and thulium) and four antiferromagnetism (cerium, neodymium, samarium, and europium).

A study of the magnetic properties of the rem is one of the most effective means of determining the nature of these metals and their compounds. Detailed studies of the ferro- and antiferromagnetism of the rem, their alloys, and their compounds, may also be of practical interest. These metals, in addition to those of the iron group, may be used in selecting new magnetic materials for the needs of industry and also other materials with special physical properties.

All the principal methods of producing single crystals may be divided into two classes: (1) the growth of single crystals from the melt, and (2) the production of single crystals from a solid polycrystalline sample by the method of recrystallization. Purification by electron-beam zone melting may also be applied.

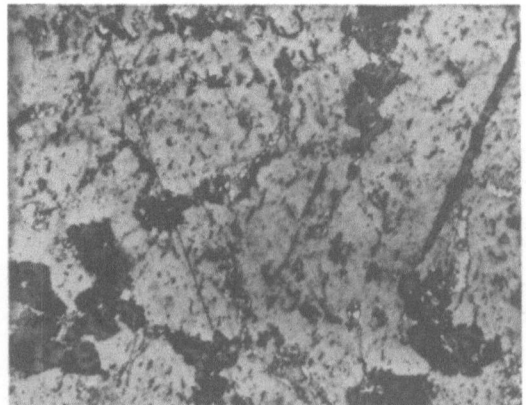

Fig. 1. Microstructure of original erbium.

118

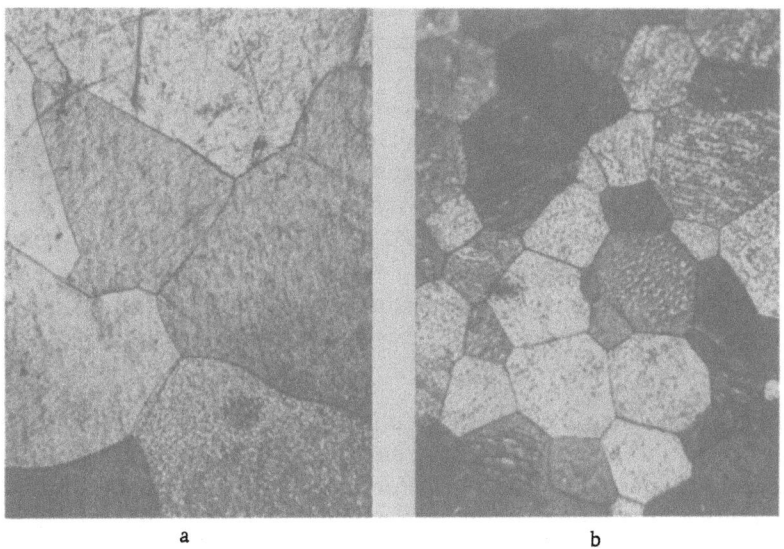

Fig. 2. Microstructures of distilled rem: a) gadolinium; b) yttrium.

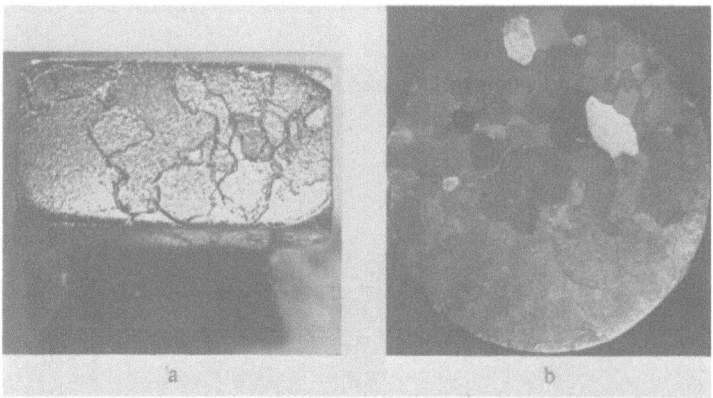

Fig. 3. Samples for growing single crystals: a) gadolinium; b) yttrium.

Fig. 4. Single crystals of yttrium grown by recrystallization annealing.

In view of the fact that methods of producing single crystals from the melt demand inactive and heat-resistant apparatus, they cannot be used for producing single crystals of the rem. In addition to this, the growth of single crystals from the melt cannot be used for polymorphic metals such as the rem. If polymorphism exists, then, on cooling, the grown single crystal may transform into a fine-crystalline aggregate of a new form [5].

We started our work on the growth of rem single crystals by adopting a suitable method of purifying the rem, since the growth of crystals in severely oxidized or contaminated metals is very difficult owing to the

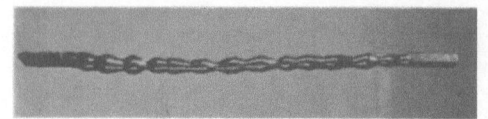

Fig. 5. Yttrium single crystal grown by zone melting.

blocking of grain boundaries. Hence, the first stage was distillation of the metals. The chief impurities in the original rem were oxygen, calcium, copper, iron, tantalum, etc. (Fig. 1). We developed conditions for the distillation of gadolinium, neodymium, scandium, yttrium, erbium, holmium, and dysprosium [4]. The purity of the distilled metals fluctuated between 99.5 and 99.8%. The distilled metals usually had coarse grains and fine boundaries (Fig. 2). In order to obtain samples for the growth of single crystals, the distilled metals were remelted in an electron-beam furnace. After remelting the metal usually contained the following amounts of impurities: oxygen, < 0.01%; nitrogen, < 0.01%; fluorine, 0.06%; and tenth parts of other rem.

In order to obtain large single crystals it is important to create such conditions as to minimize the number of nuclei developing in the sample. Usually the number of nuclei increases with increasing degree of plastic deformation. For this reason a previously worked sample, having a uniform fine structure, is subjected to a critical degree of deformation (3-6%) and then annealed. The annealing usually starts from a temperature below the recrystallization point, after which the temperature is gradually increased. In this way the crystallization centers which have already formed begin growing intensively, while no new ones develop. Final annealing is carried out at a temperature rather below that of the polymorphic transformation. We have now fully mastered the method and grown single crystals of gadolinium, neodymium, yttrium, scandium, and erbium. Figure 3 shows samples of yttrium and gadolinium from which single crystals have been grown.

The gadolinium, yttrium, and scandium were vacuum-annealed at 1250 to 1350°C for about 30 h.

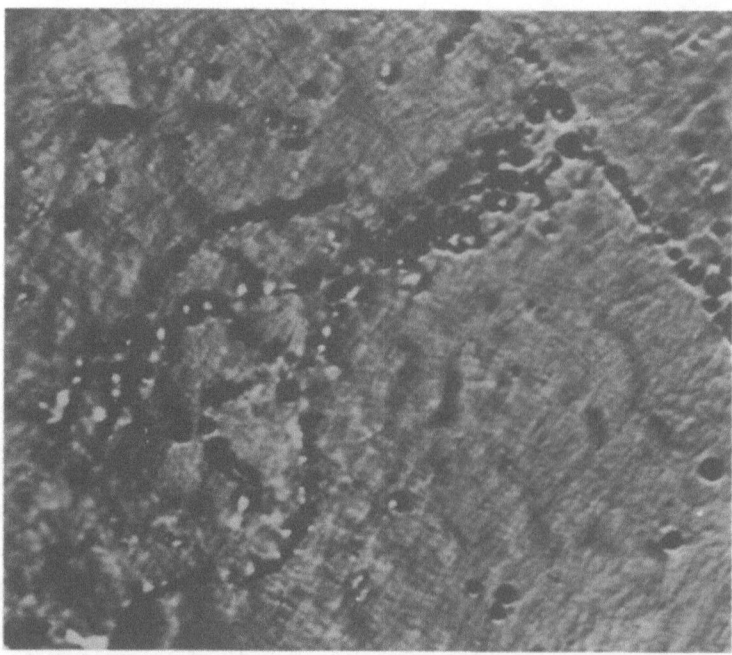

Fig. 6. Microstructure of an yttrium single crystal (× 2000).

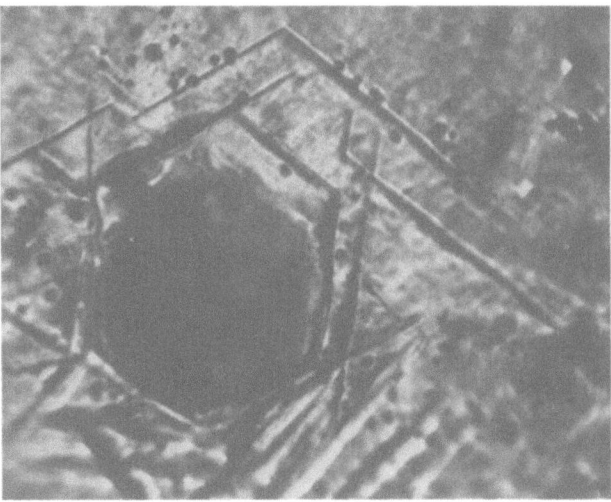

Fig. 7. Traces of twinning, forming hexagons on the basal plane of
an yttrium single crystal (× 2000).

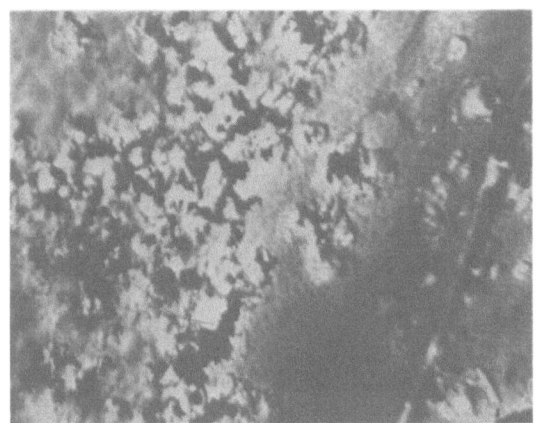

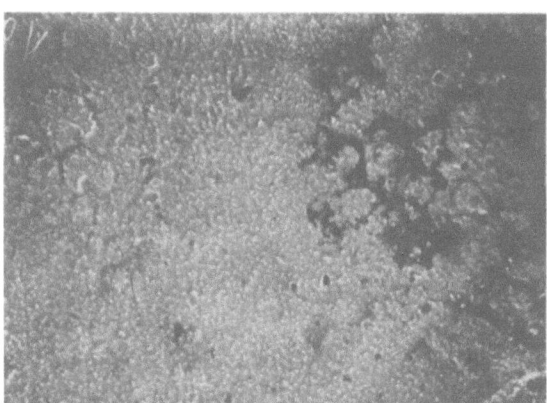

Fig. 8. Microstructure of an yttrium single   Fig. 9. Microstructure of an yttrium single
crystal (× 2600).   crystal (× 7500).

Owing to their volatility, erbium and holmium could not be annealed in vacuum, so that
the single crystals of these metals were grown in a helium atmosphere under a pressure slightly greater than 1 atm.

The annealed samples most frequently contain two or three large grains. Single crystals
were cut either on a special machine or by hand with a jeweler's saw. The work-hardened layer
is removed by etching in a solution of nitric acid, after which the single crystals are oriented
by the Laue method. Figure 4 shows prepared single crystals of yttrium growth by the method
of recrystallization annealing. By this method single crystals with dimensions of 12 × 10 × 5
mm were obtained.

In order to obtain large single crystals we have recently tried the method of producing
rem single crystals by electron-beam melting. Figure 5 shows an yttrium single crystal grown
by this method. The bar for zone refining was prepared from distilled and remelted metal.

Fig. 10. Microhardness impressions on the (1010) plane of an
yttrium single crystal.

The single crystal was obtained after three passes. After zone remelting, the sample was annealed for 1.5 h at 1000°C. Microstructures taken from the basal plane of this single crystal
are shown in Figs. 6-9.

Figure 7 shows clear traces of twinning around a defect; the traces form hexagons.
This confirms that the basal plane is obtained correctly. Figures 8 and 9 show microstructures
obtained in the electron microscope. The substructure may be seen in the photographs. More
careful research on the revelation of fine structure will be presented later.

The physical and mechanical properties were studied with oriented single crystals. It is
well known that at room temperature the majority of rem have hcp crystal structure. In
metals with hexagonal structure the variation in the properties for different crystallographical
directions is of special interest.

A considerable difference in the properties along the a and c axes was found on measuring the magnetic properties, the electrical resistance, the thermal-expansion coefficient, the
tensile strength, the ductility, the hardness, etc. We determined the thermo-emf, microhardness, and specific paramagnetic susceptibility for the single crystals.

On measuring the thermal emf relative to copper, we noted that a scandium single crystal had a high anisotropy of the thermo-emf: −8.604 μV/ deg in a direction perpendicular to
the c axis and −2.65 μV/deg in a direction parallel to the c axis. The anisotropy in the thermo-
emf of an yttrium single crystal was much smaller: 0.46 μV/ deg parallel to the c axis and
0.41 μV/ deg parallel to the a axis [2].

The anisotropy of the mechanical properties of single crystals with hexagonal structure
is widely known. We studied the microhardness of yttrium and scandium single crystals on the
(0001) basal plane and the (10$\bar{1}$0) plane. The microhardness of the yttrium single crystal on
the basal plane was 112 kg/mm^2 and that of scandium, 210 kg/mm^2. The mean microhardness
of yttrium on the (10$\bar{1}$0) plane was 61 kg/mm^2 and that of scandium, 102 kg/mm^2, i.e., only
half that on the basal plane. Of great interest is the anisotropy in the microhardness impressions on the (10$\bar{1}$0) plane, depending on the angle of rotation of the indentor. The impressions
on this plane were of irregular form and the length of the diagonals varied with varying angle
of rotation (Fig. 10). No twinning or slipping traces were observed. The difference in the

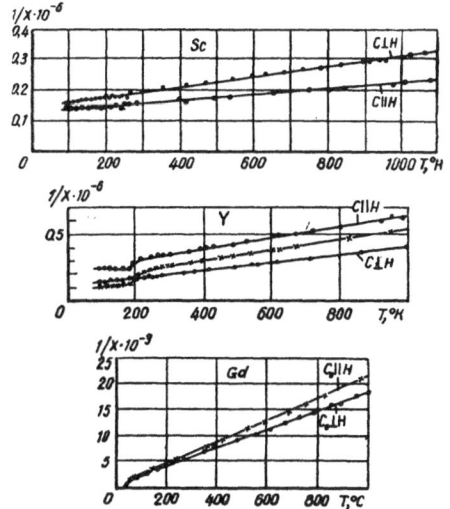

Fig. 11. Variation in the paramagnetic susceptibility of yttrium, gadolinium, and scandium with temperature.

microhardness in different planes is evidently due to a different mechanism of deformation under the impression of the indentor.

The specific paramagnetic susceptibility was measured by the Faraday—Sucksmith method from 77 to 1000°K. The sample dimensions were $2 \times 3 \times 6$ mm. Measurement was made in two directions, with the field parallel or perpendicular to the c axis.

Experiments showed that the magnetic susceptibility for yttrium was much greater for the direction in which the field was perpendicular to the c axis. In gadolinium the anisotropy appeared especially sharply for high temperatures, the magnetic susceptibility being greater in the perpendicular direction. At low temperatures up to 100°, the difference was not so sharp. In scandium the anisotropy of the paramagnetic susceptibility at room temperature reached 30% (Fig. 11).

Conclusions

We have perfected methods of growing rem single crystals (scandium, yttrium, gadolinium, erbium, etc.) (a) by recrystallization annealing, which enables more perfect single crystals to be obtained, and (b) by electron-beam zone remelting with subsequent high-vacuum annealing, which yields very large single crystals (up to 6 or 8 mm in diameter and 30-40 mm long).

We have studied the anisotropy of the physicochemical properties: magnetic susceptibility, thermal emf, and microhardness. We have found that the thermal emf along the a and c axes differs by a factor of 3 for scandium single crystals and by 30% for an yttrium single crystal. The values of microhardness obtained with respect to different planes in gadolinium, scandium, yttrium, and other single crystals differ by a factor of 2. The temperature-dependence of the paramagnetic susceptibility is distinguished by the fact that, for gadolinium, yttrium, and scandium the reciprocal of the paramagnetic susceptibility of the polycrystalline metal occupies a middle position between the values obtained for the a and c axes of single crystals.

The development of research into rem single crystals may make a considerable contribution to the theory of solid-state structure and be of practical importance in developing new materials for modern technology.

Literature Cited

1. Behrendt, D. R., Legvold, S., and Spedding, F. H., Phys. Rev., 106(4):723 (1957).
2. Naumkin, O. P., Terekhova, V. F., and Savitskii, E. M., Fiz. Met. i Metalloved., 16:5 (1963).
3. Savitskii, E. M., Terekhova, V. F., Naumkin, O. P., and Burov, I. V., Tsvetn. Metally (1963), p. 5.
4. Collection: Theory and Use of Rare-Earth Metals. Izd. Nauka, Moscow (1964).
5. Kuznetsov, V. A., Crystals and Crystallization. GITTL, Moscow (1954).

GROWTH OF RHENIUM SINGLE CRYSTALS
BY ELECTRON-BEAM ZONE MELTING

M. V. Pikunov, N. P. Koroleva,
K. V. Marunova, and E. I. Pavlova

State Institute of Rare Metals

The electron-beam melting of rhenium has been mentioned in a number of papers [1-7]; various properties of the single crystals were studied in [1-3], while the effects of melting conditions and other factors on the quality and degree of purity of the metal were considered in [6-7]. A number of important laws indicating the need for further research were found.

In the present investigation we studied the influence of technological factors (state and composition of the original material, velocity and number of passes) on the quality and purity of rhenium single crystals obtained by electron-beam zone melting, the predominant direction of growth of the crystals, and some of their other properties.

The experiments on zone melting were carried out on the apparatus constructed by the design and planning bureau of the State Institute of Rare Metals with a pressure of 1 to $5 \cdot 10^{-5}$ mm Hg in the working chamber. The leakage was $0.3 \, \mu/\text{min}$ for a chamber volume of 1 liter. Melting was carried out with tungsten and rhenium ring cathodes. The original material was cermet bars of rhenium of square cross section 5×5 to 7×7 mm and 150-250 mm long. The bars were sintered at 2300°C (in hydrogen or vacuum) from powder obtained by the hydrogen reduction of ammonium perrhenate by two methods, the differences between which are indicated

Table 1

Type of technology	Method of obtaining rhenium powder from ammonium perrhenate	Sintering atm.	Impurity content of bar			Density, g/cm³	Losses in melting for four passes (zone velocity 3 mm/min),%
			O_2	H_2	Σ, metallic		
A	Crushing the ammonium perrhenate and reducing in a current of hydrogen	hydrogen	$1 \cdot 10^{-3}$	$3 \cdot 10^{-4}$	$2.12 \cdot 10^{-2}$	18.6	30-35
		vacuum	$2 \cdot 10^{-3}$	$3 \cdot 10^{-4}$	$2.04 \cdot 10^{-2}$	18.6	30-35
B	Roasting the ammonium perrhenate of the original coarseness in nitrogen and then reducing in a current of hydrogen	hydrogen	$1 \cdot 10^{-2}$	$2.6 \cdot 10^{-3}$	$2.11 \cdot 10^{-2}$	17.5	15-20
		vacuum	$9 \cdot 10^{-3}$	$8 \cdot 10^{-4}$	$1.82 \cdot 10^{-2}$	17.5	15-20

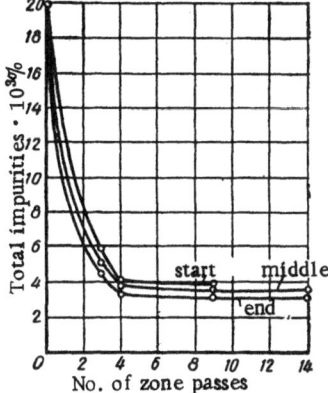

Fig. 1. Degree of purification as a
function of number of passes.

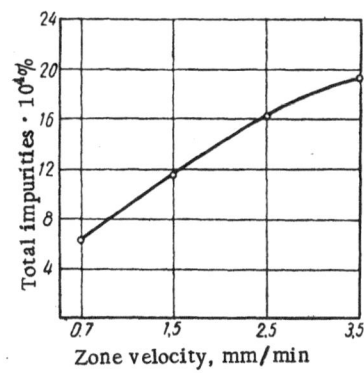

Fig. 2. Degree of purification as a
function of the zone velocity.

Table 2

Element	Content before melting, %	Content after melting, % (3 passes, v = 2 mm/min)
Fe, Mo	1·10⁻³	$5 \cdot 10^{-4}$
Al, Mg, Si, V, Ca	each	$1 \cdot 10^{-4}$
Ni, Cr		$< 1 \cdot 10^{-5}$
Cu, Mn	$1 \cdot 10^{-4}$	$< 1 \cdot 10^{-5}$
Cd, Bi, Sb, F, S, P, Hg, Ba, Sr	$1 \cdot 10^{-5}$	$< 1 \cdot 10^{-5}$
Ta, B	$1 \cdot 10^{-4}$	$1 \cdot 10^{-4}$
W	$1 \cdot 10^{-3}$	$1 \cdot 10^{-3}$
K, Na	$1 \cdot 10^{-3}$ $1 \cdot 10^{-4}$	$1 \cdot 10^{-3}$ $1 \cdot 10^{-5}$
Σ	1-$2 \cdot 10^{-2}$	2-$3 \cdot 10^{-3}$
O_2	$1 \cdot 10^{-2}$	$1 \cdot 10^{-3}$
H_2	$1 \cdot 10^{-3}$	$1 \cdot 10^{-4}$

in Table 1. GaS impurities were determined by vacuum heating with a sensitivity of $1 \cdot 10^{-3}\%$ for oxygen and $1 \cdot 10^{-4}\%$ for hydrogen. Metallic impurities were analyzed with a mass spectrometer. The total was given for 26 impurities: Al, Mg, Mn, Fe, Mo, Si, V, Pt, Ca, Na, Cr, Cu, Cd, Bi, Sb, F, S, P, Ag, Hg, Sn, Ta, B, W, K, Ni.

The sensitivity of the determination was $2 \cdot 10^{-4}\%$ Si; $2 \cdot 10^{-5}\%$ F, Na, Al, P, K, Ca; remainder 2-$4 \cdot 10^{-6}\%$.

We see from Table 1 that the sintering atmosphere (hydrogen or vacuum) has little effect on the purity of the bars of the loss of metal on melting. The greatest losses (30-35%) occur on melting rhenium obtained by method A. In this case it was only possible to create a molten zone over the whole cross section of the sample after three or four preliminary passes, during which the bar was heated up and partially melted. Even after these preliminary passes there was considerable "splashing" of the metal during the melting of such material, leading to fluctuations in the size of the molten zone and to the development of local contractions and thickenings in the samples. On melting samples prepared by method B, the process was far more stable,

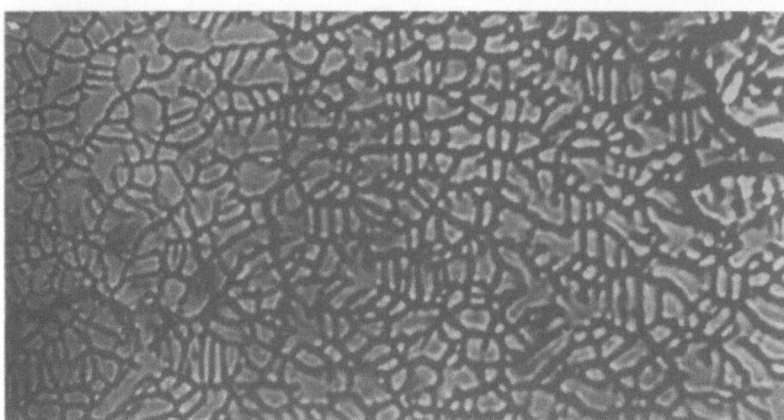

Fig. 3. Cell structure after deep electropolishing (× 400).

and after melting the rods had the same diameter along their entire length. Only one preliminary pass was needed in preparation for melting with a complete zone. The amount of metallic impurities in the original bars was roughly the same in both cases, although the amount of oxygen and hydrogen was an order greater in the bars obtained by the method giving the lower density.

A study of the effect of the number of passes (1-14) on the purification showed that after the first few zone passes the oxygen and hydrogen content fell to the limit of analytical sensitivity.

Figure 1 shows the amount of total metallic impurities in the rhenium as a function of the number of zone passes for a molten-zone velocity of 2.5 mm/min. We see that the limiting purification sets in after three or four passes, in agreement with the results of Lawley and Maddin [6]. The impurity distribution along the sample was quite uniform, with only an insignificant decrease near the end of the sample.

The effect of the zone velocity on the degree of purification was studied in the range 0.7 to 3.5 mm/min for four zone passes. The results of an analysis of 26 impurities are indicated as averages in Fig. 2. Reducing the zone velocity by five times reduces the content of metallic impurities by a factor of about three. The loss of metal, however, rises sharply, from 19% for 3.5 mm/min to 36% for 0.7 mm/min (four passes). A more complete picture of the extent of the purification is obtained on comparing analyses for individual metal impurities before and after melting (Table 2).

Clearly, the degree of purification is not always in direct relation to the vapor tension of the impurity. For example, iron and molybdenum are removed to approximately the same extent for an equal initial concentration, whereas the vapor pressure of iron at the melting point of rhenium is some 1000 times that of molybdenum. Magnesium, aluminum, silicon, and calcium are removed less than would be expected as compared with copper, manganese, nickel, and chromium. The results for the removal of potassium and sodium were rather unstable. Impurities of elements less volatile than or of similar volatility as rhenium (tantalum and tungsten) are not removed, as might be expected. For these elements it was also impossible to follow the redistribution resulting from zone melting, because the zone velocities were too small and the equilibrium partition coefficients close to unity.

After melting with two or three passes, the rhenium samples acquired single-crystal form. The rhenium bar grew as a single crystal after only one pass when obtained from the more porous original material with a zone velocity less than 1 mm/min.

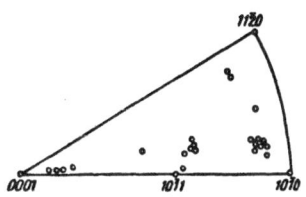

Fig. 4. Growth orientation of rhenium single crystals.

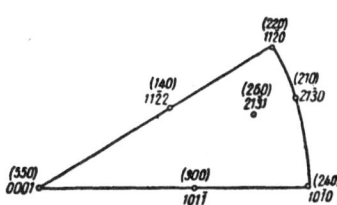

Fig. 5. Microhardness of rhenium in different planes.

The single crystals proved to be very ductile. In order to obtain perfect single crystals, it was important not to allow deformation when taking them out of the clamps or on subsequent study. In view of this the only acceptable method of cutting rhenium single crystals was the electrolytic method. Attempts to cut single crystals for metallographical study with a corundum disc not only led to extremely severe distortion of the metal structure near the cutting point but also caused deformation along the whole length of the bar. Electrolytic cutting of the samples was effected in the manner proposed in [8], using as electrolyte a mixture containing 1 g of chromic anhydride, 1 g of water, and 4 parts of glacial acetic acid. The same electrolyte was used for electropolishing the microsections and revealing the substructure of the single crystals. It was found that the most severe conditions of electropolishing were required for the basal (0001) plane: 40 V with a current density of 5-6 A/cm^2. The first-order prismatic plates ($10\bar{1}0$) were polished most easily: 20 V with a current density of 2.5-3 A/cm^2. The other planes were polished under intermediate conditions. After electrolytic etching (voltage 10-15 V, current density 1 A/cm^2) on planes close to the basal plane, etch figures in the form of hexagons were obtained. These figures did not form closed blocks but were arranged in the form of rows. On increasing the etch time, the number and mutual arrangement of these figures remained unaltered. After deep electropolishing at a voltage of 50-60 V, the cellular structure of Fig. 3 appeared on the ($11\bar{2}0$) plane of a single crystal grown in two zone passes at a velocity of 3 mm/min.

The direction of growth of some of the single crystals was determined by the Laue backphotography method. The results are shown on the stereographic triangle in Fig. 4. The growth axes are grouped mainly perpendicular to the $\{10\bar{1}0\}$, $\{10\bar{1}1\}$, and $\{0001\}$ faces. In the paper of Geach and Jones cited above [4], no preferred directions of spontaneous growth were found.

The perfection of four single crystals obtained under different conditions was studied by the Laue method, using a sharp-focus tube (diameter of focus 40 μ) giving a divergent beam. For single crystals grown with a zone velocity of 3-3.5 mm/min in three passes, there was a coarse block structure with block size 0.5 mm and disorientation of the order of 1-2°. Single crystals obtained with a zone velocity of 0.7 mm/min had a structure with a block disorientation of under 5-10'.

The microhardness was carefully measured with a PMT-3 hardness tester on the end surfaces of single crystals having different crystallographic orientations. The result presented in Fig. 5 shows the considerable influence of orientation on the microhardness, the difference in which reaches 400 units for different faces. It should be mentioned that the difference in microhardness measured on one plane but with different orientations of the indentor was no greater than 20-30 units, i.e., in this case there was no confirmation of the view expressed in [9] that the orientation of the indentor on the plane measured affected the magnitude of the microhardness more than the orientation of the plane itself.

Literature Cited

1.   Alekseevskii, N. E., Egorov, V. S., and Kozak, B. N., Zh. Éksperim. i Teor. Fiz., 44:3 (1963).
2.   Hauser, J. J., J. Appl. Phys., 33(10):3074-3077 (1962).

3.   Geach, G. A., Jeffery, R. A., and Smith, E., Rhenium, p. 84. Amsterdam-New York (1962).
4.   Geach, G. A., and Jones, F. O., J. Less-Common Met., 1(1) : 56 (1959).
5.   Chuprikov, G. E., Author's Abstract of Dissertation. Izd. IMET, Moscow (1963).
6.   Lawley, A., and Maddin, R., Acta Met., 8 : 12 (1960).
7.   Soden, R. R., Brennert, G. F., and Buchler, E., J. Electrochem. Soc., 1 : 112 (1965).
8.   Strutt, P. R., Rev. Sci. Instr., 4 : 32 (1961).
9.   Patridge, P. G., and Roberts, E., J. Inst. Met. (Oct. 1963).

SECTION II

CRYSTAL IMPERFECTIONS
AND THEIR STUDY

ON THE NATURE OF THE FORMATION OF DISLOCATION STRUCTURE IN METALLIC CRYSTALS DURING GROWTH FROM THE MELT

D. E. Ovsienko

Institute of Metal Physics of the Academy of Sciences
of the Ukrainian SSR

Real crystals as a rule contain various kinds of defects: pores, inclusions, macro- and micromosaics, dislocations, vacancies, displaced atoms, etc.; these have a considerable effect on a number of physical and mechanical properties. The recent vigorous development of new branches of science and technology, together with modern experimental research into the theory of strength and the electron structure of metals, demand the development of methods of producing crystals with perfect structures. The solution of this problem cannot be considered without a deep understanding of the processes of crystal growth and the nature of the imperfections arising in crystals.

In this short review we shall simply consider existing concepts regarding the formation of the mosaic structure of crystals growing from the melt, as well as imperfections on an atomic scale (dislocations, vacancies), to the extent to which they determine this structure. In addition to the well-known data already generalized in [1-3], we shall consider the results of work carried out in recent years which were not included in the earlier summaries, but have nevertheless greatly extended our knowledge of the subject.

Discovery and First Research on Mosaic Structure

In 1870 the Russian crystallographer Erofeev [4] suggested the idea that real crystals constituted a grouping of many indivisible crystallites. This definition was very similar to the concept of mosaic structure introduced by Darwin [5] after observing [6] the lack of agreement between the spacing and intensity of x-ray reflections obtained from real crystals and the corresponding quantities indicated by the theory of scattering based on an ideal lattice. The proposition of Darwin that a crystal consisted of indivisible regions easily disoriented relative to one another not only proved to be a convenient mathematical model for the description of imperfections in the crystal lattice, but was also accepted by many physicists as being close to reality. The intensity of research into crystal imperfections increased accordingly. Even in the early period (1930-1934) several hypotheses were advanced in order to explain the nature of the development of mosaic structure. These hypotheses may be divided into two main groups.

The first contains hypotheses explaining the origin of mosaic structure as an inherent property of thermodynamically stable crystals, and the second contains those associating its formation with disruptions in the regularity of crystal growth, the presence of impurities, etc.

The progenitor of the first point of view was Swicky [6], who considered that an imperfect crystal was more stable than an ideal one. Zwicky came to this conclusion on the basis of his own calculations and those of Lennard-Jones and Dent [7], from which it followed that the lattice parameter on the surface of a crystal, e.g., NaCl, was 5% larger than well inside it; this, in his view, must lead to the compression of the lattice and the appearance of cracks with a definite periodicity, or to the so-called secondary structure. The untenability of this theory was demonstrated even in 1934. Despite this, certain authors [10-12] continued to defend Zwicky's idea up until 1946, attempting to prove the existence of secondary structure. Without stopping to criticize these views, we must point out that the very fact that perfect silicon and germanium crystals have been obtained completely refutes them.

One of the authors of the second group of theories was Smekal [13], who showed that, in addition to the existence of a block structure, that of a perfect crystal was a real possibility. The block structure, in Smekal's view, arose in the course of growth or during subsequent deformation as a result of the penetration of pores, cracks, impurities, etc., into the body of the crystal, these constituting "weak" places and affecting structurally sensitive properties. Certainly, Smekal did not propose a specific model of mosaic structure, but he deserves great credit for actually emphasizing the necessity of separating crystal properties into structure-sensitive and structure-insensitive classes.

A more specific model, in the form of branched or dendritic structure (lineage), was proposed by Burger [8], who explained its origin as follows. During growth, a crystal nucleus, originally ideal, splits into separate branches with a slightly inclined orientation relative to the nucleus. On reaching large dimensions, the crystal will exhibit a "branched" structure, i.e., a collection of branches or blocks. The more the growth conditions depart from equilibrium, the more imperfect will the crystal be, as indeed is found experimentally. Burger, however, did not propose any mechanism for the formation of the branches in the ideal crystal, nor did he describe the boundaries of this structure. In addition to this, the actual concept of branched structure is too general. Despite this, Burger's hypothesis is more reasonable than those proposed earlier, since it associated the formation of the actual structure with the conditions of growth.

Considerable progress in explaining the character of the real structure of crystals and understanding the nature of its formation has been made in the last 10-15 years as a result of the theory of dislocations and modern methods of study. The concept of dislocations introduced in 1934 by Taylor [14] and Orowan [15], the dislocation model of low-angle boundaries, and also Frank's proof [17] of the possibility of describing any boundary by means of an appropriate distribution of edge and screw dislocations, enabled the hitherto existing indeterminacy in the description of substructural disorientations to be overcome. This has led to more purposeful investigations, so that in the last few decades it has been possible to discover several types of substructure and partly to explain the nature of their formation.

Formation of a Macromosaic Structure

Even to the unaided eye, metal crystals grown from the melt exhibit a coarse, so-called macromosaic structure. This includes such forms of substructures as dendritic structure, striations, corrugations (columnar structure), cellular structure, etc. These structures are evidently modifications of that originally described by Burger.

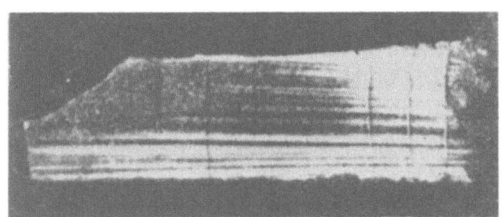

Fig. 1. Striated structure on the side surface of a tin single crystal [18].

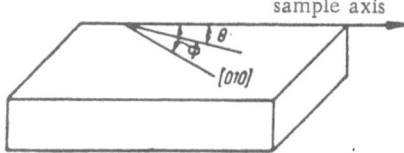

Fig. 2. Schematic arrangement of the boundary of a striated structure and the preferential growth direction [010], respectively, making angles of θ and Φ with the axis of an aluminum sample.

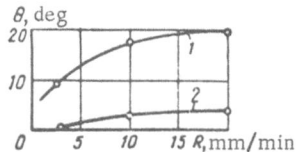

Fig. 3. Variation of θ with growth rate for aluminum crystals with different silver contents: 1) Al + 0.18 at.% Ag; 2) Al + 0.021 at.% Ag; $\Phi = 21$ deg.

Striated Structure. This type of structure (Fig. 1) was observed in many metals (Sn, Cu, Pb, Ag, Ni, Al, etc.) and studied in detail by Teghtsoonian and Chalmers [18]. A characteristic of this was an approximately parallel arrangement of bands or striations about 1 mm wide, the disorientation between which increased gradually from 0.5 to 5° on passing from the beginning of the crystal to the end. The total angle of rotation between neighboring striations in one direction approximately equals that in the opposite direction. As the growth rate increases, the transverse size of the blocks and the disorientation between them increases. The formation of this substructure requires a certain incubation period of about 1–3 cm, this being the longer, the slower the rate of growth. The direction of the striations depends greatly on the growth rate; for low growth rates they tend to arrange themselves in the direction of the thermal flux (parallel to the axis of the sample), and for high rates in the preferential growth direction, which is the [110] in tin and the [010] in aluminum. Impurities have a considerable effect on striated structure. In very pure Pb, Sn, Al (99.999%) the boundaries of the striations are corrugated or sinuous and very indistinct, but on adding 0.02–0.1% impurities [19, 20] they become straight and more regular. As the impurity concentration rises, the striations tend to arrange themselves parallel to the preferential growth direction even for low growth rates (Figs. 2 and 3). The effect of impurities depends on the magnitude and sign of the partition coefficient, for example, in aluminum the addition of 0.02 at.% Fe produces the same effect as that of 0.2 at.% of Ag, Mg, or Zn.

On the basis of the characteristics of striated structure just described, Aust and Chalmers [21] proposed a method for eliminating striations by setting up a thermal flux leading to the formation of an inclined boundary. The crystal has to be grown at a low velocity so that the distance along the normals to the inclined surface of separation should be smaller than the incubation length. Tin crystals grown under such conditions contained no striations, whereas, in the presence of a perpendicular surface of separation striations remained. It is an interesting fact that Noggle and Koehler [22], using the method of soft melting, obtained aluminum crystals containing a small number of striations with a dislocation density (0.5 cm^{-2}) two orders lower than in the experiments of Teghtsoonian and Chalmers. Teghtsoonian and Chalmers proposed a vacancy mechanism for the formation of striated structure, and we shall discuss this later.

Columnar or Cellular Structure. Both these are manifestations of one and the same structure, constituting, as it were, a bundle of pencils parallel to the growth axis. The cross section perpendicular to the growth axis has a honeycomb or cellular appearance (Fig. 4). This structure is found in metals containing impurities and is known as Smialovski structure

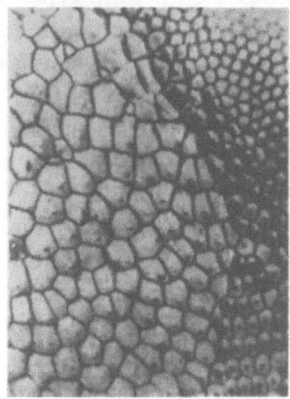

Fig. 4. Cellular structure of a decanted polycrystalline surface of separation, showing the effect of orientation on the size and shape of Pb cells [30].

[23], after the worker who first indicated its connection with impurity segregations. This structure was also studied by Pound and Kressler [24] on the crystal—melt surface of separation of polycrystalline metals obtained by rapid removal (decantation) of the liquid metal after partial solidification. Cells were observed in 99.897% Sn, 99.998% Bi, 99.999% In, 99.997% Al, etc. Hulme [25] showed that the disorientation between the fibers in zinc crystals was about 5', i.e., much less than in striated structure.

The results of [23, 24] were then confirmed and considerably expanded by Chalmers and his colleagues. Rutter and Chalmers [26] gave a qualitative explanation of the formation of this type of substructure, while Tiller et al. [27] theoretically considered the conditions required for its appearance.

These authors connected the formation of cellular structure with the concentration supercooling in front of the crystallization front discovered by Ivantsov [28] and called by him diffusion supercooling. Owing to the existence of concentration supercooling produced by the displacement of impurities in front of the crystallization front (for k < 1), the plane front loses its stability and decomposes into a series of cells, the vertices of which project into the melt while their boundaries are depressed and enriched with impurities. The vertices of the projections move across the impurity enriched layer into a region of higher supercoolings, where they grow more rapidly. Latent heat is released and the impurities are displaced; owing to diffusion flows parallel to the front, the impurities are carried away from the projections to the boundaries. Hence, the impurity concentration at the tip of a projection diminishes and that along the boundaries increases, correspondingly reducing the crystallization temperature.

If the temperature gradient in the liquid phase is so high that there is no structural supercooling, the surface of separation will be plane. Clearly, the condition for the changeover from a plane to a cellular surface of separation will be that the equilibrium temperature of the phase should equal the true temperature at the crystallization front. Hence, the condition for the stability of a plane front will have the form

$$G/R = -m(1-k)c_0/D_k, \qquad (1)$$

where G is the temperature gradient in the liquid phase, R is the growth rate, m is the slope of the liquidus line, k is the partition coefficient, c_0 is the original impurity concentration, and D is the diffusion coefficient of the impurity in the melt. The possibility of satisfying this condition was checked for tin containing traces of Pb [29] and lead containing Sn and Ag [30]; a linear relation between c_0 and G/R was obtained (Fig. 5).

It was shown in these papers and others that even a slight amount of impurity was sufficient for the formation of cellular structure in metals; for example, in zinc this structure occurred [31] in the presence of 0.05 wt.% Cd. At the same time, semiconductor crystals grown under the same conditions required a much higher impurity content. In germanium crystals in particular [32], cellular structure only started appearing after the addition of 0.2 at.% Ga; before this only twinning was observed. This is apparently associated with a different growth mechanism or with an increased energy of the surface of separation in germanium as compared with ordinary metals. Tiller and Rutter [30], studying Pb with added Sn, Ag, and Cu, observed

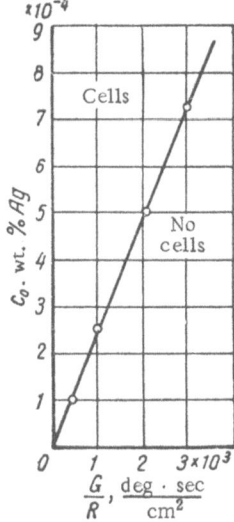

Fig. 5. Critical conditions for the transformation from a smooth to a cellular crystallization front in Pb−Ag [30].

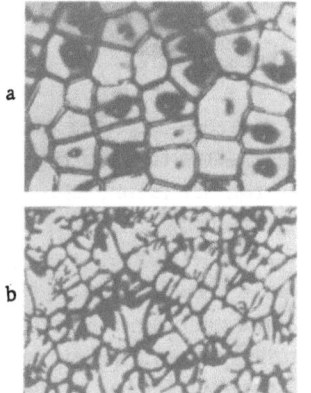

Fig. 6. Transformation from a cellular separation boundary to a dendritic one for Sn−0.2 at.% Pb alloy: a) G/R = 200; b) G/R = 31. × 150 [36].

that the transformation from a smooth to a cellular front, as well as the actual cell size, depended on the crystallographic orientation of the front (see Fig. 4). Bolling and Tiller [33] attempted a theoretical explanation of the observed variation in cell size with solidification conditions and obtained the following relation:

$$a = \frac{PD}{\pi R'} \{1 + [1 + (4\pi^2 R'/PD)(0.6\gamma/\Delta SG')]^{\frac{1}{2}}\}, \qquad (2)$$

where P is a quantity depending on the distribution of impurity in front of the cell, ΔS is the specific entropy of melting, and γ is the surface tension at the crystal−melt boundary. It follows from this expression that the cell size increases with rising c_0 and falling G and R, and also depends on γ, i.e., the crystallographic orientation. These theoretical conclusions agree qualitatively with the results of [34, 35] relating to zinc. It follows from the experiments of [30] that at a polycrystalline surface of separation the cell size is smaller for grains with small indices than for those with large. Hence, a boundary with small indices should absorb more impurities than one with large indices, so that it will find itself at a higher temperature, leaving behind the surface with large indices; finally, it may supplant the latter. As indicated by Tiller [37], this explains the preferential [100] orientation in the columnar zone of ingots of metals with face-centered lattices.

Assuming this mechanism, one may prevent the formation of cellular structure either by ensuring a slow growth rate, so that impurities may diffuse more effectively in the liquid, or by creating a sharp temperature gradient in the melt, thus removing the concentration supercooling.

If the concentration supercooling is substantial, the cellular form of the boundary begins converting itself into branched form (Fig. 6); this is called dendritic [36], although in the initial stage it has no sharp crystallographic structure. For a fairly large extent of the supercooled zone development into ordinary dendritic form may occur. Since the value of the structural supercooling depends on c_0, R, and G, the transformation from cellular to dendritic structure also depends on these factors. Tiller and co-workers [30, 38] considered lead containing traces of tin (Fig. 7) and showed that the formation of dendritic structure demanded larger concentrations and growth rates R but smaller gradients G than cellular structure. Since the critical conditions for the transformation of cells into dendrites depend on the orientation, a region exists (Fig. 7) in which some separation boundaries (with small indices) have cells and others dendritic structure. Outside this region, the boundary has one or the other structure for any orientation. Tiller [39] considers that the changeover from cellular to dendritic structure will take place on satisfying the conditions

$$G/R^{\frac{1}{2}} = A_\theta c_0/k, \qquad (3)$$

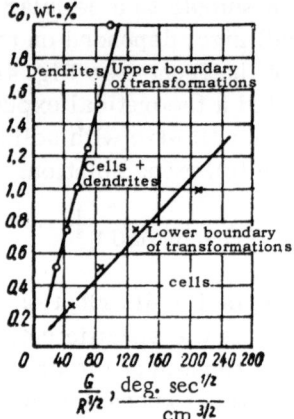

Fig. 7. Critical conditions for the transformation from cellular to dendritic form for Sn–Pb alloys [30].

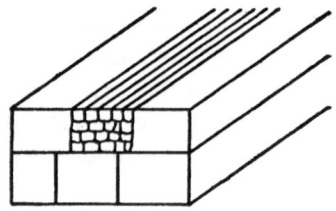

Fig. 8. Schematic representation of a crystal containing striated (large elements) and columnar (small elements) structure.

where A_θ is a constant depending on orientation. However, the experiments described in [36] showed that this condition was well satisfied in the case of lead containing Ag and Sn, while in tin containing Bi or Pb the relation between c_0/k and $GR^{1/2}$ was nonlinear. Hence, Eq. (3) requires further checking for other systems and a more rigorous theoretical derivation.

Relation Between Striated and Columnar Structure. It was noted in [19, 20] that a certain interrelation existed between the arrangement of striated and columnar (cellular) structure. It was found in particular when studying Sn, Pb, and Al crystals, that when the two structures existed simultaneously the boundaries between the striations and the columns always coincided as indicated schematically in Fig. 8. In the opinion of the authors of [19], this may be explained either by the fact that the elastic energy of the dislocation structures forming the boundaries between the columns may be reduced to a minimum as a result of impurities lying in the walls of the columns (cells), or else by the falling of a moving boundary into a "groove" (impurity wall of the cells), where it may become fixed and follow the motion of the forming cell.

This proposition was confirmed by Hunt and Smith [40], who, also studying tin, showed that the mutual arrangement of the two types of sub-boundaries depended on the form of impurity (Fig. 9). In the case of the addition of 0.1% Pb (k < 1) the boundaries of the striation, as a rule, coincide with the impurity walls, whereas, for the addition of 0.5% Sb (k > 3), the boundaries of the striation pass through the cells, intersecting their boundaries. It is true that in this case there is sometimes a coincidence of boundaries, but the striation boundaries easily migrate from the walls to the middles of the cells with high impurity concentration, where they are fixed. Unfortunately, these are so far the only experiments carried out on a decanted surface. Further development of these, we feel, should improve our understanding of the nature of the boundaries themselves. It would also seem interesting to trace the change in the arrangement of the sub-boundaries at various distances from the separation boundary, where (as shown for zinc in [41]) annealing may lead to a considerable rearrangement of the primary structure with the formation of a secondary coarser one [42].

Dendritic Growth. We noted earlier that for a sufficiently large extent of the region of structural supercooling the initial forms of dendrites might develop into ordinary dendrites. This process may be briefly described as follows. Some of the projections formed on the surface of separation may move forward into a region of high supercooling, leading to an increase in the growth rate of the projection and to its further motion into the liquid. The dissipation of the latent heat evolved in the volume of the liquid will also promote this. The rise in temperature and the displacement of impurity in the neighborhood of the projection will prevent the formation of new projections and retard the growth of projections-formed layer. As shown by Saratovkin [43], the side surface of the growing needles also contains toothlike asperities from which secondary branches are then formed. Since the branches of the dendrites

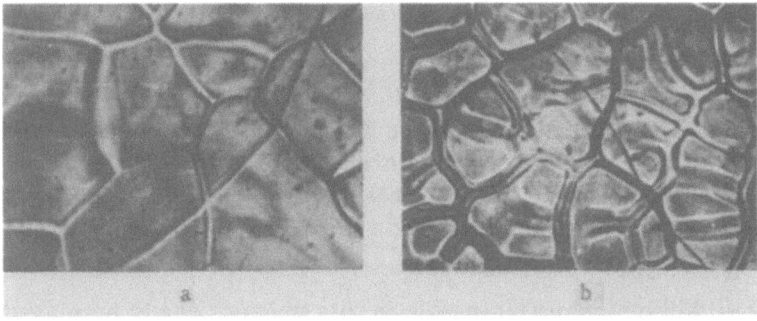

Fig. 9. Mutual arrangement of striated structure (indicated by arrow) on the decanted surface of tin containing: a) 0.5% Sb; b) 0.1% Pb.

contain less-dissolved substances, we find that, as solidification proceeds, the liquid remaining between the branches becomes richer and richer in impurities. These impurity-enriched parts will solidify last, filling the interdendrite space. In pure metals dendritic growth will also occur for fairly large (thermal) supercoolings, but without impurity inclusions.

An important feature of dendritic crystallization is the high growth rate, which may be several times the velocity of the principal front. It was found experimentally [44] that on changing the supercooling from 3 to 10° the growth rate R of tin dendrites increased from 3 to 25 mm/min, the relation between R and ΔT being described by an equation of form $R = A(\Delta T)^{1.8}$, which agrees closely with the expression $R = A(\Delta T)^{l}$ ($1.2 < l < 2.4$), theoretically obtained by Temkin [45]. Another important feature is the preferential direction of dendrite growth for each kind of crystal structure. Thus, in fcc and bcc metals, dendrites grow predominantly in the [100] directions, in hcp metals in the [1010] directions, and in tetragonal metals (tin) in the [110]. It is true that in the presence of impurities, as shown in [46], the growth direction may change. In pure metals the directions in question are also the axes of the dendrites, which should be parallel to each other. It is possible, however, that asymmetry may appear as a result of side flows of liquid due to convection, or as a result of vibration. This may lead to an incorrect growing together of the branches and to the appearance of dislocation-type defects at the joints. This is probably responsible for the disorientations of ~3° between the dendrites noted in [47].

Development of Dislocations During the Growth

of Crystals from the Melt

As has been shown, the sub-boundaries between neighboring, slightly disoriented elements of substructure may be described by means of a network of dislocations. Hence, a knowledge of the mechanism underlying the development of dislocations is also important in order to understand the nature of the formation of various kinds of substructure.

Experiments show that the density of dislocations (arranged at random or situated in walls) formed in crystals during growth from the melt depends greatly on the conditions of growth. There are grounds for considering that dislocations are formed both directly in the process of forming the crystal, i.e., at the boundary between the solid and liquid phases, and after the formation of the crystal while cooling from the melting point to room temperature. Different, recently proposed mechanisms may be involved in this. The growth mechanisms, the local capture of impurities, and the continuation of dislocations from the seed may take part in the first case, and a vacancy mechanism together with plastic deformation associated with nonuniform cooling of the crystal in the second.

Let us consider each of these mechanisms in more detail.

Development of Dislocations as a Result of Thermal Stresses. During the growth of a single crystal from the melt, cooling usually takes place in a nonuniform manner both along the axis of the sample and in the radial direction; this produces a nonuniform expansion of the lattice parameter in different parts of the crystal, leading to the development of stresses and hence strain. If these stresses are small, the strain may be elastic, but if they exceed a certain critical value, this leads to plastic yield with the formation of dislocations and other defects.

Billig [48] considered that the main role in this process would be played by the radial temperature gradient $\partial T / \partial r$, and proposed the following expressions for estimating the dislocation density:

$$\varrho = \frac{\alpha}{b} \frac{\partial T}{\partial r},\tag{4}$$

where α is the thermal expansion coefficient, and b is the Burgers vector. Billig also made some special experiments on silicon and germanium single crystals grown by Czochralski method with various temperature gradients, and observed an increase in dislocation density with increasing temperature gradient. Calculation by formula (4) for silicon gives $\rho \approx 10^4$ cm^{-2}.

Indenbom [49] considered that the axial rather than the radial gradient corresponded to the appearance of dislocations, since it was always the greater. According to Indenbom's views, owing to the existence of the axial gradient, the density of the growing layer is smaller than that of the underlying layer, and some of its atomic planes must break off, forming edge dislocations. The density of the dislocations formed by this mechanism in a lamellar crystal is determined by the relation

$$\varrho = \frac{\alpha}{b} \frac{\partial T}{\partial n},\tag{5}$$

differing from (4) in respect of the axial temperature gradient $\partial T / \partial n$; this may not produce the plastic deformation (strain) which is a necessary condition in (4). Calculations made for aluminum ($\alpha = 2.8 \cdot 10^{-6}$, $b = 2.85 \cdot 10^{-8}$ cm) for $\partial T / \partial n = 100$ deg/cm, give $\rho = 10^5 \text{ cm}^{-2}$. These estimates, however, have not been verified experimentally.

Examining the role of the axial and radial gradients, Rosi [50] established that an appreciable increase in the dislocation density of germanium (from $2 \cdot 10^3$ to $4 \cdot 10^3 \text{ cm}^{-2}$) only took place after reaching an axial gradient of 200-300 deg/cm; before this only the radial gradient (varied by varying the sample diameter) had an effect. For example, on changing the diameter from 6 to 29 mm, ρ increased from 10 to $7 \cdot 10^4 \text{ cm}^{-2}$ for an axial gradient of about 150 deg/cm. Wagner [51], however, only observed an increase in the dislocation density of silicon after thermal shock, i.e., sharp cooling. Dash [52] came to the conclusion, on the basis of experiments with germanium, that the existence of a temperature gradient merely promoted the multiplication of existing dislocations, since, in dislocation-free crystals, even thermal shocks produced no new dislocations.

Unfortunately, no similar experiments have been made for metals. We can thus only suppose that the formation of new and the multiplication of old dislocations in metals should take place more easily than in silicon and germanium, since the critical yield stresses for these are lower.

Vacancy Mechanism of the Formation of Dislocations. In a growing crystal there are always a certain number of vacancies, the equilibrium concentration of which

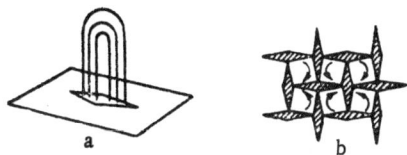

Fig. 10. System of dislocation loops [57].

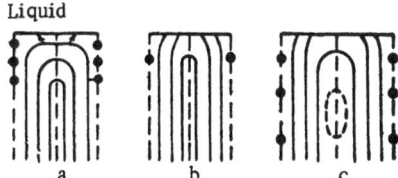

Fig. 11. Successive positions (a, b, c) of dislocation loops attracted to the crystal—liquid surface of separation [57].

falls off exponentially with temperature. Since the growth of crystals always involves a temperature gradient, the parts of the crystal at some distance from the crystallization front will have an excess of vacancies, which has to be eliminated by some mechanism or other. One such mechanism was proposed by Frank [53] and Seitz [54]. These workers supposed that in the presence of saturation the vacancies tended to gather together into groups in the form of discs or spheres, which then served as sinks, increasing in size with falling temperature. After reaching some critical size, such discs have to collapse, forming rings of dislocations, which are then able to grow by creeping [55].

Teghtsoonian and Chalmers used this idea in order to explain the striated structure mentioned earlier. In contrast to Seitz, Chalmers [56] considered that a half-disc of vacancies (bounded on one side by a surface of separation) might form, reaching critical size more rapidly than a whole disc generated within the crystal. After collapsing, this half-disc forms a half-loop of edge dislocations with its two ends (dislocations of opposite signs) coming out on the surface of separation. As the front moves (the authors suppose), these loops will lengthen as a result of excess vacancies joining them. At temperatures close to the melting point, these dislocations will be quite mobile and, if there are many of them, they will be able to join up into stable rows, forming a mosaic structure of the striated type. Since the number of rows of opposite sign is the same, this should lead to uniform rotations of the blocks in opposite directions. The rate of forming dislocations by this mechanism will increase with increasing growth rate and temperature gradient, since both factors promote supersaturation of vacancies.

Developing these ideas further, Frank [57] proposed that the dislocations formed by the vacancy mechanism would build themselves up into rows in the manner shown in Fig. 10a. The existence of such a system of loops is equivalent to the removal of a narrow rhomb of the lattice. Since such systems are arranged in mutually perpendicular planes, prisms should build up as in Fig. 10b, forming a striated mosaic. The removal of rhombs of material will rotate individual parts of the crystal lying between them in opposite directions. Calculations showed, however, that, if the striated structure were formed by this mechanism, then for the experimentally observed disorientations of about 2-3° the vacancy concentration would have to be 2-3 orders higher than was in fact possible.

In view of this, Frank had to suppose that the dislocation loops approaching the dislocation front were attracted to it by image forces and came out at the surface (Fig. 11). As a result of this, the number of dislocations and the degree of disorientation would continuously increase until the rate of forming new loops was balanced by the rate of annihilation associated with the meeting of opposite pairs.

On the basis of this mechanism, Frank calculated that the width of the striations varied in inverse proportion to the square root of the growth rate R and temperature gradient G ($\sqrt{RG}$), which in general agrees with experimental data. The existence of an incubation period for the formation of striated structure is here explained by the necessity of creating the appropriate supersaturation of vacancies, sufficient to initiate the condensation of these into discs.

Despite the fact that many characteristics of striated structure may be qualitatively explained by the theory in question, some of its assumptions, as indicated by Schock and Tiller [58], are none too well founded, so that the reality of this mechanism is subject to doubt.

In order to check the possibilities of the formation of dislocations by Frank's mechanism, these authors calculated the critical sizes of the vacancy discs, the corresponding critical supersaturation of vacancies, and the velocity of the dislocations. Assuming that the collapse of a spontaneously generated disc took place when its energy exceeded that of a dislocation ring, they obtained the following expression for the critical radius of the disc:

$$\frac{b}{r_{\text{к}}}\left[\ln\left(\frac{r_{\text{к}}}{b}\right)+1.2\right]=4\pi\left(\frac{\nu-\gamma_{\text{f}}}{3Eb}\right), \tag{6}$$

where ν is Poisson's ratio, E is the shear modulus, and γ_{f} is the energy of a stacking fault formed inside the dislocation loop. Calculation with this formula gives r_{K} almost equal to a few atomic distances. The critical supercooling at which the vacancy supersaturation required to keep a dislocation ring in equilibrium is reached may be calculated from the expression

$$\Delta T/T_{\text{m}}=3Eb^4/8\pi U_0 r\left[\ln\left(\frac{r}{b}\right)+2.8\right], \tag{7}$$

where U_0 is the energy of vacancy formation.

An estimate for Al, Cu, and Ag with $r_{\text{K}}=10$ V gives $\Delta T/T_{\text{m}}\approx 0.3$. The creeping velocity of the dislocation loops v, assuming that this may be determined from the rate of exhaustion of the surrounding regions of excess vacancy concentration $(c-c_0)$, is given in the form

$$v=0.5\frac{D_v}{b}\left(\frac{c}{c_0}-1\right), \tag{8}$$

where D_V is the vacancy diffusion coefficient.

After analyzing the temperature dependence of the velocity of the dislocations, Schock and Tiller showed that this passed through a maximum as a result of the interaction of two factors depending on temperature in different ways, viz., the mobility of the vacancies and their supersaturation. According to this estimate, the velocity maximum occurs at $\Delta T/T_{\text{m}}=0.1$, and for the ordinary temperature gradient of approximately 20 deg/cm (for copper) lies at a distance of 5 cm from the crystallization front. For a dislocation to experience the effects of image forces from the surface of separation, the latter must approach to within a distance of ≤0.1 mm from it. However, at this distance, $\Delta T/T_{\text{m}}=0.00025$, and the dislocation creep velocity equals 10^{-5} cm/sec, i.e., two orders smaller than the ordinary crystal growth velocity. From this the authors conclude that the dislocations cannot come out at the surface of separation, but will follow it at such a distance (~5 mm) that their velocity will equal the velocity of solidification. This conclusion contradicts the Frank mechanism, which cannot therefore explain the striated structure. This is also supported by the fact that striated structure occurred in Sn and Pb crystals grown at quite high temperatures (20° below the melting point), at which the critical supersaturation of vacancies necessary for the formation of dislocation rings was certainly not reached. The formation of striated structure in the opinion of these authors was associated with micro-segregations of impurities, while the condensation of vacancies only promoted the formation of dislocation loops.

Through rejecting the Frank explanation for the origin of striated structure, Schock and Tiller acknowledged the role of the vacancy mechanism [54] in the formation of dislocations. They considered that dislocation loops formed by the collapse of vacancy discs, 10-15 atoms in radius, would grow to visible dimensions of about 10 μ, introducing a certain contribution

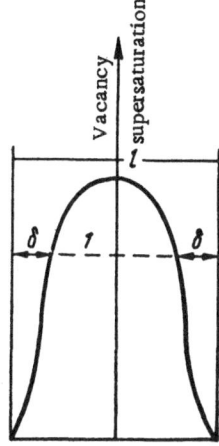

Fig. 12. Distribution of vacancy supersaturation over the cross section of the sample. 1) Proposed critical supersaturation.

into the total dislocation density. The maximum dislocation density obtained by this mechanism in the absence of other vacancy sinks is

$$\varrho_{max} = \frac{c - c_0}{2rb}. \qquad (9)$$

Calculation with this formula gives $\rho_{max} \approx 10^6$ cm^{-2} for a tenfold supersaturation, which may be achieved at a cooling rate of ~0.1 deg/sec. If the crystal contains other growth dislocations on which vacancies may condense, ρ may fall to 10^4 cm^{-2}. It was not considered, however, whether these loops could grow to visible dimensions. This question was considered by Elbaum [59], who refined expression (8) for v [v = $(\pi/b)(c - c_0)D_v$], and obtained an expression for the maximum radius of a loop r_{max} to which it could grow from the critical size r_K over a period (t) of cooling from the critical temperature T_K to room temperature

$$r_{max} = r_{\kappa} + \frac{\pi}{b}(c - c_0)D_v(T_{\kappa})\frac{RT_{\kappa}^2}{WU_{k}}\left[1 - \exp\left(-\frac{WU_{k}t}{RT_{\kappa}}\right)\right]. \quad (10)$$

Here, U_k is the kinetic energy of the vacancies and W is the cooling rate.

The results of the calculations show that v and ρ_{max} depend substantially (other conditions being equal) on the material of the crystal. Thus, for R = 10^{-3} cm/sec and G = 20 deg/cm, for Al and Cu the growth rate of the dislocation loops v at T_K is 64 and 0.6 cm/sec, respectively, and r_{max} is $3.7 \cdot 10^4$ and $3.2 \cdot 10^3$ cm, while for Si and Ge, for which v is much smaller (10^{-7} and 10^{-9} cm/sec), the value of r_{max} is only $1.45 \cdot 10^{-4}$ and $2.7 \cdot 10^{-6}$ cm. On the basis of these calculations, Elbaum concludes that, since the r_{max} of Si and Ge is very small, while the growth rate of the loops is much smaller than the over-all growth rate R, the loops never reach visible dimensions. Elbaum uses this to explain the fact that Dash [52] failed to observe the formation of dislocations as a result of the vacancy mechanism, even under the most favorable conditions. In the case of Al and Cu, however, the true size of the loops is not limited by r_{max}, since the growth rate of the loops is greater than the crystal growth rate.

Elbaum [60] considers that the vacancy mechanism plays the principal part in metals, basing this view on the fact that, using the Czochralski method, he was able to grow dislocation-free crystals under conditions in which their radius (about 0.1 mm) was smaller than the distance δ (Fig. 12) necessary for the diffusion of vacancies from the side surface during the period of crystal growth. Elbaum explained the formation of dislocations observed in samples of large diameter (i.e., for which $l > 2\delta$) by a dislocation mechanism, without, however, obtaining any direct proof of this.

Recently Jackson [61] again considered the conditions for the formation of dislocations by the vacancy mechanism, calculating the critical supersaturations S (supercoolings ΔT_K) necessary for the formation of 10^7 vacancy aggregates with various possible configurations in 1 cm^3. The results of these calculations, carried out on the assumption that the equilibrium concentration at the melting point was preserved on cooling to room temperature, are shown in Table 1.

We see from these data that even spherical aggregates, having the lowest energy for formation, may only be generated in the complete absence of other vacancy sinks at very low temperatures and correspondingly large supercoolings ΔT_K. Such generation cannot take place during the period of crystal growth, since, for the ordinary temperature gradient of about 10

Table 1

Metal	Sphere			Circular disc			Partial dislocations			Dislocation		
	T_K, °C	ΔT_K	S	T_K, °C	ΔT_K	S	T_K, °C	ΔT_K	S	T_K, °C	ΔT_K	S
Al	366	294	76.7	147	513	$1.05 \cdot 10^5$	282	388	631	55	605	$3 \cdot 10$
Cu	701	381	20.4	382	704	$3.9 \cdot 10^3$	855	228	4.8	—	—	—

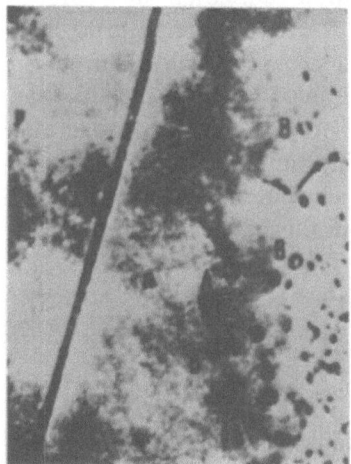

Fig. 13. Dislocation loops in quenched aluminum.

deg/cm complete solidification of the crystal occurs before any vacancy aggregates can develop. The author supposes that the formation of vacancy aggregates and correspondingly dislocations is only possible on quenching; this was also apparently found in the experiments of Hirsch and his co-workers [62, 63], who, in fact, obtained dislocation loops on electron-microscope pictures of quenched Al, Cu, and Au samples (Fig. 13), such loops not being observed after slow cooling.

It should be noted, finally, that in one of his later papers [64], Chalmers, one of the authors of the vacancy mechanism of striated structure, acknowledged the inadequacy of this hypothesis and indicated that a leading part was played by other mechanisms in the formation of dislocation structure in cast metals. The existence of striations, in particular, he linked with the amalgamation of boundaries with slight disorientations.

Thus, the role of the vacancy mechanism in the formation of dislocations during crystal growth from the melt is by no means clear, and further experimental and theoretical work is required. Meanwhile we must suppose that dislocations are formed by other mechanisms and that the presence of vacancies facilitates their creeping and formation into networks. In this some of the excess vacancies arising in the cooling process will be absorbed by the creeping dislocations, and the rest by coming out at the crystal surface, phase boundaries, and other points.

Continued Growth of Dislocations from the Seed. It was shown by Dash [65], Kokorish and Sheftal' [66], and others that the degree of perfection of the seed had a considerable influence on the perfection of the growing crystal. This may be explained as being due to the continuation of dislocations already existing in the seed, which may also be multiplied by thermal stresses. Dislocations may also propagate from surface defects of the seed (cracks, scratches, etc.), and also be formed as a result of imperfect epitaxy of the crystallizing material.

By reducing these effects to a minimum and using a dislocation-free seed, Dash [52] was able to grow silicon crystals free from dislocations. He showed, however, that dislocation-free crystals could also be grown from a seed containing dislocations, but that for this it was necessary to orient the system in such a way that the planes of least slip, the (111) or (100), should make as large an angle as possible with the growth axis. With this kind of orientation, the probability of the interface being intersected by dislocations arising from thermal stresses is reduced. It also becomes possible for dislocations to come out at the surface of the crystal, this being assisted by the presence of excess vacancies. It is interesting that Dash found no

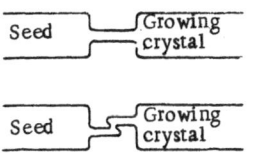

Fig. 14. Arrangement of constrictions for preventing the continued growth of dislocations from the seed into the growing crystal [67].

dislocations after subjecting dislocation-free germanium to thermal shock and thus achieving a considerable vacancy supersaturation. Dash concluded from this that vacancy saturation not only failed to create new dislocations but also promoted the emergence of existing dislocations at the surface of the crystal. No systematic investigation of this kind has been carried out for metals. Zasimchuk and Ovsienko [67], however, observed the effect of the seed on the degree of perfection of zinc crystals grown from the melt by a modification of the Bridgman method. They also proposed a method of eliminating this influence by a fine constriction (diameter ~1 mm) between the seed and the growing crystal (Fig. 14), so that a less perfect seed gave rise to more perfect crystals with the same orientation.

Formation of Dislocations During the Growth of Dendrites. It was indicated earlier that owing to convection currents in the liquid and other causes, there might be fluctuations in the branches of the dendrites during dendritic growth, leading to irregular intergrowth. As a result of this, dislocations might form at the joints, as observed by Lemmlein and Dukova [68] during the growth of toluidine crystals, as well as by other authors [2].

This mechanism should play a considerable part in growth from nuclei developing at large supercoolings and growing in the form of dendrites.

Formation of Dislocations as a Result of Impurities. When crystals are grown from the melt, there may always be a local capture of a high concentration of impurity, leading to a change in lattice parameters, this being the more pronounced, the greater the difference in the atomic radii of the impurity and solvent. This may give rise either to stressed regions, or to dislocations if these stresses are large enough. Such considerations constitute the basis of existing theories of the impurity mechanism of dislocation formation.

Frank [69] considers that, when the partition coefficient of the impurity between the solid and liquid phases is smaller than unity, the crystal will gradually become enriched with the impurity as it grows, leading to a gradual increase in the difference between the lattice parameters and hence to an increase in dislocation density from the beginning of the crystal to the end. This effect was in fact observed in the experiments of Esin and Kralina [70] on aluminum crystals, in which a regular change in the mosaic angle occurred.

Goss, Benson, and Pfann [71], studying germanium single crystals containing 0.2 at.% Sn and 6 at.% Si, observed series of dislocations at points of high impurity concentration. Assuming that all the stresses produced by a sharp change in concentration Δc led to the formation of dislocations, they proposed the following expression for the maximum linear dislocation density

$$n = \Delta c \left(\frac{da_0}{dc} \right) \frac{1}{a_0 b}, \tag{11}$$

where da_0 / dc is the change in lattice parameter a_0 for unit change in impurity concentration, and b is the Burgers vector. The authors' calculations for Ge containing Si ($da_0/dc = 2.8 \cdot 10^{11}$ for 1 at.%, $a_0 = 5.6 \cdot 10^{-8}$ cm, $\Delta c = 0.1$ at.%) gave n = 1200 cm^{-1}, which agrees closely with the observed etch-pit density of ~1400 cm^{-1}. However, this agreement is unconvincing, since in the majority of the layer lines corresponding to concentration fluctuations there were no etch pits in the photographs presented. This indicates that the assumption made in the calculations (that all the stresses were removed by dislocations) was incorrect.

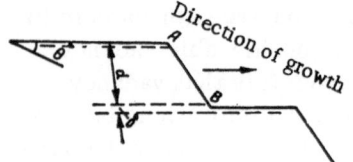

Fig. 15. Layer-like growth
illustrating the capture of
impurities.

Tiller [72] also attempted to estimate the density of the dislocations formed by the impurity mechanism in layer-like and cellular surfaces of separation. In contrast to the other authors, he also considered the actual mechanism of the local microsegregation of impurity, the essence of which reduced to the following. In the case of layer-like growth, which Tiller considered characteristic of many metals, the impurity driven back in front of the separation boundary will be carried away less readily by diffusion from re-entrant angles of the B type (Fig. 15) than from the leading edges of the plate A; hence, a higher concentration will be captured at points B. As a result of this process, as the crystallization front moves forward, a thin layer δ is formed with an increased concentration of impurity and a lattice parameter differing from that of the adjacent layers. Assuming that the removal of stresses arising in the regions of microsegregation takes place as a result of the introduction of edge dislocations, Tiller obtained expressions for the dislocation density both in the case of layer-like and cellular surfaces of separation. Later [73] these calculations were refined and a general formula for both forms of segregation was proposed:

$$\varrho = \frac{\Phi}{b}\left(\Delta c \frac{\Delta r}{r} - E_e\right), \tag{12}$$

where Φ is the form factor of the impurity structure, which is taken as about 4/d for layers and long cells and 8/d for equiaxial cells (d = width of cell or layer), b is the Burgers vector, Δr is the difference between the atomic radii of the solvent and impurity, r is the atomic radius of the solvent, E_e is the limit of elastic deformation, and Δc is the average excess impurity concentration (in atomic proportions) in the segregation region, which may be estimated from the formula

$$\Delta c = \frac{1}{W + k/(1-k_0)}\left[c_\infty + \frac{k_0 D}{m(1-k_0)} \cdot \frac{G_\tau}{R}\right]. \tag{13}$$

Here, W is the ratio of the area of the impurity boundary to the total area of the element of substructure (0.1-0.3), k_0 is the equilibrium partition coefficient, c_∞ is the impurity concentration in the volume of the liquid, G_T is the temperature gradient in the solid phase, m is the slope of the liquidus line, and R is the growth rate of the crystal.

It follows from the calculations made for zinc containing traces of cadmium [71] that the excess concentration of Cd in the boundaries of the elongated cells may reach about $9c_\infty$, and the dislocation density calculated from formula (12) on the assumption $E_e = 0$ is $\sim 10^8 c_\infty/d$. We see from (12) that the effect should depend on the form of the impurity. Thus it follows from the calculations of [72] that for layer-like growth of aluminum crystals the addition of Mn should give a much greater dislocation density ($\sim 10^6$ cm^{-2}) than the same concentration of Cu (10^4 cm^{-2}).

This theory is more perfect than the others, but it nevertheless remains qualitative, since it contains a number of simplifying assumptions, and the formulas obtained are approximate. In particular, formula (13) cannot be used for very small values of c_∞; this follows from the fact that for $c_\infty = 0$, Δc has a finite value, which has no physical meaning. The equations obtained also contain some other parameters (W, k_0), which are hard to determine, so that calculation is not really very definite.

The absence of exact calculations, however, does not raise any doubts as to the fact that sharp changes in impurity concentration may create stresses sufficient for the formation of vacancies. The contribution of this mechanism may under certain circumstances be very

Table 2

Material	R, cm/sec	δ_{max}, min	L, cm	ϱ_{exp}, cm^{-2}	ϱ_{theor}, cm^{-2}
Al (99.995%)	$1.6 \cdot 10^{-4}$	23	0.02	$5 \cdot 10^6$	—
Al (99.995%)	$1 \cdot 10^{-2}$	50	0.001	$1 \cdot 10^7$	—
Al (99.9995%)	$8.3 \cdot 10^{-4}$	15	0.1	$1.6 \cdot 10^5$	—
Al (99.9995%)+0.02 at. % Cu	$8.3 \cdot 10^{-4}$	21	0.07	$3.4 \cdot 10^5$	$2 \cdot 10^5$
Al (99.9995%)+0.02 at. % Ag	$8.3 \cdot 10^{-4}$	34	0.06	$5.6 \cdot 10^5$	$1.2 \cdot 10^4$
Al (99.9995%)+0.005 at. % Cr	$8.3 \cdot 10^{-4}$	16	0.04	$4.7 \cdot 10^5$	$1.1 \cdot 10^4$
Al (99.9995%)+0.01 at. % Mg	$8.3 \cdot 10^{-4}$	17	0.05	$4.5 \cdot 10^5$	$1 \cdot 10^5$
Zn (99.997%)	$4.1 \cdot 10^{-4}$	10	—	—	—
Zn+0.04 at. % Cd	$4.1 \cdot 10^{-4}$	13	—	$8 \cdot 10^5$	$2 \cdot 10^6$
Zn (99.997%)	$3.5 \cdot 10^{-3}$	26	—	$1 \cdot 10^6$	—
Zn+0.04 at. % Cd	$3.5 \cdot 10^{-3}$	46	—	$3 \cdot 10^6$	—

Note: According to the phase diagrams [85], the amounts of impurity introduced were in all cases smaller than the solubility limit in the solid phase at room temperature.

considerable. Thus, in Al−7.4% Mn crystals the authors of [74] found a dislocation density of about 10^7 cm^{-2}. Kratochvil [75], attempting to verify the validity of Tiller's theory [72], established that on changing the total concentration of Cu, Cd, and Pb impurities from 0.5 to 6 at.% the etch-pit density at the cell boundaries of zinc crystals grown at a rate of 1 cm/min changed by a factor of 3, from $1.3 \cdot 10^5$ to $3.7 \cdot 10^5$ cm^{-2}; on increasing the velocity from 0.6 to 2.3 cm/min, the etch-pit density increased greatly, from $1.8 \cdot 10^5$ to $1.1 \cdot 10^6$ cm^{-2}. According to the data of [76], changing the zinc concentration in cadmium crystals from 0.01 to 10 at.% produced no change in the etch pits revealed by the ion-bombardment method. This result is rather strange, since the amount of Zn introduced was much greater than the solubility in solid Cd, which should have led to the precipitation of a new phase, and hence to the appearance of an additional dislocation density. It may be that this method was insufficiently sensitive. Clearly, more accurate results could be obtained by using the x-ray method.

A considerable amount of work has been carried out in the Institute of Metal Physics [67, 77-81, 86] on the influence of growth conditions and traces of impurities on the degree of perfection of aluminum and zinc single crystals [67, 81]. The x-ray method has been used most of all, supplemented by chemical etching (for zinc containing traces of Cd). The principal results of this work are presented in Table 2, which shows the crystal growth rate R, the mean mosaic block size L, the maximum mosaic angle δ_{max}, and the dislocation density ρ_{exp} calculated from x-ray data (for Al) or etch pits (for Zn + Cd). The values of ρ_{theor} calculated from formula (12) are given for comparison.

We see from the data for Al (99.995%) that with increasing growth rate the size of the elements of substructure becomes smaller and the disorientation between them larger. Results for purer aluminum and also for zinc show that the introduction of even hundredth parts of a percent of impurities lead to a very great increase in the degree of imperfection of the crystals. As shown by the results for zinc containing traces of cadmium, the effect of impurity increases together with the growth rate. In the case of the aluminum of lower purity (99.995%) also, the effect of growth rate is associated with the presence of impurities, although aspects of the growth mechanism to be mentioned a little later may also appear here.

It thus follows from these data that changes in the degree of perfection of crystals occur as a result of the presence of impurities. The formation of dislocations as a result of thermal stresses, as indicated by estimates based on the experimental conditions (G ≈ 20 deg/cm), is negligibly small ($\sim 10^3$ cm^{-2}). On comparing the dislocation densities calculated from the x-ray data and formula (13), we see that for all the alloys except Al−Ag these are quite close to each other. However, considering the extremely approximate character of the various estimates, we can only justifiably say that the experimental data agree qualitatively with Tiller's theory. The fact that the effect of Ag was unexpectedly large in comparison with the theoretical predictions requires further study. It is possible that this is due to secondary processes associated with the precipitation of a new phase on cooling the crystal to room temperature.

There are good grounds for supposing that in metals the impurity mechanism should play a greater part than in semiconducting crystals, in which the energy of dislocation formation is much higher; for example, it is 27–28 eV for Ge and Si, and 4–5 eV for Al and Cu [59]. In fact, Alekseeva and Eliseev [82] showed that in germanium crystals, after adding bismuth within the limits of solubility, the etch-pit density remained the same as in pure germanium ($\sim 10^3$ cm^{-2}). An appreciable increase in dislocation density (up to 10^6 or 10^7 cm^{-2}) was only observed after the Bi concentration exceeded the solubility limit; this was associated with the precipitation of an aging phase created by the high stresses. Analogous results were obtained in [83] for Ge crystals containing Se, Ag, Cd, and Fe.

It seems to us that in the formation of mosaic structure the mechanism here considered may be accompanied by others associated with the influence of impurities on the structure of the melt, the mechanism of crystal growth, etc. If the crystal grows by two-dimensional nucleus formation, as in the case of gallium [84], there is always a certain supercooling in front of the crystallization front, this being the greater, the higher the growth rate. In the presence of sufficiently large supercoolings, nuclei may form on particles of active impurities which, falling on a growing face, may build up on this in a not quite regular fashion, forming defects of the dislocation type. The impurities may also change the rate of forming two-dimensional nuclei, retard their growth, and, gathering together along the periphery, lead to irregular intergrowth. Crystals may in fact grow not only by the addition of single atoms but also by that of whole regions with short-range order, the stability and coordination of which may also depend on the presence of impurities. The role of these factors in the degree of perfection should appear all the more strongly, the greater the supercooling at the front, i.e., the higher the crystal growth rate. It is as yet hard to judge the contribution of the proposed mechanism in the formation of dislocations, since the actual growth of metal crystals from the melt is not sufficiently clear, and hardly anything is known of the structure of the layer of liquid adjacent to the separation boundary. The mechanism should nevertheless not be neglected.

In conclusion it should be mentioned that impurities lead not only to the formation of dislocations but also to a limitation of their mobility, so that the effectiveness of the emergence of dislocations from the crystal becomes lower. Hence, an extremely high purity of the metal is required in order to obtain perfect crystals. In addition to this, the growth of the crystals should be carried out at a very low speed at a strictly constant temperature.

Conclusion

It follows from the foregoing discussion that a great deal of experimental material has been gathered together in recent years on imperfections in metal crystals grown from the melt. Substructures of various types and forms may arise according to the particular conditions of growth and the presence of impurities. Not all these, however, have received a satisfactory explanation. The modern views on the formation of dislocations set out in this review remain as yet on the level of suppositions, requiring proof and further development. So far there is no real certainty regarding the origin of striated structure. The Teghtsoonian−Chalmers−Frank theory, which earlier appeared to give a qualitative explanation of this structure and was generally accepted [2], now lies under some doubt, since it contains a number of unfounded assumptions and fails to explain certain important features of striated structure. The role of the vacancy mechanism in the formation of dislocations during growth is also not clear.

Considerable progress has been made in recent years in explaining the effects of growth parameters and soluble impurities. The introduction of the concept of structural supercooling and detailed consideration of the supersaturation of impurities during growth has made it possible to explain the formation of cellular (columnar) structure and (partly) dendritic growth forms. It has also been shown that the presence of very small quantities of impurities leads to the formation of dislocations. The formation of dislocations as a result of impurities and the multiplication of dislocations under the influence of thermal stresses takes place much more easily in metals than in semiconductor crystals such as germanium and silicon, in which the energy of dislocation formation is much greater and the mobility of the dislocations lower. This evidently also explains the fact that up until now it has proved impossible to produce perfect metal crystals, whereas, in the case of germanium and silicon, these are obtained quite "easily."

In order to obtain perfect crystals it is thus essential to have extremely pure metals and preserve strictly controlled conditions of growth.

It should be noted, finally, that so far the study of dislocation growth structure has been limited mainly to metals with low melting points, and little attention has been paid to the more refractory type; this is because of methodical and other difficulties associated with the characteristics of the latter. Meanwhile, the extension of such research to refractory metals is extremely important in view of the high prospects of their practical use and the influence which various types of imperfections may have on the physical and mechanical properties.

Literature Cited

1. Hirsch, P. B., in: Advances in Metal Physics. Vol. 3, p. 383. [Russian translation]. Metallurgizdat, Moscow (1960).
2. Elbaum, C., Usp. Fiz. Nauk, No. 3, p. 545 (1963).
3. Hurl, D. T., Growth Processes and the Growing of Single Crystals [Russian translation]. p. 303. IL, Moscow (1963).
4. Popov, T. M., and Shafranovskii, I. I., Crystallography, p. 142. Moscow (1955).
5. James, R. Optical Principles of X-Ray Diffraction [Russian translation]. IL, Moscow (1950).
6. Zwicky, F., Z. Krist, 24:131 (1923).
7. Leonard-Jones, J. E., and Dent, B. M., Proc. Roy. Soc., A121:247 (1928).
8. Burger, M. J., Z. Krist, 89:195 (1934).
9. Orowan, E. Z., Z. Krist., 89:242 (1934).
10. Kljatschko, J., Koloid. Beih., 44:386 (1936).
11. Born, M., Proc. Meth. Phys. Soc. Egypt, 3:35 (1946).
12. Ioffe, V. S., Usp. Khim., 2:144 (1944).
13. Smekal, A., Z. Krist., A89:386 (1934).
14. Taylor, G., Proc. Roy. Soc., A145:362 (1934).
15. Orowan, E., Z. Phys., 89:605, 614, 634 (1934).
16. Bürgers, J. M., Proc. Koninkl. Akad. Met. Amst., 42:293 (1939).
17. Frank, F. C., Conf. on Plastic Deform., Washington (1950), p. 150.
18. Teghtsoonian, M., and Chalmers, B., Can. J. Phys., 29:370 (1951); 30:388 (1952).
19. Atwater, H. A., and Chalmers, B., Can. J. Phys., 35:208 (1957).
20. Grath, J. T., and Creig, G. V., Can. J. Phys., 40(7):850 (1962).
21. Aust, K. T., and Chalmers, B., Can. J. Phys., 36:977 (1958).
22. Noggle, T. S., and Koehler, J. S., Acta Met., 3:260 (1955).

148   D. E. OVSIENKO

23. Smialovski, M., Z. Metallk., 29:133 (1937).
24. Pound, R. B., and Kressler, S. W., J. Met., 3: 1156 (1951).
25. Hulme, K. F., Acta Met., 2:818 (1954).
26. Rutter, J. W., and Chalmers, B., Can. J. Phys., 31:15 (1953).
27. Tiller, W. A., et al., Acta Met., 1:428 (1953).
28. Ivantsov, G. P., Dokl. Akad. Nauk SSSR, 81(2): 179 (1951).
29. Walton, D., et al., Trans. AIME, 7: 1023 (1955).
30. Tiller, W. A., and Rutter, J. W., Can. J. Phys., 34:96 (1956).
31. Domiano, V., and Herman, M., Trans. Am. Inst. Min., Met. Petrol., 215:196 (1959–1960).
32. Bolling, C. F., Tiller, W. A., and Rutter, J. W., Can. J. Phys., 34:234 (1956).
33. Bolling, C. F., and Tiller, W. A., J. Appl. Phys., 31(11): 2040 (1960).
34. Boček, M., Kratochvil, P., and Valouch, M., Czech. J. Phys., 8: 557 (1958).
35. Kratochvil, P., Lukac, P., and Valouch, M., Czech. J. Phys., 10:48 (1960).
36. Plasket, T. S., and Winegard, W. S., Can. J. Phys., 38: 1077 (1960).
37. See collection: Liquid Metals and Their Solidification. Metallurgizdat, Moscow (1962).
38. Moris, W., et al., Trans. Am. Soc. Met., 47:403 (1955).
39. Tiller, W. A., Can. J. Phys., 34(7): 729 (1956).
40. Hunt, M. D., and Smith, R. W., Can. J. Phys., 40: 9 (1962).
41. Domiano, V. V., and Tint, Q. S., Acta Met., 9(3): 177 (1961).
42. Biloni, H., and Bolling, G. F., Trans. Met. Soc., 227: 6 (1963).
43. Saratovkin, D. D., Dendritic Crystallization. Metallurgizdat, Moscow (1957).
44. Rosenberg, A., and Winegard, W. C., Acta Met., 2:242 (1954).
45. Temkin, D. E., Dokl. Akad. Nauk SSSR, 6:1307 (1960).
46. Esin, V. O., Fiz. Met. i Metalloved., 17(2): 298 (1964).
47. Weinberg, F., and Chalmers, B., Can. J. Phys., 29:382 (1957).
48. Billig, E., Proc. Roy. Soc., A235:37 (1956).
49. Indenbom, V.L., Kristallografiya, 2:294 (1957).
50. Rosi, E. D., RCA Rev., 13:349 (1958).
51. Wagner, R. S., J. Appl. Phys., 29:1679 (1958).
52. Dash, W. C., Dislocations and the Mechanical Properties of Crystals [Russian translation], IL, Moscow (1960).
53. Frank, F. C., Symp. Deform. Cryst., p. 89. Pittsburgh (1950).
54. Seitz, F., Phys. Rev., 50:890 (1950).
55. Cottrell, A. H., Dislocations and Plastic Flow in Crystals [Russian translation]. Moscow (1958).
56. Chalmers, B., J. Met., 6:519 (1954).
57. Frank, F. C., Verformung und Fliessen des Festkörpers (1956).
58. Schock, B. C., and Tiller, A. W., Phil. Mag., 5:49, 43 (1960).
59. Elbaum, C., Phil. Mag., 5:55 (1960).
60. Elbaum, C., J. Appl. Phys., 32(4): 742 (1961).
61. Jackson, K. A., Phil. Mag., 7(73): 1117 (1962).
62. Hirsch, P. B., et al., Phil. Mag., 3:877 (1958).
63. Hirsch, P. B., and Silcox, J., Grow a Perf. Cryst., Wiley (1958), p. 262; Phil. Mag., 4: 72 (1959).
64. Chalmers, B., J. Phys. Soc. Japan, 3(18): 64 (1963).
65. Dash, W. C., J. Appl. Phys., 30:459 (1959).
66. Kokorish, K. Yu., and Sheftal', N. N., Growth of Crystals, Vol. 3, p. 388. Izd. Akad. Nauk SSSR, Moscow (1961). [English translation: Consultants Bureau, New York (1962).]
67. Zasimchuk, I. K., and Ovsienko, D. E., Ukr. Fiz. Zh., 10:1092 (1964).
68. Lemmlein, G. G., and Dukova, E. D., Kristallografiya, 1:3 (1956).
69. Frank, F., Phil. Mag. Suppl., 1:1 (1957).

70. Esin, V. O., and Kralina, A. A., Fiz. Met. i Metalloved., 13(4) : 557 (1962).

71. Goss, A., Benson, K., and Pfann, W., Acta Met., 4 : 113 (1956).

72. Tiller, W., J. Appl. Phys., 29 : 611 (1958).

73. Tiller, W., Acta Met., 10 : 618 (1962).

74. Jmura, T., and Suzuki, T., J. Japan Inst., 1: 19 (1955).

75. Kratochvil, P., Czech. Phys. J., 12 : 927 (1960).

76. Lyapunina, N. A., et al., Fiz. Met. i Metalloved., 4 : 582 (1962).

77. Ovsienko, D. E., and Sosnina, E. I., Fiz. Met. i Metalloved., 2 : 375 (1956).

78. Ovsienko, D. E., and Sosnina, E. I., Fiz. Met. i Metalloved., 3 : 516 (1956).

79. Sosnina, E. I., and Ovsienko, D. E., Fiz. Met. i Metalloved., 3 : 527 (1956) .

80. Sosnina, E. I., Meleshko, L. I., and Ovsienko, D. E., In: Study of Imperfections in Crystal Structure. Naukova Dumka, Kiev (1965).

81. Zasimchuk, I. K., and Ovsienko, D. E., Kristallografiya, No. 11, p. 325.

82. Alekseeva, V. G., and Eliseev, P. G., Fiz. Tverd. Tela, 1(8) : 1309 (1959).

83. Belyaev, A. D., Vasilevskaya, V. N., and Miselyuk, E. G., Growth of Crystals, Vol. 3, p. 380. Izd. Akad. Nauk SSSR (1961). [English translation: Consultants Bureau, New York (1962).]

84. Alfintsev, G. A., and Ovsienko, D. E., Dokl. Akad. Nauk SSSR, 4 : 792 (1964).

85. Vol., A. E., Structure and Properties of Binary Metal Systems. Moscow (1961).

86. Sosnina, E. I., and Meleshko, L. I., This collection, p. 165.

THEORETICAL ANALYSIS OF THE CONDITIONS AFFECTING IMPURITY DISTRIBUTION DURING DIRECTIONAL CRYSTALLIZATION

B. I. Birman and B. L. Timan

All-Union Scientific-Research Institute of Single Crystals

The theoretical problem of the distribution of impurities in a crystal during directional crystallization and the diffusion mechanism of transfer in the melt has been solved in a number of papers [2-4]. In all these papers it has been assumed that the initial impurity distribution in the melt is constant and the flow of impurities at the upper end of the ampoule is zero.

At the same time the diffusion equation for the melt

$$\frac{1}{D}\frac{\partial c}{\partial t} = \frac{\partial^2 c}{\partial x^2} \tag{1}$$

was solved for initial and boundary conditions

$$c(x, 0) = c_0 = \text{const}, \tag{2}$$

$$j_L = -D\frac{\partial c}{\partial x}\bigg|_{x=L} = 0, \tag{3}$$

$$(1-k)\frac{dy}{dt}c\bigg|_{x=y(t)} + D\frac{\partial c}{\partial x}\bigg|_{x=y(t)} = 0, \tag{4}$$

and the impurity concentration in the solid phase was found from the equation

$$c_s(x) = kc(x, t)\big|_{t=y^{-1}(x)}, \tag{5}$$

where $c(x, t)$ is the concentration of impurity in the melt, $c_f(t)$ is the impurity concentration at the crystallization front on the melt side, D is the diffusion coefficient in the melt, $y(t)$ is the distance to the crystallization front from the initial point of the ampoule, $j_L(t)$ is the axial diffusion flow of impurity at the upper free surface of the melt, and $V = dy/dt$ is the crystallization velocity.

The law governing the motion of the crystallization front, known from the solution of the thermal problem, was taken as linear in [2-4] (constant crystallization rate) and either linear or proportional to the square root of the crystallization time in [3].

It follows from these treatments that the manner in which impurity enters into the crystal is not uniform. For large values of the parameter VL/D, i.e., for comparatively large crystal lengths and growth rates, in the initial stages of growth there is a buildup of impurity at a distance of the order of $\delta \approx D/V_K$ in the crystal in front of the crystallization front on the melt side (formation of a zone of concentration packing), and the impurity concentration in the crystal increases. In the middle of the growth process, when a stable zone of concentration packing has been formed at the crystallization front, the impurity concentration in the crystal is approximately equal to c_0. At the final stage of growth there is a rise in the concentration of the impurity built up in the melt as a result of the absence of any removal process; the intensity with which the impurity enters into the crystal rises sharply, and the upper part of the crystal is severely contaminated. For small values of the parameter VL/D, the manner in which the impurity enters into the crystal resembles the law represented by Pfann's formula. This means that intensive diffusion transfer may lead to a sharp increase in the impurity content at the end of the crystal even in the absence of fully developed convection.

Since the initial distribution of the impurity in the melt must affect the manner in which it passes into the crystal, as must the flow of impurity from the other end of the ampoule, the rate of crystallization, and the variation in growth conditions, we must solve Eq. (1) under conditions more general than those formulated in (2) and (3). The direct solution of the problem in this case involves serious mathematical difficulties. In order to solve the problem, it is therefore more convenient to adopt an approach inverse to that employed in [2-4].

Let us formulate the inverse problem thus: we have to solve the diffusion equation (1) with the impurity balance condition (4) satisfied at the crystallization front and with a known law of variation for $c_f(t)$. The latter condition is analogous, having an advance prescription of the impurity distribution in the crystal. With this presentation of the problem, the initial distribution of impurity in the melt $c(x, 0)$ and the diffusion flow $j_L(t)$ enabling this distribution to be set up, are unknown quantities.

In this form the problem may be solved by the method of successive approximations [5] used in [6] for determining the thermal field in the crystallization of bodies of the simplest shapes. According to the method of successive approximations, the solution of Eq. (1) may be sought in the form

$$c(x, t) = \sum_{n=0}^{\infty} c_n(x, t). \tag{6}$$

As zero approximation $c_0(x, t)$ we take the stationary solution of the diffusion equation

$$\frac{\partial^2 c_0}{\partial x^2} = 0, \tag{7}$$

and in order to find the successive approximations we solve the recurrence differential equations

$$\frac{\partial^2 c_n}{\partial x^2} = \frac{1}{D} \frac{\partial c_{n-1}}{\partial t}. \tag{8}$$

The boundary conditions for the zero approximation are given in the form

$$c_0(x, t)\big|_{x=y(t)} = c_f(t), \tag{9}$$

$$D\frac{\partial c_0}{\partial x}\bigg|_{x=y(t)} = -(1-k)\frac{dy}{dt}c_f(t), \tag{10}$$

and for the successive approximations in the form

$$c_n(x, t)\big|_{x=y(t)} = 0 \qquad n > 1, \tag{11}$$

$$\frac{\partial c_n}{\partial x}\bigg|_{x=y(t)} = 0 \qquad n > 1. \tag{12}$$

The zero approximation satisfying boundary conditions (9) and (10) has the form

$$c_0(x, t) = c_f(t) + \frac{1}{D}[y(t) - x](1 - k)\frac{dy}{dt}c_f(t). \tag{13}$$

By successively integrating Eq. (8), preserving boundary conditions (11) and (12), we obtain the following expressions for the n-th approximations $(n \geq 1)$:

$$c_n(x, t) = \frac{1}{D}\int_y^x\left[\int_y^\eta \frac{d}{dt}c_{n-1}(\xi, t)\,d\xi\right]d\eta = \frac{1}{D^n}\frac{d^{n-1}}{dt^{n-1}}\left\{\frac{dc_f}{dt}\frac{[y(t) - x]^{2n}}{(2n)!}\right\}$$

$$+ \frac{1}{D^{n+1}}\frac{d^n}{dt^n}\left\{[1 - k(t)]\frac{dy}{dt}c_f(t)\frac{[y(t) - x]^{2n+1}}{(2n+1)!}\right\}. \tag{14}$$

Collecting together the terms of the zeroth and n-th approximations, we obtain the desired solution to the diffusion equation for the problem considered (7):

$$c(x, t) = \sum_{n=0}^{\infty}\frac{1}{D^n}\frac{d^n}{dt^n}\left\{\frac{[y(t) - x]^{2n}}{(2n)!}c_f(t)\left[1 - \frac{k(t)[y(t) - x]}{(2n+1)D}\frac{dy}{dt}\right]\right\}. \tag{15}$$

This solution enables us to analyze the conditions for the impurity to pass into the crystal in accordance with any desired law of variation along the length of the crystal, for an arbitrary law v(t) of growth velocity as a function of time, and also makes it possible, in principle, to consider any relationship between the partition coefficient, the growth velocity v(t), and the concentration at the crystallization front $c_f(t)$.

The solutions contained in formula (15) strictly satisfy Eq. (1) and boundary condition (4). The solutions may serve to determine the initial distributions of impurity in the melt necessary to ensure any required penetration of impurity into the crystal for any one of the several forms of variation of the diffusion flow at the opposite end of the ampoule; they are thus useful in practical applications. However, there are clearly still further solutions of the diffusion equation which satisfy (1) and (4) and ensure any required penetration of impurity into the crystal for other laws of variation of the diffusion flow $j_L(t)$. These solutions correspond to rather different initial distributions of impurity in the melt; they cannot be obtained by the method based on using the stationary solution of the diffusion equation as zero approximation.

Let us consider some particular solutions of (15) representing considerable interest in the practice of crystal growth.

1. For a constant partition coefficient but an arbitrary law of variation of crystallization velocity, the expressions for c(x, t) and j(x, t) in the melt will be (to an accuracy of terms of the order of $1/D^3$) as follows:

$$c(x, t) = c_f(t)\left\{1 - (1 - k)\left[\left(z - \frac{z^3}{2} + \frac{z^3}{6}\right) + W_2\left(\frac{z^3}{6} - \frac{z^4}{8}\right) + \right.\right.$$

$$+ W_3 \frac{z^5}{120} \Big] \Big\} + \frac{D}{\dot y^2} \frac{dc_f}{dt} \Big\{ \Big[\frac{z^2}{2} - \frac{(2-k) z^3}{6} + \frac{(3-2k)z^4}{24} \Big]$$

$$- W_2 \frac{(3-2k) z^5}{120} \Big\} + \frac{D^2}{\dot y^4} \frac{d^2 c_f}{dt^2} \Big\{ \Big[\frac{z^4}{24} - \frac{(3-k) z^5}{120} \Big] \Big\} + \frac{D^3}{\dot y^6} \frac{d^3 c_f}{dt^3} \frac{z^6}{720} + \cdots, \qquad (16)$$

$$j(x, t) = -\dot y \Big\{ c_f(t)(1-k) \Big[\Big(1 - z + \frac{z^2}{2} \Big) + W_2 \Big(\frac{z^2}{2} - \frac{z^3}{2} \Big) + W_3 \frac{z^4}{24} \Big] \Big\}$$

$$+ \frac{D}{\dot y} \frac{dc_f}{dt} \Big\{ \Big[z - \frac{(2-k) z^2}{2} + \frac{(3-2k) z^3}{6} \Big] - W_2 \frac{(3-2k) z^4}{24} \Big\}$$

$$+ \frac{D^2}{\dot y^3} \frac{d^2 c_f}{dt^2} \Big\{ \Big[\frac{z^3}{6} - \frac{(3-k) z^4}{24} \Big] \Big\} + \frac{D^3}{\dot y^5} \frac{d^3 c_f}{dt^3} \frac{z^5}{120} + \cdots. \qquad (17)$$

Here we have introduced the following dimensionless parameters:

$$z = \frac{x - y(t)}{D} \frac{dy}{dt}, \qquad W_2 = D \frac{\dfrac{d^2 y}{dt^2}}{\Big(\dfrac{dy}{dt} \Big)^3}, \qquad W_3 = D^2 \frac{\dfrac{d^3 y}{dt^3}}{\Big(\dfrac{dy}{dt} \Big)^5}. \qquad (18)$$

Formulas (16) and (17) are suitable for calculations with small z, i.e., for small growth rates or at small distances from the crystallization front. We see from the structure of formulas (16) and (17) that the power series with the coefficients W_2, W_3, etc., characterizing the change in growth rate with time, begins with a power of higher order than the power series depending only on the growth rate. This means that the change in growth rate with time only slightly affects the nature of the impurity distribution in the melt immediately next to the crystallization front.

2. Of greatest interest in the practical growing of single crystals is the case of constant growth rate.

It follows from formulas (16) and (17) with $W_2 = W_3 = \ldots = 0$ that the expressions for the impurity concentration in the melt and the diffusion flow of impurity will be as follows:

$$c(x, t) = c_f(t) \Big[1 - (1-k) \Big(z - \frac{z^2}{2} + \frac{z^3}{6} \Big) \Big] + \frac{D}{V^2} \frac{dc_f}{dt} \Big[\frac{z^2}{2} - \frac{(2-k) z^3}{6}$$

$$+ \frac{(3-2k) z^4}{24} \Big] + \frac{D^2}{V^4} \frac{d^2 c_f}{dt^2} \Big[\frac{z^4}{24} - \frac{(3-k) z^5}{120} \Big] + \frac{D^3}{V^6} \frac{d^3 c_f}{dt^3} \frac{z^6}{720} + \cdots, \qquad (19)$$

$$j(x, t) = -V c_f (1-k) \Big[1 - z + \frac{z^2}{2} \Big] +$$

$$+ \frac{D}{V} \frac{dc_f}{dt} \Big[z - \frac{(2-k) z^2}{2} + \frac{(3-2k) z^3}{6} \Big] + \frac{D^2}{V^3} \frac{d^2 c_f}{dt^2} \Big[\frac{z^3}{6} - \frac{(3-k) z^4}{24} \Big] + \frac{D^3}{V^5} \frac{d^3 c_f}{dt^3} \frac{z^5}{120} + \cdots. \qquad (20)$$

Let us study these expressions in the case of a small parameter VL/D. If we call

$$\frac{Vt}{L} = \tau, \qquad \frac{VL}{D} = \lambda, \qquad (21)$$

and suppose that

$$c_f(\lambda, \tau) = A(\lambda, \tau)(1 - \tau)^{k-1}, \qquad (22)$$

then after certain transformations the expression for the diffusion flow at the end of the ampoule (21) may be put in the form

$$j_L = - V (1-\tau)^k \left\{ \frac{dA}{d\tau} + \lambda \left[\frac{(1-k)(1+k)}{3} A - \frac{(4-k)}{6} (1-\tau) \frac{dA}{d\tau} + \frac{(1-\tau)^2}{6} \frac{d^2A}{d\tau^2} \right] - \right.$$

$$- \lambda^2 \left[\frac{(1-k)(2+k)(3+k)}{30} (1-\tau) A - \frac{(12-7k)(3+k)}{12} (1-\tau)^2 \frac{dA}{d\tau} \right.$$

$$\left. + \frac{(6-k)(1-\tau)^3}{60} \frac{d^2A}{d\tau^2} - \frac{(1-\tau)^4}{120} \frac{d^3A}{d\tau^3} \right] + \cdots . \tag{23}$$

Assuming that $A(\lambda, \tau)$ may be put in the form of a series in λ:

$$A(\lambda, \tau) = A_0 + A_1 \lambda + A_2 \lambda^2 + \cdots, \tag{24}$$

equating terms with equal powers of λ, and supposing that $j_L{}_{x=L} = 0$ at any moment of time, we find the following equations for determining A_i:

$$\frac{dA_0}{d\tau} = 0, \tag{25}$$

$$\frac{dA_1}{d\tau} + \frac{(1-k)(1+k)}{3} A_0 = 0, \tag{25a}$$

$$\frac{dA_2}{d\tau} + \frac{(1-k)(1+k)}{3} A_1 - \frac{(4-k)}{6} (1-\tau) \frac{dA_1}{d\tau} - \frac{(1-k)(2+k)(3+k)}{30} (1-\tau) A_0 = 0, \tag{25b}$$

$$\text{etc.}$$

On successive integration we find

$$A_0 = c_0, \tag{26}$$

$$A_1 = c_0 \frac{(1-k)(1+k)}{3} (1-\tau) + c_1, \tag{26a}$$

$$A_2 = c_0 \frac{(1-k)(2+k)(3+k-5k^2)}{90} (1-\tau)^2 + c_1 \frac{(1-k)(1+k)}{3} (1-\tau) + c_2, \tag{26b}$$

where c_0, c_1, c_2, etc., are constants of integration.

Thus, for small λ, the solution for $c_f(t)$ satisfying the condition of the absence of a diffusion flow at the end of the ampoule may be put in the form

$$c_f(\lambda, \tau) = c_0 (1-\tau)^{k-1} + \lambda \left[c_1 (1-\tau)^{k-1} + c_0 \frac{(1-k)(1+k)}{3} (1-\tau)^k \right] +$$

$$+ \lambda^2 \left[c_2 (1-\tau)^{k-1} + c_1 \frac{(1-k)(1+k)}{3} (1-\tau)^k + c_0 \frac{(1-k)(2+k)(3+k-5k^2)}{90} (1-\tau)^{k+1} \right] + \cdots . \tag{27}$$

We see that as $\lambda \to 0$ the solution of (27) passes into the Pfann solution (1).

In growing crystals at rates of the order of 1–4 mm/h with an ampoule length of 100–200 mm and a diffusion coefficient of the impurity in the melt of the order of $3 \cdot 10^{-5}$ cm^2/sec, the parameter $\lambda = VL/D$ lies in the range 10–80, i.e., much greater than unity, and Eqs. (19) and (20) are inapplicable.

In this case it is better to use the following expressions for c(x, t) and j(x, t), suitable for any z:

$$c(x,\ t) = \sum_{s=0}^{\infty} \frac{D^s}{V^{2s}} \frac{dc_f^s}{dt^s}\ L_s(z) + k \sum_{s=0}^{\infty} \frac{D^s}{V^{2s}} \frac{d^s c_f}{dt^s}\ K_s(z), \tag{28}$$

$$j(x,\ t) = - \sum_{s=0}^{\infty} \frac{D^s}{V^{2s-1}} \frac{d^s c_f}{dt^s}\ L_s^{\bullet}(z) - k \sum_{s=0}^{\infty} \frac{D^s}{V^{2s-1}} \frac{d^s c_f}{dt^s}\ K_s^{\bullet}(z), \tag{29}$$

in which we have the functions

$$L_s(z) = H_s(z) - K_s(z), \qquad L_s^{\bullet}(z) = H_s^{\bullet}(z) - K_s^{\bullet}(z), \tag{30}$$

$$H_s(z) = \sum_{l=0}^{\infty} \frac{(l+s-1)!}{l!\,(s-1)!} \frac{(-1)^l z^{l+2s}}{(l+2s)!}, \tag{31}$$

$$K_s(z) = \sum_{l=0}^{\infty} \frac{(l+s)!}{l!s!} \frac{(-1)^l z^{l+2s+1}}{(l+2s+1)!}, \tag{32}$$

$$H_s^{\bullet}(z) = K_{s-1}(z),$$
$$K_s^{\bullet}(z) = L_s(z). \tag{33}$$

Here

$$z = \frac{V(x - Vt)}{D} = \lambda\left(\frac{x}{L} - \tau\right). \tag{34}$$

Figures 1-4 present graphs of these functions for $s \leq 4$ in the range 0-10; these may be used for practical calculations.

If it is necessary to grow crystals with a uniform distribution of impurity over the length, it is necessary that c_f should remain constant in the course of growth.

It follows from formulas (29) and (30) that, in this case, the distribution of impurity concentration and the impurity diffusion flow in the melt have the form

$$c(x,\ t) = c_f\ [k + (1-k)\,e^{-\frac{V(x-Vt)}{D}}], \tag{35}$$

$$j(x,\ t) = - Vc_f(1-k)\,e^{-\frac{V(x-Vt)}{D}}. \tag{36}$$

We see from this that, in order to ensure uniform penetration of the impurity into the crystal for constant growth rate, the initial impurity distribution in the melt should vary according to

$$c(x,\ 0) = c_f[k + (1-k)\,e^{-\frac{Vx}{D}}], \tag{37}$$

i.e., the impurity concentration in the melt should reach a maximum below the ampoule equal to $c_f = c_s/k$, and, moreover, it should fall exponentially at a slope depending on the growth rate and diffusion coefficient of the impurity in the melt.

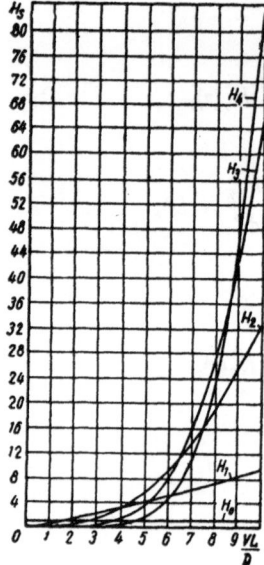

Fig. 1. Graph of the function $H_S(z)$.

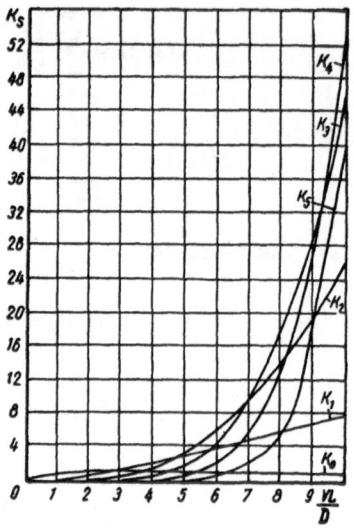

Fig. 2. Graph of the function $K_S(z)$.

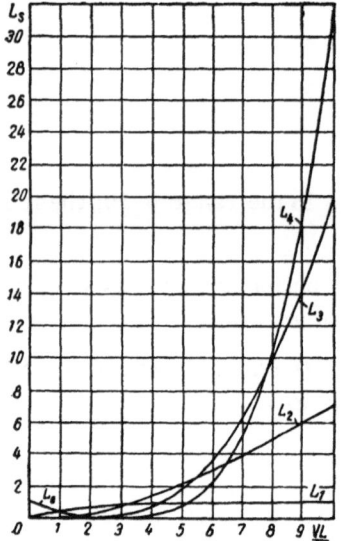

Fig. 3. Graph of the function $L_S(z)$.

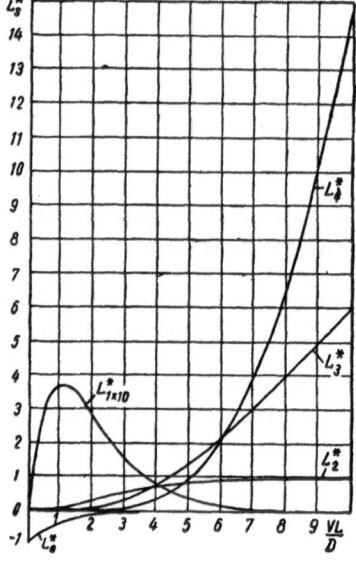

Fig. 4. Graph of the function $L_S^*(z)$.

Thus, a zone of concentration thickening of the impurity must be created artificially in the melt. The higher the growth rate, the narrower must the zone be, and the more sharply must it vary.

A second condition emanating from this solution is the necessity for the impurity to be carried away from the upper part of the ampoule in accordance with the law

$$j_L(t) = -c_f V(1-k)e^{-\frac{VL}{D}}e^{\frac{V^2 t}{D}}. \tag{38}$$

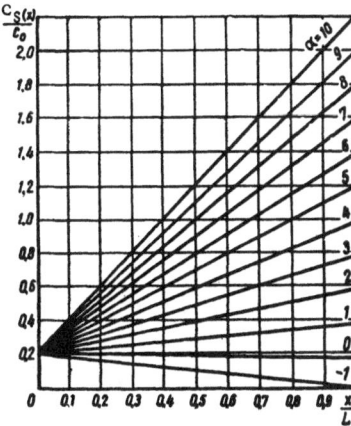

Fig. 5. Graphs of the necessary initial impurity distribution in the melt for a linear variation of impurity concentration in the solid phase $c_S = kc_0(1 + \alpha x/L)$; $k = 0.2$.

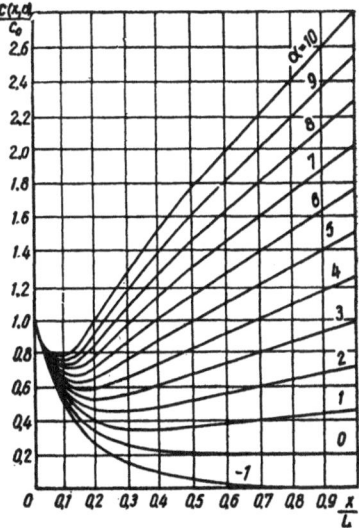

Fig. 6. Graphs of various impurity distributions in the solid phase corresponding to an initial impurity distribution in the melt of the type shown in Fig. 5: $VL/D = 10$, $k = 0.2$, $c_S = kc_0(1 + \alpha x/L)$.

For large VL/D, the intensity of the outflow of impurity required to ensure uniform penetration into the crystal at the beginning and in the middle of the growth process is very close to zero and only at the end of growth does it reach an appreciable value, rising exponentially with time.

If the condition for the outflow of impurity on the required law at the end of growth, for example, is omitted, this will lead to an increase in the intensity with which the impurity enters into the crystal. However, this increase in the impurity content of the crystal at the end of growth for an original impurity distribution in the melt obeying law (37) will be much less than that which is obtained for an original uniform distribution in the melt, since the total amount of impurity used in growth is greatly reduced. We must thus mention one further advantage in the new method of growing crystals: the great economy in the use of the additive ("impurity") deliberately introduced into the crystal.

In order to control the manner in which the impurity enters into the crystal, producing an intentionally nonuniform distribution, we must create conditions differing from those described by formulas (37) and (38).

If we suppose that the impurity distribution required in the crystal may be interpolated by a power polynomial of the form

$$c_s(x) = \sum_{m=0}^{r} \alpha_m \frac{x^m}{L^m}, \qquad (39)$$

then the expression for $c_f(t)$ for a constant growth rate is written in the form

$$c_f(t) = \frac{1}{k} \sum_{m=0}^{r} \alpha_m \left(\frac{Vt}{L}\right)^m, \qquad (40)$$

and the derivatives of $c_f(t)$ will be

$$\frac{d^s c_f(t)}{dt^s} = \frac{1}{k} \sum_{m=s}^{r} \frac{m!}{(m-s)!} \frac{V^m t^{m-s}}{L^m}. \qquad (41)$$

We shall now present some specific calculations from this formula:

   1. For a linear law of variation of impurity content in the solid phase,

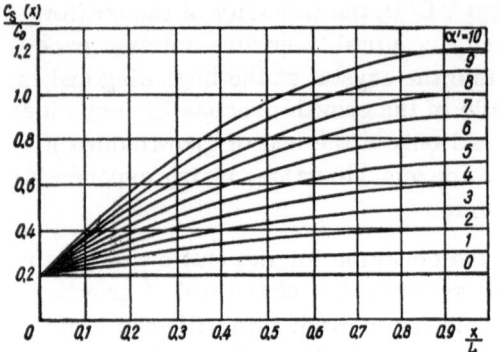

Fig. 7. Graphs of the initial impurity distribution in the melt required to produce a quadratic law of impurity concentration in the solid phase: k = 0.2, $c_S = kc_0[1 + \alpha'(x/L - x^2/2L^2)]$.

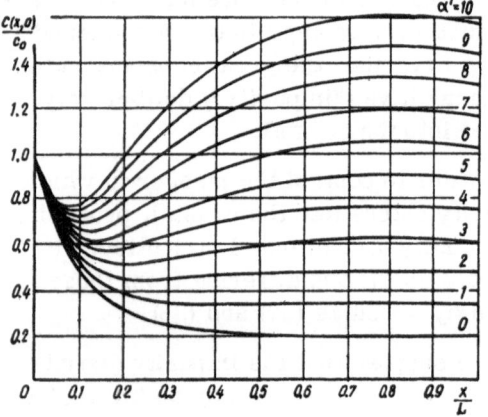

Fig. 8. Graphs of various impurity distributions in the solid phase corresponding to an initial impurity distribution in the melt of the type shown in Fig. 7: VL/D = 10, k = 0.2, $c_S = kc_0[1 + \alpha'(x/L - x^2/2L^2)]$.

$$c_s(x) = kc_0\left(1 + \alpha\frac{x}{L}\right). \qquad (42)$$

The results of calculating the necessary initial impurity distribution in the melt are presented in Fig. 5, and the desired distribution in the crystal in Fig. 6.

2. For a quadratic law of variation of impurity content in the solid phase,

$$c_s(x) = kc_0\left[1 + \alpha'\left(2\frac{x}{L} - \frac{x^2}{L^2}\right)\right]. \qquad (43)$$

The results of calculating the necessary initial impurity distribution in the melt are shown in Fig. 7, and the desired distribution in the crystal in Fig. 8.

These results show that, in the presence of a suitable diffusion mechanism, the desired impurity distribution in the crystal may be obtained by regulating the initial distribution in the melt.

Literature Cited

1. Pfann, D., Zone Melting [Russian translation]. IL, Moscow (1960).
2. Smith, V., Tiller, W., and Rutter, J., Can. J. Phys., 33:723 (1955)
3. Lyubov, B., and Temkin, D., In: Growth of Crystals, Vol. 3, pp. 59-67. Izd. Akad. Nauk SSSR, Moscow (1961). [English translation: Consultants Bureau, New York (1962).]
4. Kulik, I., and Zil'berman, G., In: Growth of Crystals, Vol. 3, pp. 85-89. Izd. Akad. Nauk SSSR, Moscow (1961). [English translation: Consultants Bureau, New York (1962).]
5. Kreith, F., and Romie, F., Proc. Phys. Soc., 68:277 (1955).
6. Borisov, V., Lyubov, B., and Temkin, D., Dokl. Akad. Nauk SSSR, 104(2):223 (1955).
7. Birman, B., and Timan, B., Dokl. Akad. Nauk SSSR, 161(1):78 (1965).

STUDY OF CONCENTRATION PACKING AT THE CRYSTAL—MELT INTERFACE

K. M. Rozin

Moscow Institute of Steels and Alloys

The interface between the original medium and the growing crystal justifiably forms the basis of many papers aimed at describing the principal laws underlying the mechanism of crystallization. In essence, the main responsibility for the process of crystal formation is borne by a comparatively thin layer of the mother solution, in which a variety of phenomena, not all understood in detail, take place. The difficulties besetting the direct experimental study of these phenomena are accentuated by the comparatively sharp changes in many physical properties within this narrow region. Especially characteristic are certain special hydrodynamic conditions, which in turn produce physical and chemical inhomogeneities in the mother solution.

In growing a crystal from the melt, which always contains a greater or smaller amount of dissolved impurities, a phenomenon known as "concentration thickening" develops in front of the crystallization front; this constitutes an impurity-enriched layer of melt (partition coefficients of the impurities smaller than unity), the properties of which depend both on the conditions of the crystallization process and on the rate of diffusion of the impurities from the crystallization front into the main volume of melt. There is really no difference in principle between the concepts of "concentration thickening," "diffusion layer," and "crystallization courtyard," since these all refer to a comparatively narrow region of liquid phase in front of the crystallization front in which there is a great difference in chemical composition from the main bulk of the liquid phase. In fact, whereas in crystallization from the melt the layer of liquid phase in front of the crystallization front is enriched with impurities which cannot, generally speaking, enter into the crystal, in crystallization from solution there is a region in front of the crystallization front with a concentration of dissolved material lower than that of the main bulk of the solution (and, correspondingly, an increased concentration of solvent).

In order to describe the region of concentration thickening it is usual to employ a spatial parameter, the "thickness" of the region of concentration thickening in the growth direction, which depends on the following factors: the intensity of circulation of the liquid phase, its viscosity, and the diffusion coefficient of the impurity. Other conditions being equal, the "thickness" decreases with increasing rate of circulation of the mother medium. Remembering the various ways of defining the thickness of the region of concentration thickening, or, as it is sometimes called, the thickness of the diffusion layer, we shall understand this quantity to mean the distance from the crystallization front (at which the impurity content in the melt is a maximum) to the beginning of the region of liquid phase in which the impurity content may be regarded as equilibrium.

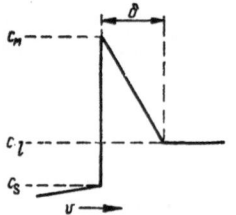

Fig. 1. Distribution of impurity in the direction of growth near the crystallization front.

The inadequacy of this parameter in describing the concentration thickening is obvious, since it only characterizes the linear dimensions of a region of chemical inhomogeneity at the crystallization front, without giving any indications of the impurity content within this region or of the degree of chemical inhomogeneity itself. Yet, the existence of even an approximate idea of the character of impurity distribution and impurity concentration would lead to a number of very interesting conclusions regarding both the actual mechanism of crystallization and the choice of optimum conditions for the process. In addition to this, the popular concept that the thickness of the region of concentration thickening is independent of the crystallization rate is not very well founded, since the region in question depends on the kinetics of the process and represents the relation between the rate of impurity buildup in front of the crystallization front (which is determined by the crystallization rate) on the one hand, and the rate of impurity outflow (which ultimately also depends on the crystallization rate) on the other.

Thus, the foregoing consideration indicates the desirability of studying the characteristics of the region of concentration thickening and their dependence on the parameters of the crystallization process.

In analyzing concentration thickening we shall, in accordance with the scheme presented in Fig. 1, use the following characteristic concepts:

a. The fall in impurity concentration in the region of concentration thickening $c_M - c_l$.

b. The thickness of the region of concentration thickening δ.

c. The average impurity concentration gradient in the region of concentration thickening in the growth direction, defined as the ratio of the fall in impurity concentration to the thickness of the region of concentration thickening $\overline{G}_c = (c_M - c_l)/\delta$.

d. The limiting (minimal) average temperature gradient in the region of concentration thickening, characterizing the transition into a state of concentration supercooling and defined as the product of the average impurity concentration gradient and the reciprocal of the temperature coefficient of impurity solubility in the main component

$$\overline{G}_s = \overline{G}_c \frac{\partial T}{\partial c}.$$

In order to determine the fall in impurity concentration in the region of concentration thickening we may use the well-known model based on the principle that the equilibrium impurity partition coefficient k_0 is realized at the actual crystallization front [1]. In this case, the effective partition coefficient is defined as the ratio of the impurity concentrations in the solid phase (at the crystallization front) and in the main bulk of the liquid phase, $k = c_s/c_l$. Thus, by using this model, we determine the unknown fall in impurity concentration in the region of concentration thickening:

$$c_M - c_l = c_s \left(\frac{1}{k_0} - \frac{1}{k} \right).$$

Of course, this relation may also be successfully used for solving the inverse problem of determining the effective partition coefficient from known experimental values of c_s and c_M.

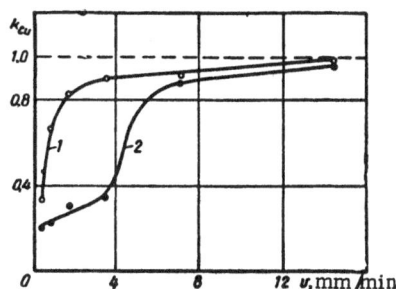

Fig. 2. Effective partition coefficients of copper impurity in aluminum as functions of crystallization rate. 1) For simple zone recrystallization; 2) for zone recrystallization with forced circulation.

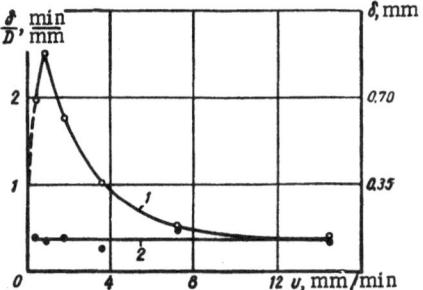

Fig. 3. Reduced thickness of the region of concentration thickening as a function of crystallization rate, using the notation of Fig. 2.

In order to determine the thickness of the region of concentration thickening δ, we use the relation between the equilibrium and effective impurity partition coefficients

$$k = \frac{k_0}{k_0 + (1-k_0)\exp\left(-v\dfrac{\delta}{D}\right)},$$

where D is the diffusion coefficient of the impurity in the melt and v is the crystallization velocity [2]. In this theory we consider the case of growing a crystal of constant cross section (so-called one-dimensional growth), no restrictions being imposed on the value of δ (the thickness of the region of concentration thickening); this also enables us to assume the dependence of the thickness of the region of concentration thickening on the rate of crystallization.

In order to study concentration thickening we used the method of zone recrystallization, which enabled us to control the crystallization velocity. Aluminum bars about 500 mm long (made from metal of the AV000 type) were placed in graphite boats and subjected to single-zone recrystallization in quartz tubes evacuated to $2 \cdot 10^{-5}$ mm Hg at various crystallization rates (from 0.5 to 14.5 mm/min). In order to study the influence of forced circulation on the crystallization process, we used a movable magnetic field created by a system of resistance heating elements. The heaters were fed from stabilized voltage sources. The effective partition coefficients of copper impurity in aluminum were determined from the results of emission spectral analysis by the method of [3].

The results of the experiments are shown in Fig. 2. The curves relating the effective partition coefficients to crystallization rate are of a monotonic character. With forced circulation of the melt (curve 2) there is a displacement in the sense of increased crystallization rates, which indicates the considerable possibilities of intensifying the process. Thus, the use of circulation makes it possible to increase the crystallization rate by a large factor with the same value of impurity partition coefficient. In the same way, with the help of circulation, we can control the effective partition coefficient over fairly wide limits for a fixed crystallization rate, and this is also of considerable practical interest.

The results obtained also indicate the serious difficulties arising in the case of simple zone recrystallization of aluminum at velocities of 2 mm/min or less. In fact, in view of the sharp slope of the curve relating the effective partition coefficient to crystallization rate, fluctuations in crystallization rate will, in this case, lead to considerable variations in the effective partition coefficient. The use of forced circulation enables us to carry out the crystallization process on the slanting part of the curve at crystallization rates up to 4 mm/min, which corresponds to a sharp reduction in the deviations of the instantaneous values of the effective partition coefficient from the nominal value.

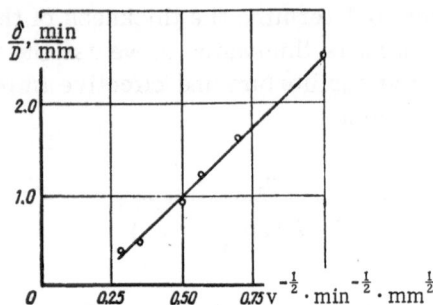

Fig. 4. Thickness of the region of concentration thickening as a function of the square root of the reciprocal of the crystallization rate for simple zone recrystallization.

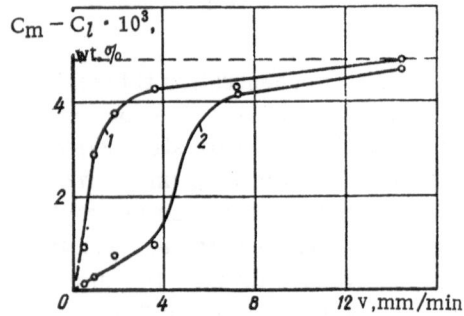

Fig. 5. Concentration drop in the region of concentration thickening as a function of the crystallization rate; notation as in Fig. 2.

The values of effective partition coefficients obtained enable us to analyze the behavior of the thickness of the region of concentration thickening as the crystallization rate varies. Figure 3 shows values of the so-called reduced thickness δ/D, which differs from the actual thickness in respect of the factor $1/D$. The thickness of the region of concentration thickening falls monotonically on increasing the crystallization rate from 1 to 14.5 mm/min (for simple zone recrystallization, without forced circulation), reaching some limiting value. In this range of velocities the thickness changes by a factor of six. Such behavior may be associated with the kinetic origin of the concentration thickening. In fact, if we trace the behavior of an artificial region of concentration thickening at a stationary melt−crystal interface, then, as a result of equalizing diffusion in the melt, the thickness will increase with time (no doubt as a result of a fall in the impurity concentration in the original region of concentration thickening) for a certain period until the impurity concentration becomes uniform over the whole volume of the melt. Thus, an increase in the crystallization rate should be considered in analogy with a reduction in the period of equalizing diffusion. If we then present the experimental data as a relation between the thickness of the region of concentration thickening and the square root of the reciprocal of the crystallization rate (by analogy with the relation between diffusion distance and time), we obtain a linear relationship (Fig. 4), which indicates the validity of the comparison.

The existence of intensive forced circulation enables us to reduce the thickness of the region of concentration thickening to a minimum (as compared with simple zone crystallization); in this case the thickness of the region will be constant over the range of crystallization rates studied. The maximum diffusion length is here evidently limited by the thickness of the region of concentration thickening, which depends on the intensity of the forced circulation of the melt. Here we should note that the movable magnetic field in our experiments had a value of ~120 Oe.

It is interesting to compare this behavior of the thickness of the region of concentration thickening with that of the drop in concentration. Figure 5 shows the drop in concentration for various crystallization rates. The behavior of the drop in concentration is very much the same as that of the effective partition coefficients; both the shape of the corresponding curves and their individual details are very similar. The drop in concentration rises monotonically from zero to a limiting value, which is determined by the approach of the effective partition coefficient to unity for fairly high crystallization rates. For all values of crystallization rate the concentration drop is lower in the case of forced circulation than in the case of simple zone recrystallization, the difference being especially great in the range of 4-5 mm/min. The concentration drop is linearly related to the effective partition coefficient (Fig. 6). It should be noted

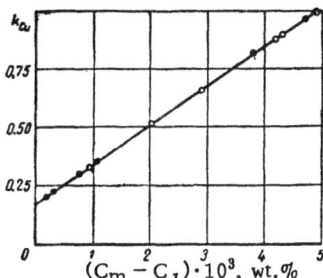

Fig. 6. Effective partition coefficient as a function of the concentration drop in the region of concentration thickening: ○ − simple zone recrystallization; ● − zone recrystallization with forced circulation.

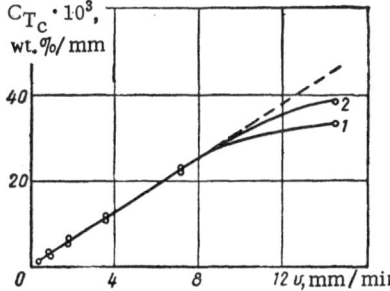

Fig. 7. Average concentration gradient in the region of concentration thickening as a function of crystallization rate; notation as in Fig. 2.

that there is no clear correspondence between the thickness of the region of concentration thickening and the value of the concentration drop.

The values of the average concentration gradient in the region of concentration thickening were determined from the experimental results (Fig. 7). For crystallization rates up to 8 mm/min all the experimental points lie neatly on a straight line passing through the origin of coordinates. Thus, the effect of forced circulation may be considered as a reduction in the thickness of the region of concentration thickening (or a reduction in the concentration drop). As regards the character of the relation between the concentration gradient and the crystallization rate, this may be qualitatively explained by the fact that, as crystallization rate increases, there is an increase in the buildup of impurity in front of the crystallization front owing to the difference in the solubilities of the impurities in the solid and liquid phases (here, as in the rest of the article, we consider the case in which the effective partition coefficient is less than unity).

Thus, the foregoing data lead to certain conclusions regarding the mechanism underlying the control of the composition of the solid phase. For a given amount of impurity in the melt, the composition of the solid phase at the melt−crystal interface is determined at any moment of time by the limiting impurity concentration in the melt c_M at the crystallization front; this limiting concentration may be varied either by regulating the thickness of the region of concentration thickening without changing the concentration gradient (by means of forced circulation), or by regulating the concentration gradient without changing the thickness of the region of concentration thickening (by varying the crystallization rate).

The results here obtained may be used for estimating the value of the limiting (minimal) average temperature gradient in the melt at the crystallization front characterizing the boundary of the region of concentration supercooling. For an original concentration of copper impurity in aluminum equal to $1 \cdot 10^{-3}$ wt.%, we obtain (after substituting the usually recommended value for the diffusion coefficient of impurity in the melt at the crystallization temperature, $D = 5$ cm^2/day [4]) a series of rough values for the limiting temperature gradient in question: $4 \cdot 10^{-3}$ deg/mm for a crystallization rate of 0.5 mm/min, $1.7 \cdot 10^{-2}$ deg/mm for a rate of 1.8 mm/min, $6 \cdot 10^{-2}$ deg/mm for a rate of 7.2 mm/min, etc. Comparing these values with the extremely rough results of our direct measurements of temperature gradient in the melt at the crystallization front, carried out during the zone melting of aluminum by means of the 0.2-mm junction of a KhA thermocouple immersed in the melt and an automatically recording potentiometer of the KVT type with a recording scale of 45 mm/mV (giving a gradient of the order of 1-2 deg/mm), we note that the actual temperature gradient in the case considered, even on a fairly coarse estimate, greatly exceeds the limiting temperature gradient at the crystallization front. Thus, in the particular case of the crystallization of aluminum with the parameters indicated there is a distinct possibility of substantially reducing the temperature gradient in the region of the crystallization front without hazarding the concentration supercooling.

The methods here used for studying the region of concentration thickening will be extended to other systems.

Literature Cited

1.   Mullin, J. B., and Hulme, K. F., J. Phys. Chem. Solids, 17:1 (1960).
2.   Burton, J. A., Prim, R. C., and Slichter, W. P., J. Chem. Phys., 21:1987 (1953).
3.   Shilkin, A. I., and Kuliev, A. A., Izv. Akad. Nauk AzerbSSR, Ser. Fiz.-Mat. i Tekhn. Nauk, 1:57 (1961).
4.   Frenkel', Ya. I., Kinetic Theory of Liquids (1945).

EFFECT OF THE CONDITIONS OF GROWING ALUMINUM SINGLE CRYSTALS FROM THE MELT ON THEIR DISLOCATION STRUCTURE

E. I. Sosnina and L. I. Meleshko

Institute of Metal Physics of the Academy of Sciences
of the Ukrainian SSR

In an earlier paper [1] we studied the effect of small additions of copper, silver, magnesium, and chromium, soluble at room temperature, on the formation of dislocations in aluminum single crystals of type V3 (99.9994%) grown from the melt at a rate of $8.3 \cdot 10^{-4}$ cm /sec. It appeared interesting to study this influence for a higher (by one order) growth rate. Since the substructure of the crystals was quite coarse, so that its dimensions could not be determined very accurately from the total intensity of double-reflection x-ray curves, in the present investigation we supplemented this by metallographic observations of etch pits and the Schulz method of x-ray crystal topographical study.

Method

The single crystals had the form of parallelepipeds $2.5 \times 0.7 \times 0.5$ cm^3 in size. The sides coincided with the (100) and (110) planes to an accuracy of 1-2°. The end surface, normal to the growth direction of the crystals, approximately corresponded to the (110) plane. Since dislocations are revealed better on planes of low indices, this type of boundary for the crystals is most convenient when observing dislocations by means of etch pits.

In fcc crystals with high-energy defect packing (sodium, aluminum) dislocations are properly arranged in the {111} planes. There are four different planes of this type, forming a tetrahedron [2]. The Burgers vector of the complete dislocations has twelve <110> directions, according to the number and arrangement of nearest neighbors in the lattice. These directions coincide with the directions of the six edges of the tetrahedron mentioned.

If we suppose that dislocations of one type (edge or screw) with twelve different Burgers vectors are arranged with equal probability in the lattice, we may estimate the proportion of dislocations in the crystal observable by means of etch pits in various crystallographic planes. Estimates of this kind showed that the (100) and (110) planes contained up to half of the dislocations in the crystals. The inclinations of the dislocation lines to the (100) plane are such that, other conditions being equal, the points at which the lines emerge on this plane should etch better than in the case of the (110) plane. The number of such dislocations, however, is smaller. Hence the etch pits on the (100) plane of the crystal should be larger but their density smaller.

165

The above discussion relates to the revelation of fresh dislocations, not surrounded by clouds of impurities. However, in metals, the etching of fresh dislocations can only be achieved in a limited number of cases. Most frequently only dislocations decorated with impurities can be observed. The appearance of such dislocations depends on the amount of decorating impurity [3] and the time for the formation of impurity clouds around them. For a small amount of impurity in the metal or an inadequate aging time, not all the dislocations emerging on the plane under examination can be revealed. In addition to this, we must remember that the etchability of dislocations decorated to different degrees may also vary. In view of this we made no attempt to determine the dislocation density by the metallographic method. We simply tried to follow the variation in dislocation distribution as a function of the conditions of crystal growth.

The single crystals to be studied were obtained by the method of [4], etched in order to remove oxide films, and polished electrochemically. The polishing was carried out in a solution of 68% H_3PO_4, 13% H_2SO_4, and 19% H_2O with a current density of 0.24 A/cm^2, a voltage of 8-9 V, and an electrolyte temperature of 85°C. The current was fed to the single crystals by means of a "soft" electrode in the form of an aluminum foil stuck to one of the ends of the crystal. The polishing time required in order to obtain a good surface was 0.5-1 h. After polishing, the crystals were carefully washed in boiling and cold twice-distilled water and immediately placed in the etchant. In order to reveal low-angle boundaries we used an etchant [5] quite sensitive to slight disorientations.

The etchant was first tested for the revelation of dislocations by reference to aluminum single crystals of the AV000 type (99.99%) and the etching conditions were selected accordingly. For this purpose freshly grown single crystals (worked, unworked, worked-and-annealed, and also naturally aged) were subjected to etching. The single-crystal samples 0.15 cm thick were worked by bending. Annealing for the formation of impurity atmospheres was effected at 200°C for 8 h [6]. The etch pictures on the worked samples were observed in the (110) planes parallel to the bending radius.

In worked and freshly grown single crystals, the boundaries of the substructure etched poorly. In worked-and-annealed samples, sharply expressed etch-pit walls appeared in the direction of the bending radius. The boundaries of the subgrain structure were clearly revealed in naturally aged crystals as well. Hence, subsequently, we only studied the dislocation distribution from etch pits in samples which had undergone natural aging for several months.

We also checked the reproducibility of the etch figures by deep polishing of the sample surface. We found that after the removal of a layer a few tens of microns thick, the picture of the subgrain structure was reproduced quite satisfactorily.

A comparison of the reactivity of aluminum of various purities showed that, whereas in naturally aged crystals of the purer V3 aluminum only individual pits appeared after 10-20 h etching, crystals of the less pure AV000 aluminum overetched in this period. In single crystals of V3 aluminum with the addition of impurities, a satisfactory etch picture was obtained in the etchant of [5], diluted in the ratio 1:4, after 15 h. We used this concentration for observing the substructure in the crystals of the alloys studied.

Single-crystal samples were placed in the solution in a suspended state. Etching took place without agitation of the etching solution. After etching, the samples were twice washed in distilled water, carefully dried with decalcified filters, and examined under the microscope.

Preliminary study of long single crystals showed that the density of the sub-boundaries was not uniform along the sample. At the beginning of the sample the density was low; it became larger in the middle and, as a rule, larger again at the end. Different samples were therefore compared with respect to pictures obtained from their middle sections.

We studied the arrangement of dislocations chiefly on the (100) side surface of the single-crystal samples by means of etch pits. From the same surface we obtained reflection curves [1] and Schulz photographs. In the latter case we used an ARS-4 x-ray system with a point-focus BSV-5 tube. We studied the influence of both impurities and growth rate on the substructure of V3 aluminum single crystals.

Results and Discussion

Effect of Impurities. Since there are not etchants revealing fresh dislocations in aluminum, it is quite difficult to follow the form of the substructure in single crystals of high-purity aluminum. We can only say that it has a form drawn out in the [110] growth direction. The cross section of this structure reaches dimensions of 0.1-0.15 cm for a low growth rate and 0.07 cm for a high growth rate. The substructure becomes much smaller on adding impurities, the extent of the change depending on the type of impurity. This may clearly be seen from Table 1, which gives the metallographic dimensions L of the subgrains in crystals of different alloys for two growth rates V. For a low growth rate there are transverse subgrains, normal to the growth direction. The number of these, however, is small. For high growth rates these subgrains are in the majority of cases absent. Only longitudinal subgrains occur, and these sometimes run the whole length of the sample. This kind of substructure was observed on both side surfaces, the (100) and the (110). An analogous picture was obtained on the (110) end surface.

This substructure is clearly associated with the cellular growth of crystals from the melt. Since the impurity content c_∞ in the alloys was small, only the initial stages of this kind of growth occurred. The cells had an elongated form. It seemed interesting to compare these dimensions with the values calculated from the formula

$$d = 2\pi \sqrt[3]{\frac{2D}{kV} \cdot \frac{\sigma T_0}{gQ}}, \tag{1}$$

obtained in [7]. Here k is the equilibrium impurity partition coefficient, D is the impurity diffusion coefficient in the melt (taken as being the same for all the alloys and equal to $5 \cdot 10^{-5}$ cm^2/sec); V is the growth rate (velocity); σ is the surface tension of molten aluminum; T_0 is its melting point; Q is the heat of fusion per unit volume; and g is the temperature gradient averaged over the two phases.

The latter is calculated from the expression

$$g = \frac{g_l \varkappa_l + g_s \varkappa_s}{\varkappa_l + \varkappa_s}, \tag{2}$$

where g_l and g_s are the temperature gradients in the liquid and solid phases; $\varkappa_l$ and $\varkappa_s$ are the corresponding thermal conductivities.

The values of g_l and g_s, and also the dimensions d of the subgrains, are given in Table 1. These dimensions, as may be seen from the table, are quite close to the cross sections determined metallographically.

On checking the criterion for the instability of a plane growth front given by Voronkov [7] and Sekerka [8] [formula (23)], we found that in the majority of cases the criterion was satisfied. This means that, under our conditions, owing to the presence of an impurity of concentration c_∞ in the melt, the plane growth front is unstable. The front breaks up, with the formation of elongated cells, both in the direction of the growth front and at right angles to it. The substructure observed metallographically (Figs. 1 and 2) is clearly a reflection of this kind of growth.

Table 1

Material	V, cm/sec	k	c_∞, at. parts	L, cm	g_l, deg	g_s, deg	d, cm
Al—Cu	$8.3 \cdot 10^{-4}$	0.12	$2 \cdot 10^{-4}$	$7 \cdot 10^{-2}$	33	35	$8.3 \cdot 10^{-2}$
Al—Ag	$8.3 \cdot 10^{-4}$	0.29	$2 \cdot 10^{-4}$	$6 \cdot 10^{-2}$	33	35	$6.3 \cdot 10^{-2}$
Al—Mg	$8.3 \cdot 10^{-4}$	0.33	$1 \cdot 10^{-4}$	$5.2 \cdot 10^{-2}$	33	35	$6.1 \cdot 10^{-2}$
Al—Cr	$8.3 \cdot 10^{-4}$	1.54	$1 \cdot 10^{-5}$	$4.2 \cdot 10^{-2}$	33	35	$3.6 \cdot 10^{-2}$
Al—Cu	$8.3 \cdot 10^{-3}$	0.12	$2 \cdot 10^{-4}$	$5 \cdot 10^{-2}$	24	29	$4.3 \cdot 10^{-2}$
Al—Ag	$8.3 \cdot 10^{-3}$	0.29	$2 \cdot 10^{-4}$	$3.5 \cdot 10^{-2}$	24	29	$3.2 \cdot 10^{-2}$
Al—Mg	$8.3 \cdot 10^{-3}$	0.33	$1 \cdot 10^{-4}$	$3.5 \cdot 10^{-2}$	24	29	$3.1 \cdot 10^{-2}$
Al—Cr	$8.3 \cdot 10^{-3}$	1.54	$1 \cdot 10^{-5}$	$2.6 \cdot 10^{-2}$	24	29	$1.8 \cdot 10^{-2}$

Table 2

Material	V cm/sec	B, min	β, min	L, cm	ϱ_c, cm^{-2}	ϱ_d, cm^{-2}
Al	$8.3 \cdot 10^{-4}$	15	3.4	$1 \cdot 10^{-1}$	$1.7 \cdot 10^6$	$1.3 \cdot 10^6$
Al—Cu	$8.3 \cdot 10^{-4}$	21	4.9	$7 \cdot 10^{-2}$	$3.4 \cdot 10^5$	$1.9 \cdot 10^6$
Al—Ag	$8.3 \cdot 10^{-4}$	34	7.0	$6 \cdot 10^{-2}$	$5.6 \cdot 10^5$	$2.6 \cdot 10^6$
Al—Mg	$8.3 \cdot 10^{-4}$	17	4.6	$5 \cdot 10^{-2}$	$4.5 \cdot 10^5$	$1.8 \cdot 10^6$
Al—Cr	$8.3 \cdot 10^{-4}$	16	3.9	$4 \cdot 10^{-2}$	$4.7 \cdot 10^5$	$1.5 \cdot 10^6$
Al	$8.3 \cdot 10^{-3}$	14	3.5	$7 \cdot 10^{-2}$	$2.4 \cdot 10^5$	$1.3 \cdot 10^6$
Al—Cu	$8.3 \cdot 10^{-3}$	22	5.2	$5 \cdot 10^{-2}$	$5.0 \cdot 10^5$	$2.0 \cdot 10^6$
Al—Ag	$8.3 \cdot 10^{-3}$	41	7.1	$4 \cdot 10^{-2}$	$8.6 \cdot 10^5$	$2.7 \cdot 10^6$
Al—Mg	$8.3 \cdot 10^{-3}$	18	4.9	$3 \cdot 10^{-2}$	$7.9 \cdot 10^5$	$1.9 \cdot 10^6$
Al—Cr	$8.3 \cdot 10^{-3}$	18	4.2	$3 \cdot 10^{-2}$	$6.7 \cdot 10^5$	$1.6 \cdot 10^6$

Table 3

Material	V, cm/sec	B, min	β, min	Axis of sample	Angle of sample axis to the [100], deg
Al	$8.3 \cdot 10^{-4}$	64	25	[210]	27°
Al	$8.3 \cdot 10^{-4}$	52	9	[310]	18°

    Inside the grains of the first-order impurity substructure, the etch pits are, as a rule, distributed in a disordered manner; in the (100) plane they are larger and their density smaller than in the (110) plane. Sometimes, however, the etch pits join up into the boundaries of second-order subgrains (Fig. 3). These subgrains have smaller dimensions and their shape is more equiaxial. The dislocation density, and hence the disorientation in the sub-boundaries, is smaller. It may be that these sub-boundaries are formed at not very high temperatures, when the metal is cooling after solidification. Owing to their low mobility, the dislocations cannot arrange themselves into completed sub-boundaries in the cooling period.

    The aluminum and alloy single crystals were also photographed in a diverging white x-ray beam from a sharp-focus tube in the Schulz manner. Figures 4-6 present photographs of Schulz photographs from the (100) planes of some of the samples, taken in copper radiation. The photographs show dark and light sub-boundaries, depending on the sign of the inclination of

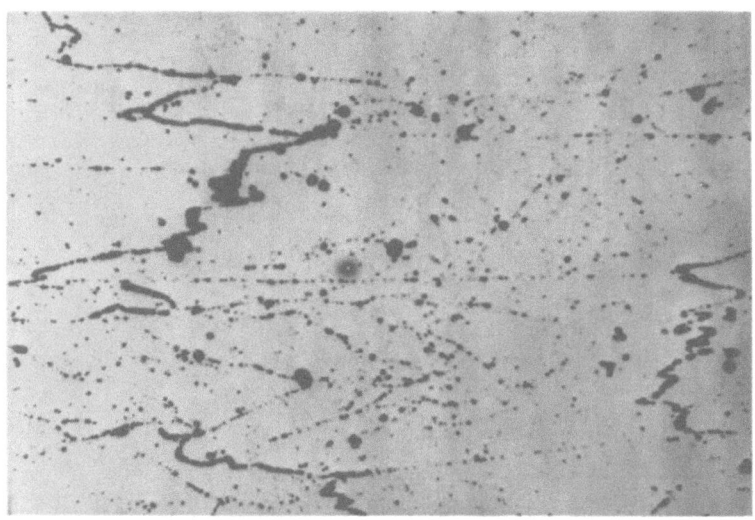

Fig. 1. Etch picture of the (100) plane of Al—Cu single crystal
No. 129. V = 8.3 · 10^{-4} cm/sec (× 70).

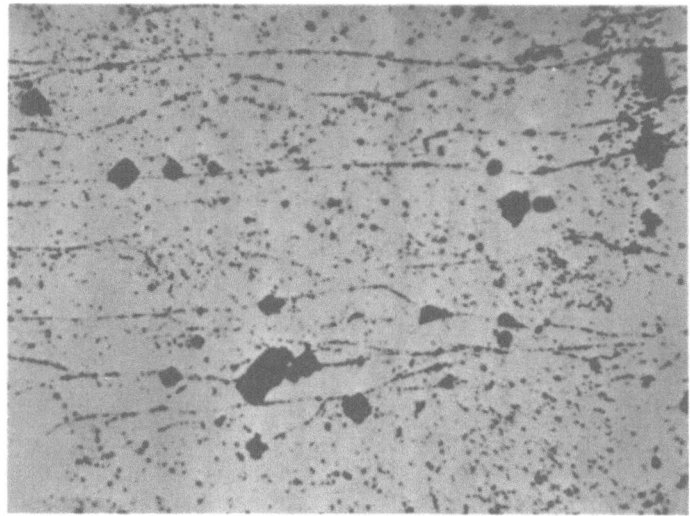

Fig. 2. Etch picture of the (100) plane of Al—Cr single crystal
No. 155. V = 8.3 · 10^{-3} cm/sec (× 70).

adjacent subgrain to one another. The disorientations of the subgrains at these boundaries are quite large, from several minutes to several degrees. In the alloys the upper limit of these disorientations is rather lower than in V3 aluminum. The boundaries are directed either in the [110] growth direction or at small angles to it.

Sub-boundaries with very small disorientation angles were not found on the Schulz photographs. The number of these over the cross section of the sample is therefore smaller than that of chemically revealed sub-boundaries, and the distance between them is greater.

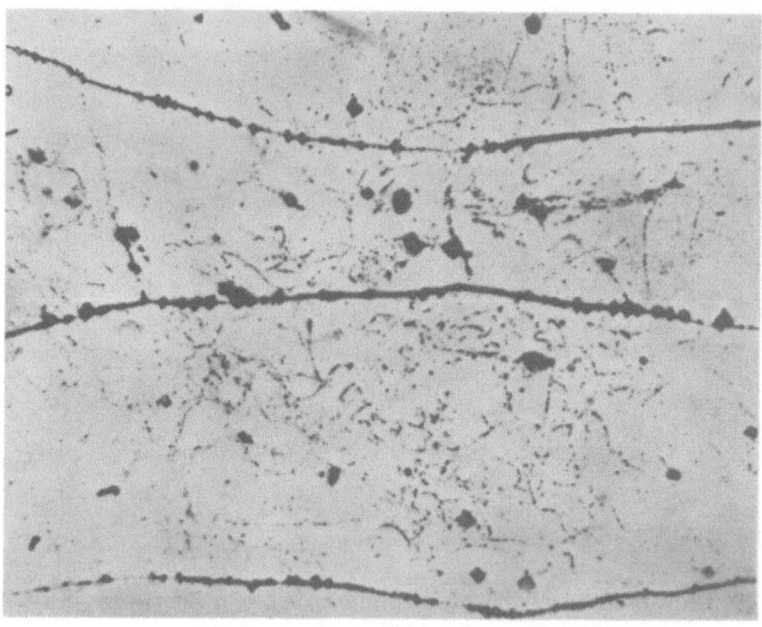

Fig. 3. Substructure of the first and second orders in the (100) plane of Al$-$Mg single crystal No. 169; V = 8.3 $\cdot$ 10^{-4} cm/sec ($\times$ 70).

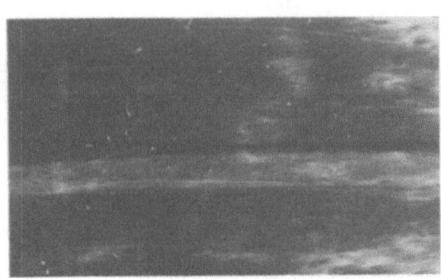

Fig. 4. Schulz photograph of the surface of V3 aluminum single crystal No. 154; V = 8.3 $\cdot$ 10^{-3} cm/sec ($\times$ 5).

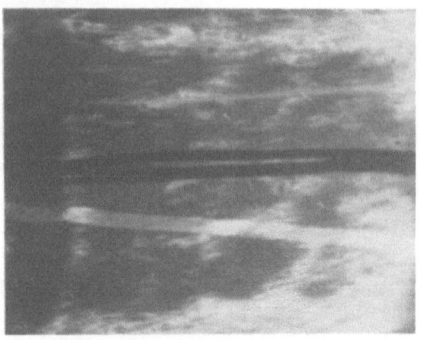

Fig. 5. Schulz photograph of the surface of Al$-$Cr single crystal No. 152; V = 8.3 $\cdot$ 10^{-3} cm/sec ($\times$ 5).

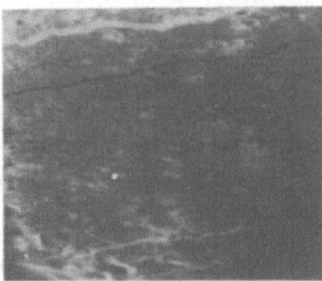

Fig. 6. Schulz photograph of the surface of Al$-$Cu single crystal No. 146; V = 8.3 $\cdot$ 10^{-3} cm/sec ($\times$ 5).

The metallographic dimensions of the subgrains were used for calculating the dislocation density in the sub-boundaries. Table 2 shows the total width B of the double-reflection x-ray curves, the half-width β, the dislocation density ρ_c calculated from the model of low-angle boundaries, and the density ρ_d calculated from the model of a disordered distribution (where $<b^2> = \frac{1}{3}a^2$ and not as in [1]). We see that for both low and high growth rates of the single crystals the impurities produce broadening of the reflection curves and an increase in dislocation density. The mechanism underlying this effect is evidently that of Tiller, as indicated in [1].

Effect of Velocity. The data of Table 2 indicate the influence of impurity on the size and disorientation of subgrains for both growth rates (velocities). This influence is represented by a reduction in the transverse dimensions of the subgrains, an increase in their disorientation, and a corresponding increase in dislocation density. As regards the influence of growth rate, an increase in this leads to a slight reduction in the cross section of the subgrains and to a lengthening of these in the direction of growth. The disorientations increase very little. The dislocation density also rises, but not so much as on adding impurities.

It should be noted that the effect of growth rate in crystals of V3 aluminum and aluminum alloys was not so great as in single crystals of AV000 aluminum [9]. One of the reasons for the different effects of the same range ($8.3 \cdot 10^{-4}$–$8.3 \cdot 10^{-3}$ cm/sec) of growth rates of seeded and unseeded single crystals on their angular range of reflection is apparently the difference in the conditions of growth. In the single crystals studied in the present investigation (grown with seeds) the direction of the growth axis was the same for different growth rates. In the unseeded crystals this direction was probably different. Table 3 presents the results of measuring the width and half-width of double-reflection curves from the (111) planes of two unseeded 99.99% aluminum crystals studied earlier in [9]. The growth rate of the crystals was the same. Determination of the orientations of the crystal axes in the growth direction showed that these were different. Hence, the angles between these axes and the direction of preferential growth, the [100], in these crystals were also different [10]. The sample with the greater angle had a much greater width, and especially half-width, in its reflection curves. Hence, the crystal with the greater angle of inclination of its growth axis from the preferred growth direction was the more perfect. This agrees with the results of [11]. It was shown in the latter that the structure of aluminum crystals is the more perfect, the more the orientation of the crystal in the growth direction deviates from the direction of preferential growth.

Thus in the unseeded crystals the severe change in substructure with growth rate may have arisen both as a result of the change in growth rate and as a result of a change in the crystallographic orientation of the growth axis. The latter effect may indeed have predominated. In the seeded crystals, no such change in the orientation of the axis took place. It is possible that the slight change in substructure with growth rate may have been due to this. However, this is not the only explanation. Others are possible as well. The method of taking the double-reflection curves, in particular, may have had an effect.

Literature Cited

1. Sosnina, E. I., Meleshko, L. I., and Ovsienko, D. E., in: Study of the Imperfection of Crystal Structure, p. 122. Naukova Dumka, Kiev (1965).
2. Thompson, N., Proc. Phys. Soc., 66B(6): 402, 481 (1953).
3. Zasimchuk, I. K., and Ovsienko, D. E., Ukr. Fiz. Zh., 9(10): 1092 (1964).
4. Ovsienko, D. E., and Sosnina, E. I., Ukr. Fiz. Zh., 8(1): 121 (1963).
5. Metzger, M., and Intrater, J., Nature, 174(4429): 547 (1954).
6. Wyon, G., and Lacombe, P., Rept. Conf. Defects Cryst. Solids, Bristol (1954); Phys. Soc. Lond. (1955), pp. 187-196.
7. Voronkov, V. V., Fiz. Tverd. Tela, 6(10): 2984 (1964).

8.　Sekerka, R., J. Appl. Phys., 36(1) : 264 (1965).

9.　Ovsienko, D. E., and Sosnina, E. I., Fiz. Met. i Metalloved., 3(2) :374 (1956).

10.　Lainer, D. I., Petrusevich, R. L., and Sollertinskaya, E. S., Fiz. Met. i Metalloved., 9(4) : 535 (1960).

11.　Esin, V. O., and Kralina, A. A., Fiz. Met. i Metalloved., 13(4) : 577 (1962).

EFFECT OF SILICON CONCENTRATION ON THE SUBSTRUCTURE OF ALUMINUM SINGLE CRYSTALS

D. E. Ovsienko, I. K. Zasimchuk, and L. I. Meleshko

*Institute of Metal Physics of the Academy of Sciences
of the Ukrainian SSR*

The formation of the dislocation structure in metal crystals grown from the melt is largely determined by the presence of impurity atoms [1-6].

Tiller [7, 8] theoretically studied one possible mechanism representing the effects of dissolved impurity on the principle of the development of microsegregations in the crystal during growth; it is nevertheless impossible to say, on the basis of existing experiments, to what extent this theory corresponds to the true facts.

In order to develop our ideas regarding the character and mechanism of the effects of impurity on the dislocation structure of crystals further, we require some experimental data relating the dislocation structure to the conditions of growth, the form of impurity, and the impurity concentration.

To this end we studied the substructure of aluminum single crystals as a function of silicon content. We used very pure V3 (99.9994% Al) samples (principal impurities $2.1 \cdot 10^{-4}\%$ Si, $1.6 \cdot 10^{-4}\%$ Fe, $1.3 \cdot 10^{-4}\%$ Cu, $0.8 \cdot 10^{-4}\%$ Mg), high-purity AV000 aluminum (principal impurities $1.4 \cdot 10^{-3}\%$ Si, $1 \cdot 10^{-3}\%$ Fe, $1.1 \cdot 10^{-3}\%$ Cu), and also alloys prepared in a vacuum furnace with induction heating, based on V3 aluminum with the addition of 0.02, 0.15, and 1 at.% of 99.9995% pure Si.

We thus had at our disposal a set of aluminum–silicon alloys not distinguished from each other by containing other impurities. An exception was the AV000 aluminum, which differed from the others by containing other impurities in addition to silicon.

Single crystals of rectangular shape, $0.8 \times 0.5 \times 2.5$ cm in size, were grown by directional crystallization of the melt in a graphite mold [9]. By means of a seed all the crystals were given the same orientation, such that the growth direction was close to the [110], and the wide side surface coincided to an accuracy of 1.5° with the (100) plane. In order to limit the possibility of imperfections propagating from the seed into the sample, the transition between them was made by way of a thin neck some 0.1 cm in diameter. The growth rate was 10^{-2} cm/sec with a temperature gradient at the interface between solid and liquid phases of about 27 deg/cm.

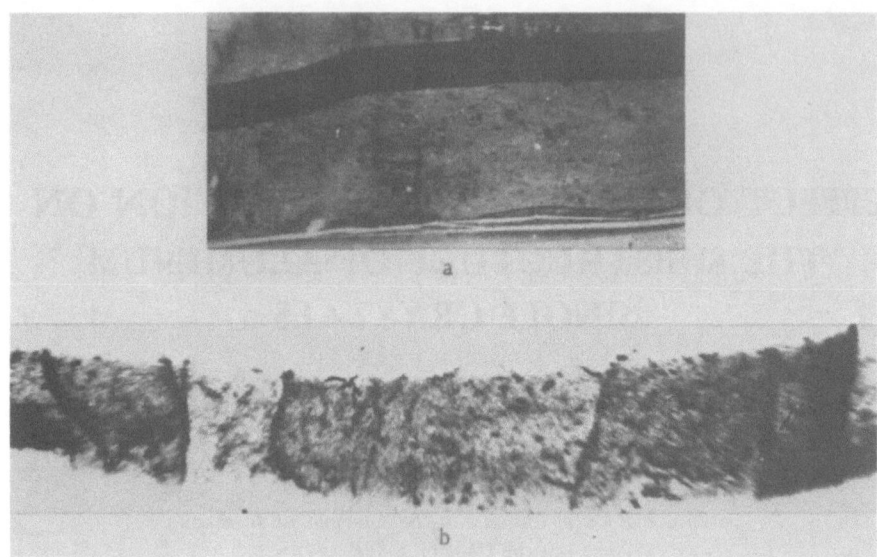

Fig. 1. Substructure of a V3 aluminum single crystal. Topographs of side
surface. a) Schulz, × 4.5; b) Berg-Barret, × 52.

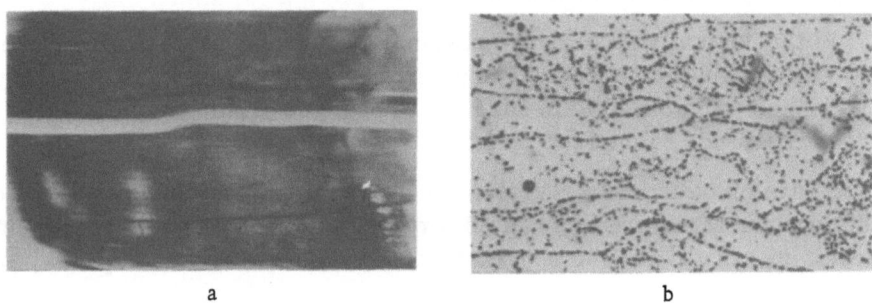

Fig. 2. Substructure of an AV000 aluminum single crystal. a) Schulz topograph, × 4.5; b) optical micrograph, × 66.

After growth, the single crystals were etched in a structure-sensitive reagent in order to
give a coarse visual idea of their imperfections; then they were chemically removed from the
seed and electrochemically polished. The substructure was studied by x-ray methods and in
some cases checked metallographically.

Precision measurement of the angular characteristics of mosaic structure was carried
out with a diffractometer based on the URS-50I and described in detail in [10]. For each
sample we recorded reflection curves from the (200) plane (on the principle of the two-crystal
spectrometer with rotation of the crystal under consideration) at different points of the surface
and for various positions of the longitudinal axis of the crystal relative to the incident beam.
The interplane distances for the (200) planes of aluminum and the (11$\bar{2}$0) planes of quartz (a perfect quartz crystal was used to obtain the parallel monochromatic beam) differ so little that it
was permissible to neglect the broadening of the double-reflection curves due to dispersion
when working with copper K$_{\alpha_1}$ radiation. In addition to this, we made the following experiments
in order to determine the sizes of subgrains disoriented relative to each other by angles greater
than 1' and also in order to obtain the distribution of disorientations between such subgrains.

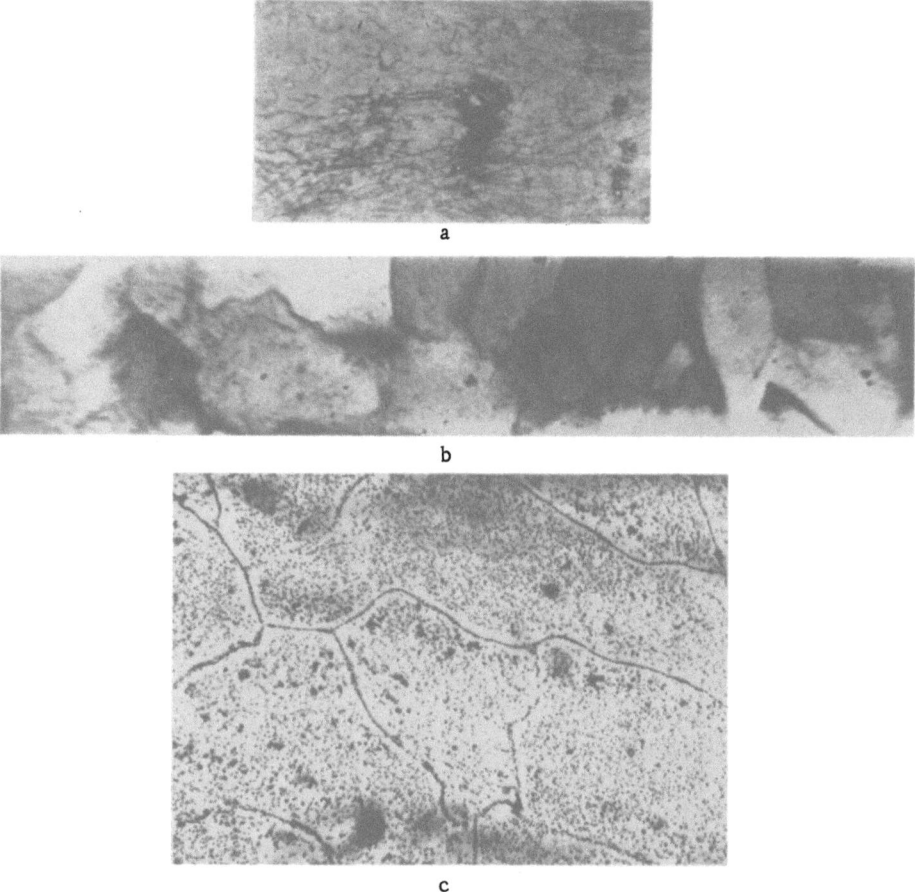

Fig. 3. Substructure of a single crystal of V3 aluminum with the addition of 0.02 at.% Si. Topographs: a) Schulz, × 4.5; b) Berg-Barret, × 52; c) optical micrograph, × 66.

By means of a micrometer screw, the sample was gradually moved in 0.01-cm steps relative to the incident x-ray beam, and at each point the angular position of the sample corresponding to maximum reflection intensity was determined (on the goniometer scale). This method was especially useful in studying the characteristics of coarse first-order substructure.

We also used the methods of x-ray topography.

A general idea of the character of the substructure of single crystals is given by x-ray photographs obtained by the Schulz method. The photographs were taken with a sharp-focus BSV-5 tube (copper anode) in white radiation. The x-ray beam illuminated the whole surface of the sample and the photograph also gave an image of the whole surface. Depending on the sign of the disorientation of the subgrains, the boundaries were revealed on the Schulz photographs as dark or light bands (Figs. 1a and 2a) of different widths, depending on the degree of disorientation.

More detailed information regarding the substructure in individual parts of the crystal surface was obtained by using x-ray diffraction topography on the Berg-Barret principle in a monochromatic parallel beam. In this case the work was carried out on the same apparatus as that used for the double-reflection curves. The topographs were taken on a fine-grain film (MR or MK NIKFI) from the wide and narrow side surfaces of the single crystals, reflection from

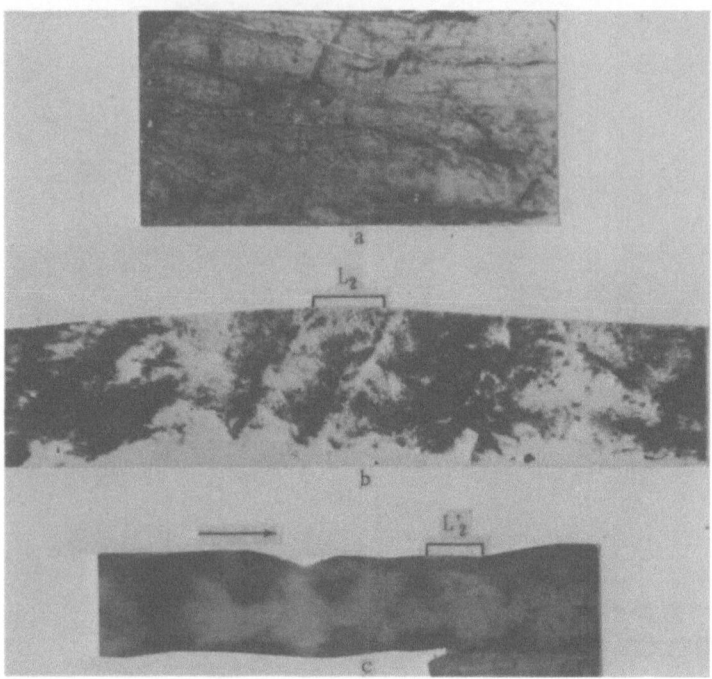

Fig. 4. Substructure of a single crystal of V3 aluminum containing 0.15 at.% Si. Topographs: a) Schulz, × 4.5; b, c) Berg-Barret photographs from the wide (b) and narrow (c) side surfaces, × 52; growth direction shown by an arrow.

Fig. 5. Substructure of a single crystal of V3 aluminum with the addition of 1 at.% Si. Schulz topograph, × 4.5.

the {311} plane being used in both cases. The distance between the sample and film was 0.1 cm for a vertical convergence of the incident beam of less than 25'.

The image on the Berg photograph is formed as a result of the different intensity of the reflection of x rays from individual parts of the irradiated surface. The intensity is determined by two factors: a) the degree of agreement between the part irradiated by the x-ray beam and the optimum reflecting position, and, b) the degree of local crystal-lattice distortion, since local disruptions of the lattice, for example, subgrain boundaries, individual dislocations, and local deformations reduce the extinction effect and lead to an increase in intensity.

The metallographic method of studying microstructure was used in addition to the x-ray methods. The use of published [10] etchants proved successful for the case of AV000 aluminum and the alloy of V3 aluminum with 0.02 at.% Si. Figures 1-5 show photographs of the substructure of single crystals in order of increasing silicon concentration, and in these we observe three orders of substructure, differing in the size and relative disorientation of the subgrains. The character of the substructure depends greatly on the amount of silicon impurity in the aluminum.

The single crystals of pure V3 or AV000 aluminum, as may be seen from the Schulz photographs (Figs. 1a and 2a), are characterized by division into several subgrains drawn out along the whole crystal and turned relative to one another through angles of 15' to 1.5° around the longitudinal axis. The boundaries between these subgrains are revealed on the Schulz photographs in the form of wide, straight bands. As a rule, the number of such subgrains (which we shall subsequently call first-order structure) is greater for the less pure AV000 than for V3 aluminum, in which there are only two or three subgrains for the whole surface. Inside these subgrains we find a finer, second-order substructure, consisting of subgrains also severely drawn out along the growth direction. Figure 1b shows a Berg photograph of part of the surface of a V3 aluminum single crystal within the bounds of one first-order subgrain. The dark lines across the photographs correspond to the boundaries of second-order grains. Inside these are dark spots, streaks, and bent lines, which form no closed contours, but only mark them out. The Berg photograph for aluminum of the AV000 type differs from that of the V3 aluminum in the greater density of such points and streaks and the smaller dimensions of the regions marked out by these. On comparing with microstructural data we may conclude that such dark spots and streaks on the topographs correspond to groups and rows of dislocations, and sometimes even to single dislocations.

The addition of 0.02 at.% Si, as may be seen from Fig. 3a, leads to a more homogeneous substructure in respect of the distribution of disorientations and the sizes of the subgrains. In these samples there are no long, straight boundaries with large disorientations, i.e., the first-order substructure is not expressed. The Berg photographs of the same surface (Fig. 3b) clearly reveal second-order substructure, which is also noticeable on the Schulz photograph. The elements of the substructure are subgrains ~200-400 μ in size, slightly drawn out along the growth direction (by ~1.5 times) and disoriented with respect to each other by angles between half a minute and several minutes. Within the bounds of such subgrains the crystal lattice is also subjected to certain bends, disorientations, and other disruptions, as may be seen from the Berg photographs; finer substructure, however, is not seen very sharply, and only rough statements may be made regarding the dimensions of the third-order subgrains.

Further increasing the silicon content (to 0.15 at.%) again leads to an increase in the degree of inhomogeneity of the distribution of disorientations, but this inhomogeneity is of a different character from that in pure aluminum and is evidently associated with the cellular character of the growth of these crystals, and hence with the periodic distribution of impurity in the course of growth. We see from Fig. 4a that the crystal consists of subgrains drawn out in the growth direction with a transverse size of ~1.5 mm. The disorientation between these is of a complex nature, containing a considerable component of torsion. Inside the large subgrains is a second-order substructure consisting of elements drawn out along the [001] direction with transverse dimensions of ~100 μ on the wide side surface. In the Berg photograph taken from the same side surface (Fig. 4b) the cross section of this element is denoted by L_2. We see from this photograph that the reflection from such a subgrain is not homogeneous, but is divided into separate parts, corresponding to elements of third-order substructure. Figure 4c shows a Berg photograph of the narrow side surface of the same sample. This includes a subgrain drawn out along the direction of growth, indicated by an arrow: L_2' corresponds to the dimension of the projection of the second-order subgrain on the photographed surface. In contrast to the cases of lower silicon contents, the Berg photographs from single crystals containing 0.15 at.% Si show no sharp boundaries between the subgrains of the second and third orders. This characteristic of the Berg photographs may be associated with the fact that the crystal lattice in the space between the subgrains suffers no sharp change of orientation, but only local bending. In addition to this, a considerable increase in the reflection intensity (degree of blackening or photographic density in the Berg photograph) may take place as a result

Table 1

Material	S, mm²	β, sec	δ, min	N	φ, min
Al V3	0.5 2	60 97	8.5 11	1.3 2.2	1 3
Al AV000	0.5 2	90 128	12.5 15.3	1.6 2.7	2.5 2.7
Al V3+0.02 at. % Si	0.5 2	57 80	5 6.4	1.5 2	1 2
Al V3+0.15 at. % Si	0.5 2	65 106	6 8	1.4 2	1 2
Al V3+1 at. % Si	0.5 2	125 194	14 23	2 5.5	3 6

Table 2

Material	L_1, cm	θ_1, min	L_2, cm	θ_2, min	L_3, cm	θ_3, sec	ρ exp, 1/cm²	ρ_{theor}, 1/cm²
Al V3	0.3	50	0.04	1.5	$8 \cdot 10^{-3}$	10	$2 \cdot 10^6$	$5 \cdot 10^4$
Al AV000	0.15	30	0.04	2.5	$5 \cdot 10^{-3}$	15	$4.0 \cdot 10^6$	$5 \cdot 10^4 - 4 \cdot 10^5$
Al V3+0.02 at. % Si	—	—	0.025	2.0	$6 \cdot 10^{-3}$	17	$3.5 \cdot 10^6$	$8 \cdot 10^5$
Al V3+0.15 at. % Si	0.13	15	0.015	1.5	$3 \cdot 10^{-3}$	30	$(4-10) \cdot 10^6$	$3 \cdot 10^6$
Al V3+1 at. % Si	0.06	20	0.01	2.0	—	—	$(5-50) \cdot 10^6$	$\leqslant 10^7$

of the increased defectiveness of the crystal lattice in individual regions, resulting from local supersaturation of these regions with silicon, this being caused by the nonequilibrium capture of impurity atoms in the course of growth. This is supported by the presence (Fig. 4c) of dark regions concentrating at the periphery of the subgrain and arranged along its length with a period equal to the projection of the cross section of the second-order subgrains.

Single crystals of aluminum containing 1 at.% Si show a substructure (Fig. 5) corresponding to dendritic growth of these crystals, with straight boundaries along the [001] direction with respect to the planes in which adjacent branches of the dendrites come together. Unfortunately, for this case it proved impossible to choose polishing conditions giving an adequately smooth crystal surface, so that the Berg-Barret method could not be used.

Table 1 shows the results of measuring double-reflection curves from the (200) plane, averaged with respect to 4 or 5 samples for each concentration. The characteristics obtained do not correspond to any definite type of substructure, but constitute averages over a large number of arbitrary parts of the crystal. In Table 1 S is the area of the irradiated section, β is the half-width of the reflection curve, δ is the angular range of the reflection at a height of 1% of the maximum intensity, N is the average number of peaks on the curve, and φ is the average angle between the peaks.

By analyzing all the experimental data, we obtained the mean sizes of the first-, second-, and third-order subgrains L_1, L_2, and L_3 for each concentration and correspondingly the mean disorientation angles between them: θ_1, θ_2, and θ_3. These data, together with the dislocation densities in the crystal obtained experimentally from the substructural data (ρ_{exp}) and the dislocation densities calculated for our crystals by the Tiller [7, 8] formula (ρ_{theor}), are shown in Table 2.

It follows from Table 2 that, as the silicon content increases, the size of the substructural elements of all orders in general becomes smaller. As regards the disorientations, here there is no such law. For substructures of the second order the disorientations vary little, for substructures of the third order they increase slightly with concentration, while for substructures of the first order the greatest disorientations occur for the purest aluminum (V3). For an intermediate concentration (0.02 at.% Si) the first-order substructure is entirely absent. Appearing again for a silicon concentration of 0.15 at.%, the first-order sub-boundaries increase their disorientation on passing to a silicon content of 1 at.%.

However, the formation of first-order substructure clearly has a different nature for pure aluminum and aluminum containing silicon. For aluminum containing 0.15 to 1 at.% Si, the existence of first-order substructure (and also that of other orders) is associated with the segregation of impurities in the course of crystal growth. In the case of pure V3 and AV000 aluminum the nature of these sub-boundaries is different. They are evidently formed as a result of the displacement and rearrangement of dislocations existing in the crystals at temperatures close to the crystallization point. The same nature may well apply to sub-boundaries of the second and third orders for these samples, but these are only formed at correspondingly lower temperatures, when the displacement of the dislocations is limited by shorter distances. The greater amount of impurity in AV000 leads to a reduction in the mobility of the dislocations and the size of the subgrains as compared with V3 aluminum. This point of view regarding the nature of the sub-boundaries in pure aluminum is confirmed by microstructural studies. Whereas in AV000 aluminum (Fig. 2b)* all the dislocation pits are collected together into the boundaries, in aluminum containing 0.02 at.% Si the etch pits within the second-order subgrains are distributed at random, which may be explained by the limited mobility of the dislocations at lower temperatures in the presence of 0.02 at.% Si. Clearly, it is with the higher degree of order in the dislocation distribution that we must associate the higher angular characteristics (Table 1) of the AV000 aluminum crystals as compared with aluminum containing 0.02 and 0.15 at.% Si.

The values of ρ_{exp} given in Table 2 were calculated as the sums of the densities of the dislocations forming the boundaries of the three orders of subgrains. As regards the third-order structure, however, we have the least accurate data; hence, the estimate of the contribution from dislocations forming this substructure and situated within it is the least accurate. Nevertheless, this contribution may be extremely substantial for cases of high silicon concentrations. In view of this, for crystals containing 0.15 and 1 at.% Si, we also estimated the maximum dislocation density from the formula $\rho = \beta^2/9b^2$ [12], obtained on the assumption of a chaotic distribution of dislocations; the values of β were taken from Table 1. The greater of the two values of ρ_{exp} given in Table 2 corresponds to this estimate. In the last column of this table we give, for comparison, the densities of the dislocations which, according to Tiller's theory, should be introduced into the crystal as a result of the segregation of impurities in the course of growth. For the cases of 0.02–1 at.% Si the densities were calculated from the formulas for cellular growth [8], the size of the cells being taken as the period of the wrinkles formed on the side surfaces of the crystals. For the AV000 aluminum single crystals, the impurity structure of which in the course of growth was not known, the upper limit of dislocation density was also calculated from the formulas relating to the case of cellular growth, the transverse dimension of the cells being taken as 20 μ. The dislocation density in V3 aluminum was determined from the formula for layer-like growth [7], just as the lower limit of the density for the AV000 type.

---

*For comparison, Figs. 1b, 3b, and 4b should be turned through 90°.

We see from Table 2 that the dislocation density in the single crystals rises slightly with increasing silicon content. After comparing these data quantitatively with the dislocation densities calculated by the Tiller theory, it hardly seems possible to draw any definite conclusion regarding the validity of this theory. For pure V3 and AV000 aluminum, the theoretical dislocation density calculated from Tiller's formulas is much smaller than the experimental value; this may be due to the imperfections of the theory itself. However, this lack of agreement may also be explained by supposing that, for the purity of the metal in question and the conditions of crystallization prevailing, the impurity mechanism is not the decisive one in the formation of dislocations, and that the role of the impurities reduces to their influence on the mobility of the dislocations. For high silicon contents, the calculated dislocation densities are quite close to the experimental values, which is a point in favor of the impurity mechanism of Tiller. However, in this range of silicon concentrations there may also be other effects associated with the increase in impurity content, also leading to a rise in dislocation density (local supersaturations of the solid solution and irregular intergrowth of the dendrite branches).

Thus we see from the results presented that the introduction of silicon into aluminum leads to an increase in the dislocation density and greatly affects the character of the substructure.

Literature Cited

1.   Sosnina, E. I., and Ovsienko, D. E., Fiz. Met. i Metalloved., 3(3) : 527 (1956).
2.   Lyuttsau, V. G., and Rovinskii, B. M., Kristallografiya, 8(5) : 742 (1963).
3.   Zasimchuk, I. K., and Ovsienko, D. E., Ukr. Fiz. Zh., 9(10) : 1092 (1964).
4.   Atwater, H. A., and Chalmers, B., Can. J. Phys., 35 : 208 (1957).
5.   Sekerka, R. F., Bolling, G. F., and Tiller, W. A., Can. J. Phys., 38 : 883 (1960).
6.   McGrath, J. T., and Craig, G. B., Can. J. Phys., 40(7) : 850 (1962).
7.   Tiller, W., Elementary Processes of Crystal Growth, p. 272 [Russian translation]. IL, Moscow (1959).
8.   Tiller, W., Acta Met., 10(7) : 681 (1962).
9.   Sosnina, E. I., Meleshko, L. I., and Ovsienko, D. E., In: Study of Imperfections in Crystal Structure, p. 122. Naukova Dumka, Kiev (1965).
10. Zasimchuk, I. K., and Ovsienko, D. E., Kristallografiya, Vol. 1 (1966).
11. Metzger, M., and Intrater, J., Nature, 174(4429) : 547 (1954).
12. Hirsch, P. B., In: Advances in the Physics of Metals, Vol. 3, p. 383 [Russian translation] Metallurgizdat, Moscow (1960).

STUDY OF THE CONDITIONS OF GROWTH OF CADMIUM AND ZINC MICROCRYSTALS AND THEIR DISLOCATION STRUCTURE

M. V. Zakharova and G. M. Zinenkova

M. V. Lomonosov Moscow State University

It is well known that microcrystals of various materials possess unusual properties; hence, great efforts are now being directed toward the study of their strength, ductility, and dislocation structure. Considerably less attention has been devoted to the conditions and mechanism of their growth [1-3]. Such data may nevertheless greatly assist our understanding of their very high strength. In this paper we shall set out some results on the conditions for the growth of Cd and Zn microcrystals and on their dislocation structure.

The Cd and Zn microcrystals were grown by condensation from the vapor in an argon atmosphere at various pressures (10 to 600 mm Hg) [4]. Under these conditions growth took place as a result of the transport of material from the hot to the cold part of the tube by diffusion through the residual gas. Three zones are clearly distinguished in the growth tube (Fig. 1): I) the region of evaporation; II) the region of diffusion transport without condensation; and, III) the region of condensation of the zinc and cadmium vapor.

The diffusion process may be described by the following system of equations:

$$\varrho_I = \text{const}, \qquad \varrho'_{II} = 0,$$

$$D\,\varrho'_{III} = \frac{\gamma v}{R}\,[\varrho_{III} - \varrho_0(x)],$$

where ρ_I, ρ_{II}, and ρ_{III} are the densities of the vapor in regions I, II, and III, respectively; $\rho_0(T)$ is the thermodynamic equilibrium density corresponding to temperature $T(x)$; R is the tube radius; D is the diffusion coefficient of the vapor in the residual gas; v is the mean thermal velocity of the vapor molecules; and γ is a coefficient characterizing the intensity of the condensation process. By solving this system of equations, we may calculate the vapor pressure (Fig. 1, curve 3) at each point along the growth tube, and hence also the supersaturation, which constitutes one of the main factors influencing the growth of microcrystals. We see from the calculations that the distribution of cadmium vapor density is almost the same for all residual gas pressures for a given temperature gradient in the tube. Hence, the supersaturation calculated, for example, for a pressure of 600 mm Hg (Fig. 2) remains intact for pressures of 100, 50, and 10 mm Hg.

181

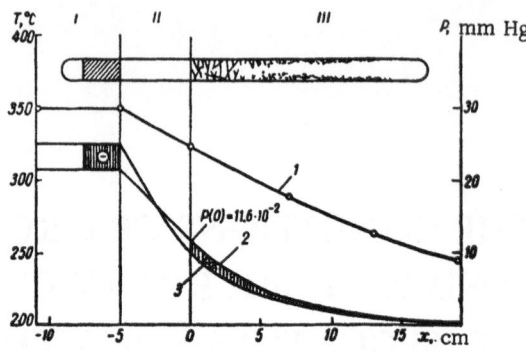

Fig. 1. Variation in the temperature, equilibrium pressure, and nonequilibrium pressure of the cadmium vapor along the growth tube. 1) Temperature distribution $T(x)$ along the tube; 2) nonequilibrium pressure of cadmium vapor $P(x)$ along the tube; 3) equilibrium pressure of cadmium vapor $P_0(x)$ corresponding to temperature $T(x)$.

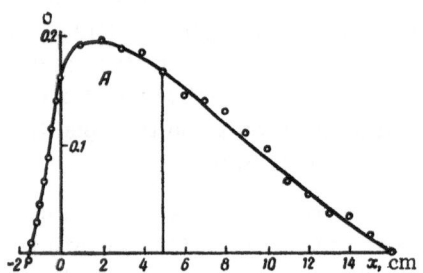

Fig. 2. Variation in supersaturation along the growth tube. A) Region corresponding to the growth of "thread-like" crystals ("whiskers").

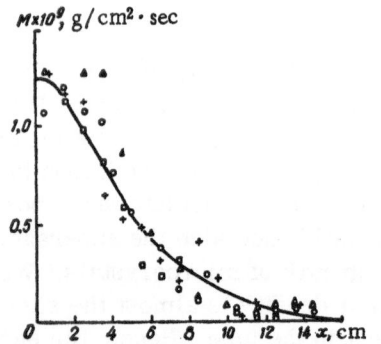

Fig. 3. Distribution of condensate along four growth tubes for an inert-gas pressure of 600 mm Hg. ○ — 100 h; △ — 82 h; □ — 100 h; + — 114 h.

Knowing the vapor pressure at each point, we may calculate the actual mass distribution of the condensate along the growth tube. In order to check such calculations we made an experimental study of the mass distribution in question. Figure 3 shows the mass distribution of condensate for four tubes at an argon pressure of 600 mm Hg. The calculated mass distribution (continuous curve) satisfactorily described the experimental data. Analogous calculations were also made for pressures of 10, 50, and 100 mm Hg. Detailed analysis of the distribution of condensate along the tube showed that for all pressures in the range 10–600 mm Hg the origin of condensation approximately corresponded to a crystallization temperature of 320°C for Cd and 420°C for Zn. The main bulk of the condensate for all pressures was concentrated in the initial region of growth. Quantitatively the deposited mass fell off sharply at the end of the tube, where condensation practically ceased. The growth region of Cd and Zn whiskers corresponded to the temperature range 295 to 320 for Cd and 395 to 420°C for Zn. The supersaturation of Cd vapor corresponding to this temperature range was 0.17 to 0.2 (see Fig. 2). In the temperature range 205 to 250°C, on varying the supersaturation from 0.17 to 0, very small whiskers (1 mm long) and other crystals of irregular shape were formed.

On studying the Cd and Zn whiskers it was observed that these comprised crystals of two types: "cylindrical," with a hexagonal cross

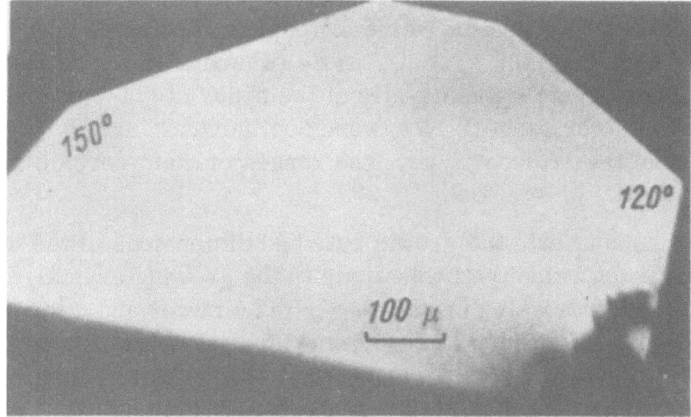

Fig. 4. Lamellar zinc single crystal.

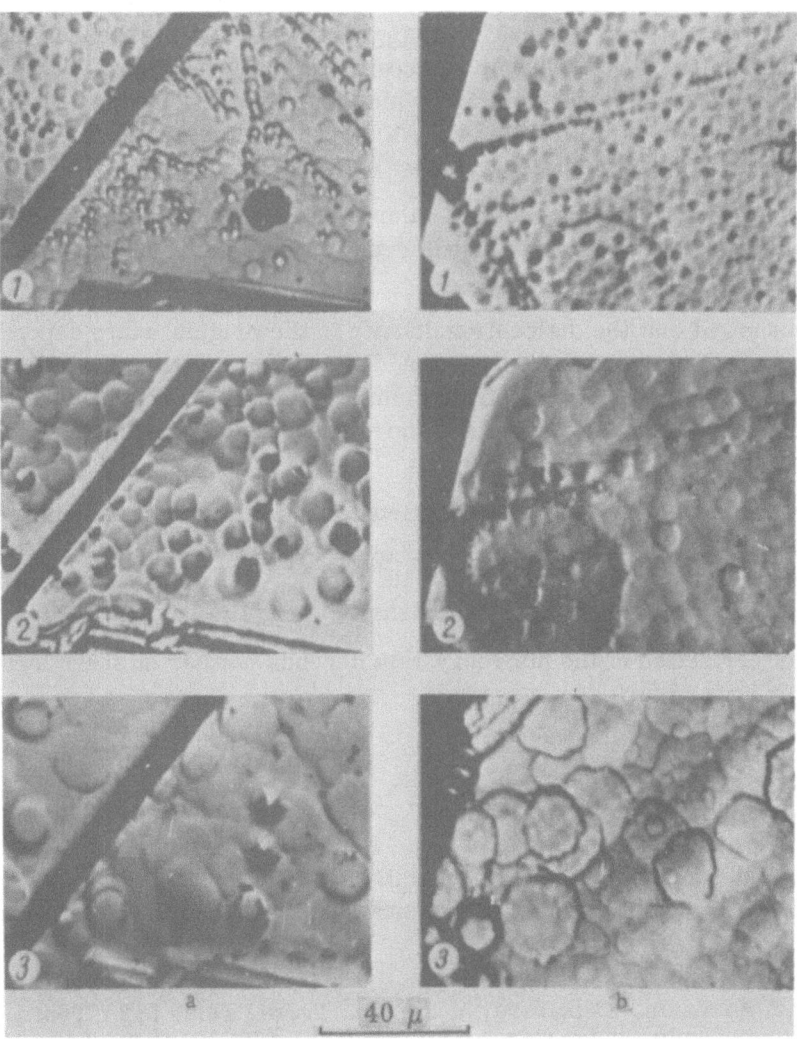

Fig. 5. Dislocation distribution with respect to thickness in (a) cadmium and (b) zinc microcrystals. 1) Original dislocation structure; 2) dislocation structure after polishing off a layer 0.04 μ thick; 3) dislocation structure after polishing off a layer 0.1 μ thick.

section, and "ribbon" with a cross section in the shape of a rectangle. The first had a growth direction [2$\bar{1}\bar{1}$3] and the second [2$\bar{1}\bar{1}$0]. The h/l ratio (where h and l are the dimensions of the cross section) in the ribbon crystals was usually of the order of 0.1-0.2. It was also found that the fine crystals in the diameter range 1-15 μ were "cylindrical" and the thick ones with diameter 10 μ and over were of the "ribbon" type. The ranges of diameters corresponding to these two types hardly overlapped.

The inert gas pressure affects the growth rate and dimensions of the whiskers by reducing or increasing the diffusion velocity of the atoms to the growing crystal. Thus, for an argon pressure of 600 mm Hg, "cylindrical" crystals 1-5 μ in diameter and up to 1 cm long tend chiefly to grow. For an argon pressure of 10-50 mm Hg a large number of "ribbon" and lamellar crystals occur in addition to the "cylindrical" whiskers. It is interesting to note that the "ribbon" crystals grew independently, while the majority of "cylindrical" Cd and Zn whiskers grew out of lamellar microcrystals. In the lamellar Cd and Zn microcrystals the large surface was the basal plane. The extent of this surface reached 0.5 mm^2, and in individual cases 1 mm^2, with a thickness of up to 100 μ. The form of the Cd and Zn single crystals is rather different. The cadmium microcrystals are hexagonal with angles of 120° and their edges are directed along the close-packing directions. The Zn microcrystals have polygonal form, in which the 120° angles are accompanied by others of 150°, which corresponds to edge directions perpendicular to the close-packing direction (Fig. 4). As a result of their orientation, the lamellar crystals are suitable for revealing dislocations by selective etching [7, 8].

Edge dislocations with Burgers vectors ±a and screw dislocations with Burgers vectors ±a±c lying in pyramidal planes of the first and second kinds emerge on the basal plane and may be observed there.

Examination showed that the dislocation density in the original state depended on the dimensions of the crystals. In order to prevent additional deformation, the microcrystals were not removed from the substrate. Plates under the critical size of about 7000 μ^2 contained no growth dislocations. In some exceptional cases there were also no dislocations in larger lamellar microcrystals.

In microcrystals exceeding the critical dimensions, the dislocation density reaches 10^3 to 10^4 cm^{-2}. Usually in the larger cadmium and zinc microcrystals there is a deformed layer about 1 μ thick with a dislocation density of 10^5-10^6 cm^{-2}. After removal of this layer (Fig. 5), growth dislocations remain in the crystals, the density being 10^3-10^4 cm^{-2}.

The degree of perfection of the microcrystals depends neither on their location in the growth chamber nor on the inert-gas pressure.

Literature Cited

1. Price, P. B., Phil. Mag., 5:473 (1960).
2. Collman, R. V., and Sears, G. V., Acta Met., 5:131 (1957).
3. Dittmar, W., and Neumann, K., Elektrochemie, 4:297 (1960).
4. Shvidkovskii, E. G., Predvoditelev, A. A., and Zakharova, M. V., Fiz. Tverd. Tela, 6:4 (1964).
5. Honig, R. B., RCA Rev., 18:195 (1957).
6. Predvoditelev, A. A., and Zakharova, M.V., Fiz. Tverd. Tela, 7:2 (1965).
7. Tyapunina, N. A., and Zinenkova, G. M., Kristallografiya, 9:6 (1964).
8. Predvoditelev, A. A., Bushueva, G. V., and Stepanova, V. M., Fiz. Met. i Metalloved., 14:5 (1962).

STRUCTURE OF THE SURFACES OF DROPS OF TIN CRYSTALLIZING ON POLISHED GLASS PLATES

O. A. Mikhno and N. L. Pokrovskii

M. V. Lomonosov Moscow State University

The conventional metallographical methods of preparing samples for structural study are sometimes unsuitable for metals with low melting points. Tin and lead samples, in particular, after grinding and polishing, exhibit an oxide film [1] and also a work-hardened layer, which easily recrystallizes, even at room temperature [2]. This gives rise to a structure greatly differing from that of the as-cast state [3].

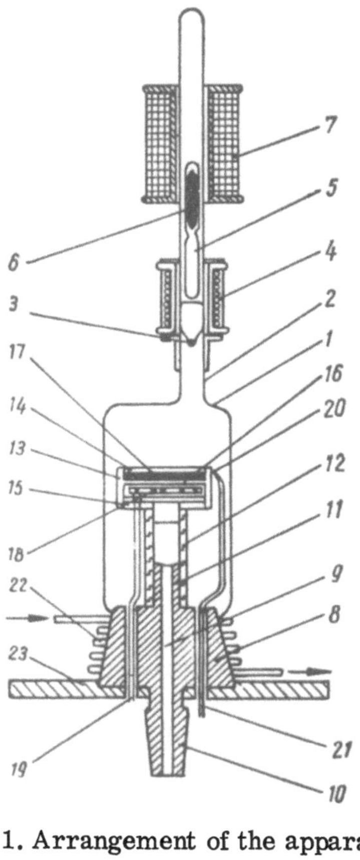

We used the method [4] of obtaining tin samples with a mirror surface in vacuum by crystallizing them on optically polished glass plates. This method, requiring no mechanical finishing or chemical etching, proved completely justified and was later used by other authors [5, 6], although without the use of vacuum.

On further development of this method we set up a special apparatus (Fig. 1) enabling us to obtain (in a vacuum of 10^{-5} mm Hg) small samples of Pb, Sn, Bi, and other metals of low melting point in the form of flat drops crystallizing on polished glass plates (Fig. 2) for any given temperature, from the melting point down to room temperature.

The apparatus was made of refractory borosilicate (molybdenum) glass, and was based on a cylindrical vacuum jacket 1; sealed to the upper end of this was a tube 2 furnished with a glass capillary 3 having an aperture of 0.3-0.8 mm. Into the tube 2 we placed the metal under examination, melting this at the instant required by means of an electric heater 4 (250 W). Inside the tube 2 was a glass float 5 containing a piece of soft iron 6 and set into motion by a solenoid 7. A Dural stopper 8 was ground into the opening of the vacuum jacket 1; this had an axial channel 9 and ended in a conical ground joint 10 for con-

Fig. 1. Arrangement of the apparatus.

185

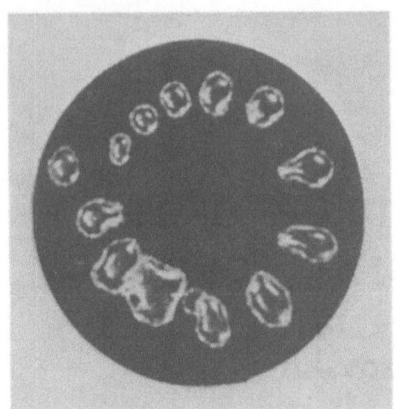

Fig. 2. Glass disc with tin samples.

nection to the vacuum system. The conical part of the stopper 8 had a cylindrical projection 11 on which was placed a porcelain tube 12 bearing a heating stand 13 (made of stainless steel) comprising two hollow cylinders (chambers) 14 and 15 separated by a screen 16. At the bottom of the upper chamber 14 was a polished glass plate 17 having the form of a flat disc of diameter 50 mm and thickness 1 mm (Fig. 2). In the lower chamber 15 was a 250-W Nichrome heater 18. The leads from the heater 18 passed through the channel 19 in the stopper 8 and were connected to the electrical supply system. For measuring the temperature of the glass substrate 17, the apparatus was furnished with a Chromel—Alumel thermocouple 20, which passed into the apparatus through channel 21 of stopper 8. The channels 19 and 21 have special vacuum sealing. The conical stopper 9 was water-cooled, the water circulating through coil 22, fixed to the lower part of vacuum jacket 1. The whole apparatus was fixed to a special stand 23.

In order to obtain samples of the metal to be examined, a small quantity of this was placed in tube 2. Then the glass float 5 was put in position and the tube was sealed at the top. The apparatus was connected to the vacuum system through the conical ground joint 10. The glass disc 17 on which crystallization of the samples was to take place was first carefully degreased. On reaching the required vacuum of $1 \cdot 10^{-5}$ mm Hg and the desired temperature of the glass substrate, the metal in tube 2 melted. The molten metal did not flow out through the narrow capillary 3 (diameter 0.3-0.5 mm), being held back by the forces of surface tension.* Arrangements were made in the apparatus, however, for forcing individual drops of metal out through the capillary 3, as follows. After melting the metal and reaching the desired melt temperature, the circuit of the solenoid 7 was closed and the glass float 5 slowly lifted up over the surface of the metal. Then the solenoid was disconnected and the flat fell back, striking the surface of the molten metal and forcing a drop through the capillary 3; the drop fell directly on the glass plate 17 and there crystallized under known temperature conditions.

The apparatus enabled us to obtain a series of identical samples without disrupting the experimental conditions. For this purpose the vacuum jacket 1 was rotated through a small angle and a free part of the glass plate presented to the capillary to receive the next drop. Thus, 10-15 samples were obtained on each plate (see Fig. 2).

The samples of tin thus obtained had a mirror surface and could be studied under the microscope without any further preparation and without removing from the glass plate. This is illustrated in Figs. 3 and 4, which show the structure of the crystallized drops. It should be mentioned that the crystallization of these samples took place in quite nonequilibrium conditions, leading to the formation of various kinds of defects. We may suppose that the samples studied contained defects of the vacancy and dislocation type, and that the regular geometric figures formed on their surfaces represent peculiar "etch pits" constituting the outcrops of dislocations, which simultaneously act as vacancy sinks. In addition to this there are other mechanisms [7-9] giving rise to regions with increased vacancy concentration, leading to the formation of pits. We also note that the surface structure of the tin samples depended greatly on the presence of impurities and was largely determined by their surface properties. In our experiments we used surface-active sodium and inactive zinc.

* The diameter of the capillary and the maximum height of the column of molten metal are determined by the surface tension of the metal in question.

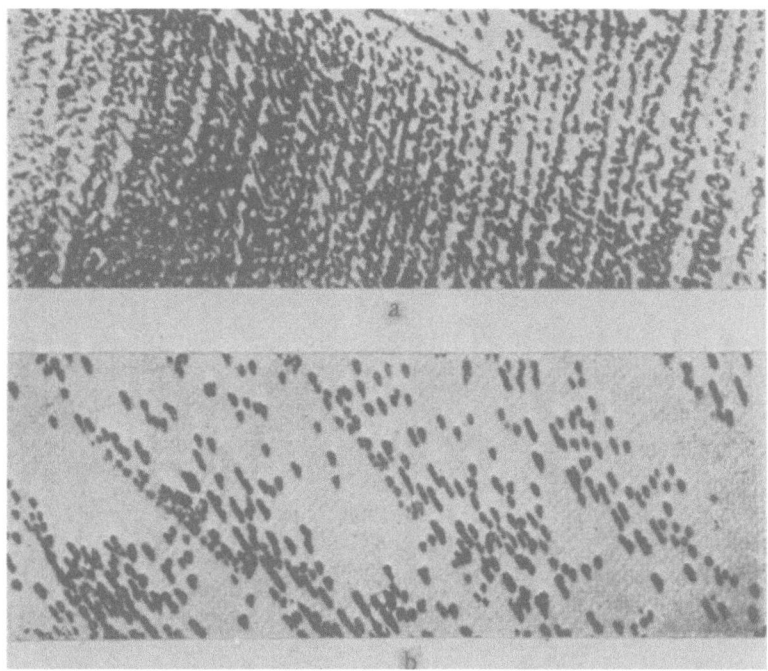

Fig. 3. Surface structure of drops of tin containing 0.05 at.% sodium; temperature of glass plate 100°C. a) × 100; b) × 300.

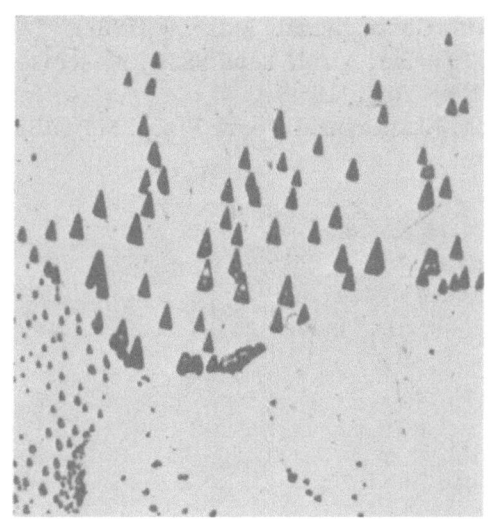

Fig. 4. Structure of the surface of a drop of tin containing 0.1 at.% zinc; temperature of glass plate 100°C. × 300.

We see from the photographs that the pits associated with specific regions of the samples not only had the same regular geometrical shape but also oriented themselves in the same sense; this was clearly associated with the crystallographic orientation of the various parts of the sample.

After annealing the tin samples (Fig. 5) we noted a change in the arrangement of the pits. This may clearly be seen by reference to the stationary reference point (shown as an arrow in the photographs). The changes in the positions of the pits took place in a specific direction as indicated by the arrows. We are making a careful study of this phenomenon at the present time.

Literature Cited

1. Richter, H., Z. Angew. Phys., 8(12) : 585 (1956).
2. Gay, P., and Kelly, A., Acta Cryst., 2 : 172 (1953).
3. Pokrovskii, N. L., and Smirnova, T. G., Fiz. Met. i Metalloved., 12(5) : 708 (1961).
4. Pokrovskii, N. L., and Galanina, N. D., Zh. Fiz. Khim., 23(3) : 324 (1949).
5. Pandya, N. S., and Shah, C. J., J. Sci. Ind., BC18(2) : 1385 (1959).

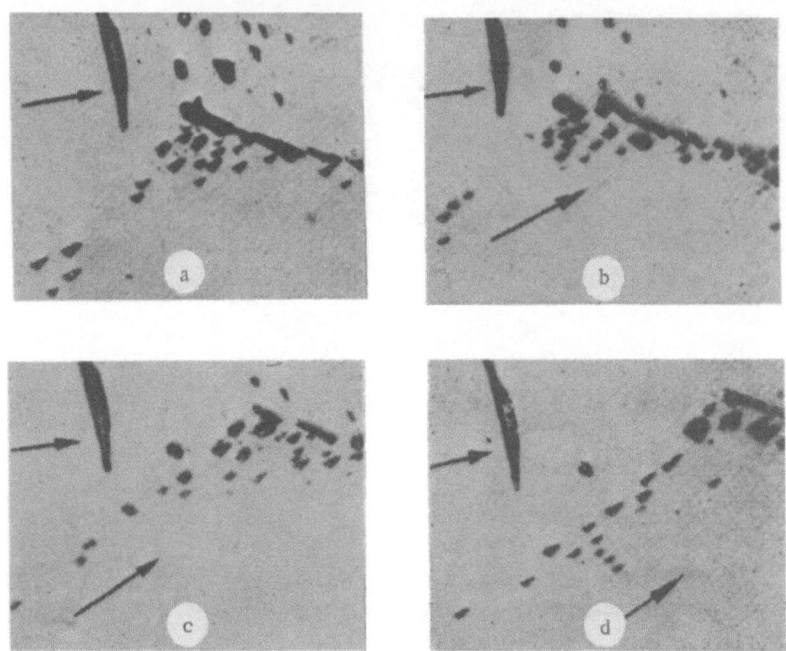

Fig. 5. Successive stages in the displacement of etch figures as a
result of annealing at 200°C. a) Initial stage; b) annealing for 7 h;
c) annealing for 17 h; d) annealing for 27 h. × 100.

6.   Takahashi, N., and Kadzato, K., Compt. Rend. Acad. Sci., 243(19) : 1408 (1950).
7.   Frenkel', Ya. I., Introduction to the Theory of Metals. Gostekhizdat, Moscow (1952).
8.   Geguzin, Ya. E., In: Growth of Crystals. Vol. 1, p. 91. Izd. Akad. Nauk SSSR, Moscow (1957). [English translation: Consultants Bureau, New York (1959).]
9.   Garber, R. I., Kogan, V. S., and Pomekov, L. M., Zh. Éksperim. i Teor. Fiz., 35(6) : 1364 (1958).

MICROINHOMOGENEITY OF IRON SILICIDE

Ya. V. Grechnyi, K. M. Zhak,
and É. N. Pogrebnoi

Dnepropetrovsk Metallurgical Institute

During the solidification of alloys both chemical and physical inhomogeneities arise. There is a definite link between these, and they exert a considerable influence on the properties of castings [1-3]. However, in contrast to chemical inhomogeneities, the structural picture of physical inhomogeneities created in castings on solidification and their influence on structural changes in iron silicide* have been little studied.

In this paper we shall consider the nature of the chemical inhomogeneity and the distribution of dislocations and microscopic breaks in continuity in cast, worked, and annealed iron silicide containing 0.020-0.043% C and 2.90-3.30% Si. The original material consisted of castings 50 mm in diameter crystallized in a preheated ceramic mold, and also samples selected at various stages in the technological conversion of transformer steel (from ingots, slabs, hot-rolled strip, and sheet). For these samples we studied the inhomogeneity in the cast state and its behavior during heat treatment and working.

The chemical inhomogeneity was revealed by electrolytic etching (current density 3-4 A/dm^2) or chemical etching in an aqueous solution of chromic anhydride, a reagent sensitive to silicon segregations. The dislocation structure was revealed by electrolytic etching in the same solution by a method similar to that of [4].

The alloys containing 0.02-0.04% C and 2.9-3.3% Si are well known to crystallize with the formation of an α solid solution. During the crystallization of such alloys, intracrystalline segregation of silicon takes place, i.e., the silicon is concentrated less in the central parts of the dendrite branches than in the peripheral parts [5, 6]. Metallographic study of cast samples showed the existence of direct intracrystalline segregation (Fig. 1a, b; Fig. 2). An increased amount of silicon occurs in the interbranch spaces and at the joints of the dendrite crystals. The dendritic character of the silicon segregations is more strongly expressed in the casting cooled in the ceramic (Fig. 2a) mold than in the ingot (Fig. 2b).

Owing to the segregation processes, austenite may be formed in the minute volumes with reduced silicon content on cooling [6]. On further cooling, the austenitic regions experience polymorphic and eutectoid transformations. We see from Fig. 2 that the pearlite columns are mainly formed in the middle of the dendrite branches with low silicon content. The pearlite regions are bordered with a low-silicon ferritic layer formed as a result of $\gamma \rightarrow \alpha$ recrystallization.

---

*We will use the terms iron silicide, transformer steel, and silicon steel interchangeably.

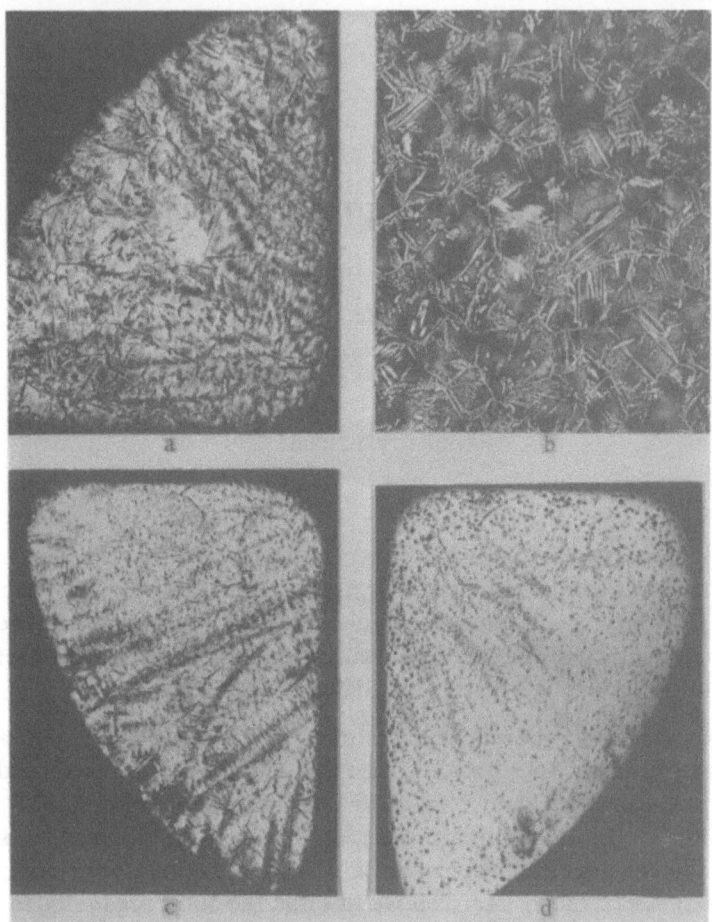

Fig. 1. Macrostructure of iron silicide in the cast (a, b) and annealed (c, d) states (× 2.5).

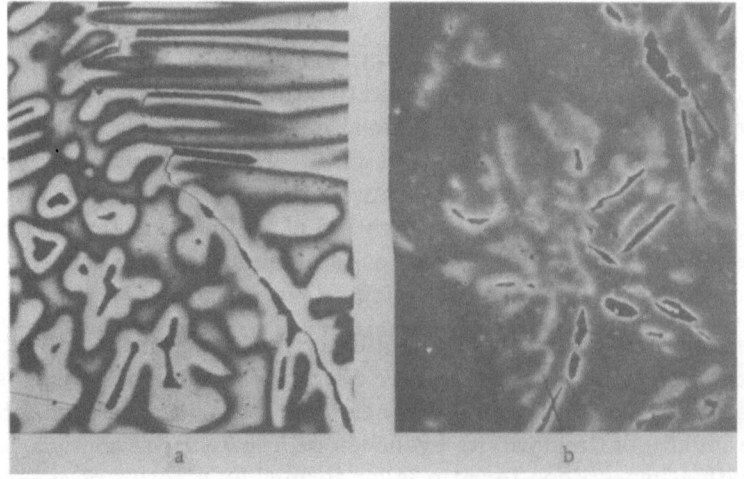

Fig. 2. Microsegregations of silicon in the cast state. a) Casting 50 mm in diameter (× 150); b) ingot (× 50).

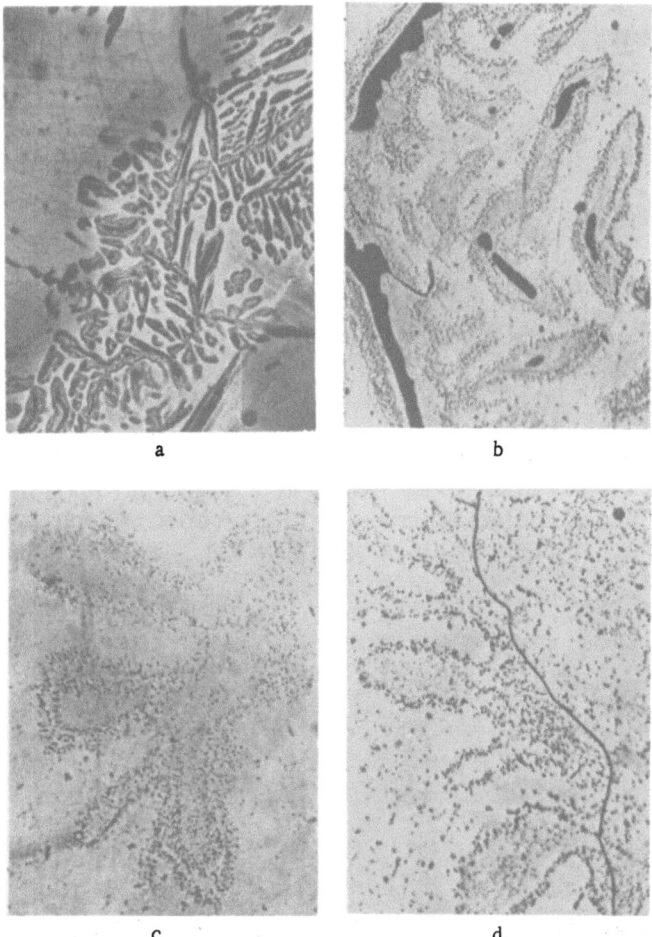

Fig. 3. Dislocation structure of iron silicide in the cast (a-c)
and annealed (d) states: a) × 15; b,c) × 150; d) × 350.

During crystallization of the alloy a physical inhomogeneity is also created; this is associated with the chemical microinhomogeneity. A study of the dislocation structure revealed by etch pits illustrates the relation between the dislocations and silicon segregations. In the original samples the arrangement of the dislocations follows the dendritic picture. Within the dendrite the dislocations are distributed nonuniformly, the dislocation density being higher along the boundaries of the concentration zones (Fig. 3a, b, c). The preferential formation of dislocations in these zones is evidently associated with the high concentration gradient of silicon and other impurities. In these regions large stresses arise as a result of the change in lattice parameter, and these are removed by the formation of dislocations [7].

The chemical and physical inhomogeneity created in castings has a great influence on structural changes in subsequent operations.

On studying samples cut from a slab subjected to hot rolling with a total reduction of about 85%, we observed a striated or banded arrangement of the segregation regions (Fig. 4a). In the course of rolling, the dendrites are distorted and drawn out in the rolling direction, giving rise to a structure in which sections of high and low silicon content alternate. In the light parts, corresponding to the dendrite branches, pearlite columns (also of extended form) occur.

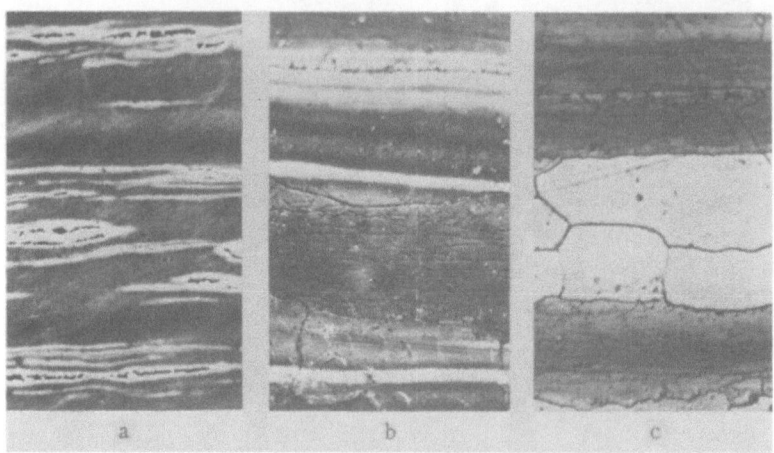

Fig. 4. Banded (striated) structure in hot-rolled silicon steel.
a) × 15; b) × 350; c) × 350.

Fig. 5. Relation between silicon segregation (a) and dislocation
density (b) in samples annealed after rolling (× 150).

This microinhomogeneity is weakened, but not eliminated, in subsequent technological operations on transformer steel. In samples cut from strip subjected to hot rolling and re-crystallization-annealing at 800-850°C (black annealing), the banded arrangement of the segregation regions is clearly revealed by reagents sensitive to silicon segregation (Fig. 4b, c). The existence of silicon segregation has a substantial influence on the recrystallization of worked ferrite. The parts impoverished with respect to silicon recrystallize more rapidly (Fig. 4c). In the silicon-rich parts the textured grain and subgrain structure is preserved for a long time

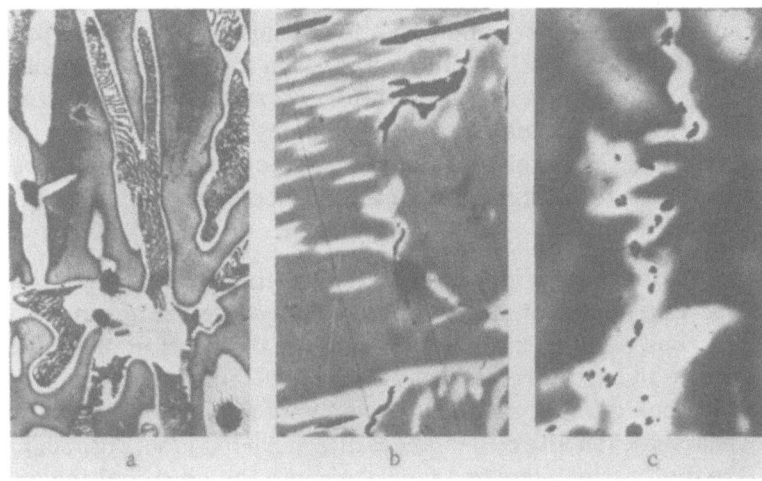

Fig. 6. Structure of iron silicide after annealing at 750°C. a, b) × 500;
c) × 100.

(Fig.4b, c), and in these, as a rule, electrolytic etching reveals a higher dislocation density (Fig. 5b).

The nonuniformity in the recrystallization of worked ferrite is due both to the microchemical and to the related physical inhomogeneity. The influence of these should appear both in the course of recrystallization treatment and during secondary recrystallization and texture formation.

We studied the behavior of the microchemical and physical inhomogeneity on annealing. Annealing at 700-800°C with a holding time of 30-35 h does not remove the intracrystalline segregation of silicon in cast (and worked) samples (see Fig. 1c). A characteristic of silicon steel is the development of microinhomogeneities with respect to silicon (associated with the dissolution of the carbide phase) on annealing (Fig. 6a, b). In the regions earlier occupied by pearlite and excess carbide, the silicon content is low (light parts). Reagents sensitive to silicon segregation clearly reveal this microinhomogeneity, together with the intracrystalline segregation. On annealing, the former is removed more rapidly than the latter, since the redistribution of silicon takes place within very small volumes.

On raising the annealing temperature, the intracrystalline segregation and the microinhomogeneity associated with the dissolution of carbides are removed more rapidly. As before, more time is needed for the elimination of the intracrystalline segregation. Traces of the dendritic segregation of silicon are visible even after 20-h annealing at 1100°C (see Fig. 1d). After high-temperature homogenization, intragranular and intergranular micropores are observed in the matrix, apparently of diffusion origin. Often these have crystalline facing and the dislocation density around the pores is higher.

In the course of low-temperature annealing (below 800°C) graphitization of the carbide phase takes place. Graphite inclusions appear inside the ferrite crystals and at the boundaries between them (Fig. 6). The precipitation of graphite occurs in regions both rich and poor in silicon.

The dissolution of the carbide phase and the homogenization of the inhomogeneous solid solution on annealing lead to the development of excess vacancies and dislocations as a result of the lack of agreement between the specific volumes of the carbide and the ferrite [8] and the unequal partial diffusion coefficients of iron and carbon [9]. The excess vacancies, interacting with the dislocations, grain boundaries, and sub-boundaries, settle on these, forming diffusion micropores. These subsequently become sinks for vacancies. The possibility of pores being formed by the supersaturation of silicon in iron—silicon alloys was demonstrated experimentally in [9, 10]. Diffusion micropores and microdefects formed on crystallization play a catalytic role during graphitization. Since the graphite inclusions are, in general, formed heterogeneously [8], the graphitization of the carbide phase taking place during low-temperature annealing makes it possible to reveal the distribution of microscopic discontinuities suitable for the formation of graphite.

On annealing, the dislocation structure of the castings changes considerably as a result of the interaction and regrouping of dislocations during structural changes and the homogenization of the inhomogeneous solid solution. However, the dislocation picture associated with the chemical inhomogeneity is hardly removed at all by low-temperature annealing, although the dislocation density is considerably reduced. Regions of dendritic dislocation distribution occur near the grain boundaries even after 10-h annealing at 900°C (see Fig. 3c). On further raising the temperature, the dislocation picture associated with the chemical inhomogeneity is removed more rapidly. After 20-h annealing at 1100°C it scarcely appears at all.

The foregoing data lead to the following conclusions:

The chemical and physical inhomogeneity created in the course of solidification in iron silicide has a great influence on structural changes taking place during working and heat treatment.

The physical microinhomogeneity is associated with the structural and chemical inhomogeneity. There is a relation between the dislocation distribution and the silicon segregations in the cast, worked, and annealed states.

The chemical inhomogeneity greatly affects the development of recrystallization in worked iron silicide. In the silicon-rich parts recrystallization is considerably retarded.

We have also studied the effect of annealing on the chemical inhomogeneity (with respect to silicon) associated with the dissolution of the carbide phase in iron silicide containing carbon.

Literature Cited

1. Movchan, B. A., Microscopical Inhomogeneity in Cast Alloys. Gostekhizdat, UkrSSR, Kiev (1961).
2. Rutter, J. W., In: Liquid Metals and Their Solidification [Russian translation], p. 297. Metallurgizdat, Moscow (1962).
3. Hurl, D. T., In: Growth Processes and the Growing of Single Crystals [Russian translation], p. 386. Metallurgizdat, Moscow (1963).
4. Shestak, B., Czech. Phys. J., 9:3 (1959).
5. Malinochka, Ya. N., Liteinoe Proizv., 1:22 (1963).
6. Mironov, L. V., et al., Phase Transformations and Properties of Electrotechnical Steels, p. 4. Metallurgizdat, Moscow (1962).
7. Tiller, W. A., J. Appl. Phys., 29:611 (1958).
8. Bunin, K. P., Baranov, A. A., and Pogrebnoi, É. N., Graphitization of Steel. Izd. Akad. Nauk UkrSSR, Kiev (1961).
9. Bunin, K. P., Dokl. Akad. Nauk SSSR, 95(1):97 (1954).
10. Fitzer, Arch. Eisenhüttenw., 25:9/10 (1954).

ON THE NATURE OF THE DEFORMATION ARISING IN THE MATRIX DURING THE GROWTH OF GRAPHITE

É. N. Pogrebnoi and K. M. Zhak

Dnepropetrovsk Metallurgical Institute

During phase transformations accompanied by a change in the specific volume of the phases, stresses and strains may arise. Graphite formation in iron alloys is one of such transformations. It was shown in [1] on the basis of an analysis of contact pressures that relaxation of the stresses created during the growth of graphite may be effected by way of the dislocation-creep mechanism.

The deformation (strain) structures created in the matrix during the growth of graphite have not been studied. It is hard to reveal the dislocation structure of graphitizing high-carbon steels and cast irons with a large carbide content. This problem is more easily solved for silicon steel, containing a small quantity of easily graphitizing carbide phase [2, 3]. The dislocation structure may be revealed by the selective-etching method [4, 5]. In this paper we shall present the results of such a study carried out on samples of cast transformer steel containing 0.03% C and 3.20% Si.

The original structure of a casting cooled in a ceramic mold consisted of large ferrite crystals, fine boundary veins of carbide, and pearlite at the joints and branches of the dendrites (Fig. 1a). Parallel and intersecting twins occurred in the ferrite crystals.

A graphitizing anneal of the samples was carried out at 750°C for 25 h. Graphite inclusions were formed in the ferrite crystals, at the intergranular and twin boundaries (Fig. 1b), and at the meeting points of twins and obstacles (Fig. 1d, e). Some graphite inclusions had radial structure (Fig. 1c). The inhomogeneity in the dislocation distribution around the graphite (Figs. 1e, 2) was revealed by electrolytic etching, resulting in the appearance of decorated dislocations [4]. In shape and character, the regions with high dislocation density around the graphite (Fig. 2a) were reminiscent of deformation rosettes formed in crystals of ordinary and silicon iron by a concentrated load [6-9]. The existence of developed substructure and the appearance of new ferrite grains near the graphite (Fig. 2b) indicates that, during the growth of the graphite, polygonization and recrystallization processes controlled by the self-diffusion velocity of the iron atoms takes place in the matrix.

In the dislocation rosettes the dislocation density is not always the same (Figs. 3a-c). Along certain directions the dislocation density is higher, and the rosettes have a radial or petal structures (Figs. 2 and 3). The number of directions is determined by the anisotropy of

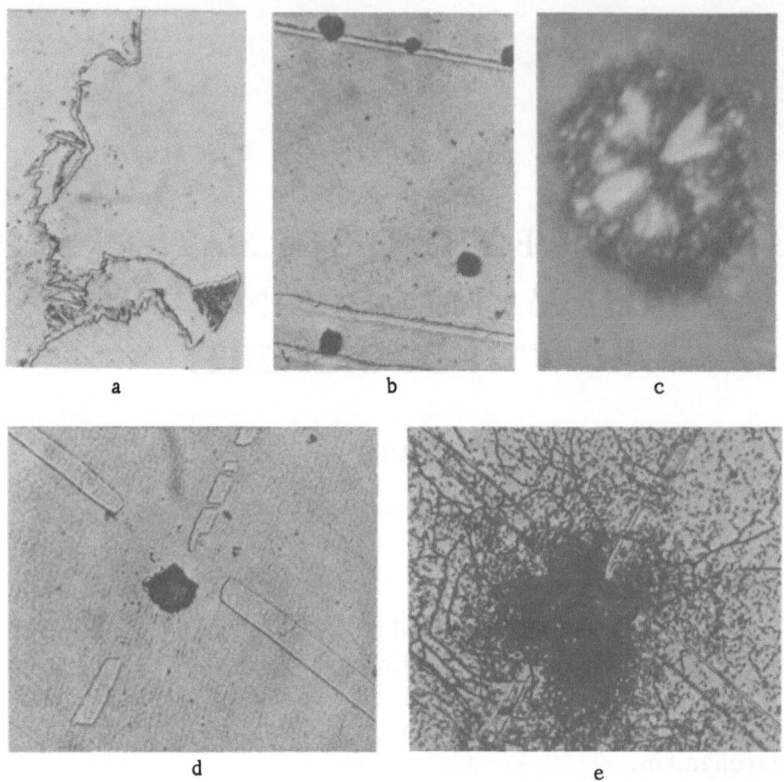

Fig. 1. Structure of the original (a) and annealed (b-e) iron silicide.
a,b) × 350; c) × 2000; d, e) × 600.

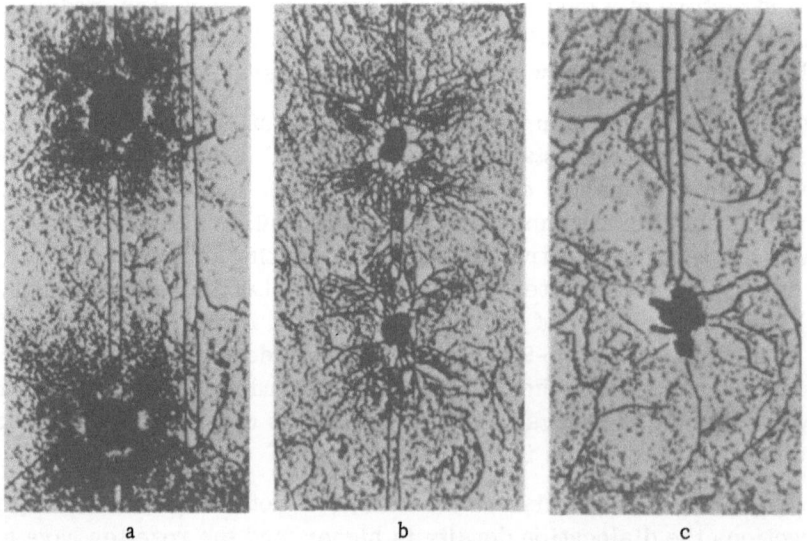

Fig. 2. Nonuniform dislocation density around graphite. a , b) × 500;
c) × 850.

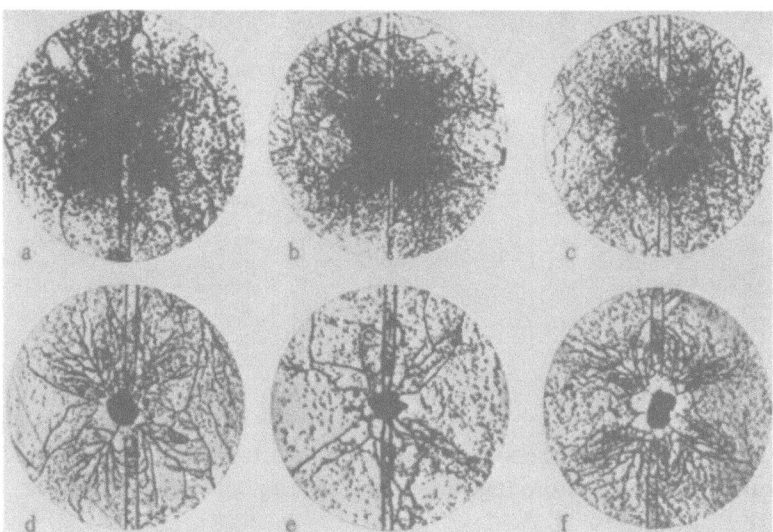

Fig. 3. Deformation rosettes around graphite inclusions (× 850)
(reduced by a factor of 2).

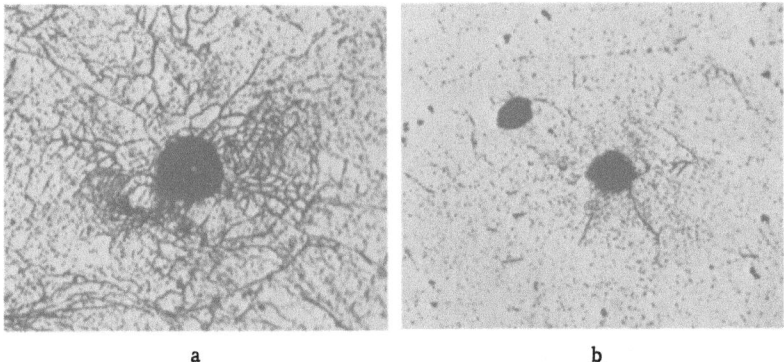

a   b

Fig. 4. Nonuniformity of dislocation density around nonmetallic
inclusions in annealed iron silicide (× 350).

the deformation of the ferrite crystals. Across the plane of the microsection there are usually either four or six principal directions of localized deformation (Fig. 3); two-, three-, and five-rayed rosettes occur less frequently. From the nature of the deformation of the ferrite around the graphite particles we may conclude that the evacuation of the matrix atoms is effected not only by the supply of vacancies and by creep, but also by acts of slip.

The radial structure of the rosettes is not always very sharp. This is because of the character of the deformation and softening processes taking place in the deformed matrix on annealing. Recovery, polygonization, and recrystallization change the structure of the rosettes. The redistribution of dislocations and their formation into rows and chains leads to the development of elements of subgranula and granula structure (Figs. 2 and 3). At certain stages of the development of subgranular structure, the petal (radial) structure of the rosettes is revealed even more sharply (Fig. 3d-f), and it is easier to estimate the nature of the deformation in the matrix during the growth of the graphite.

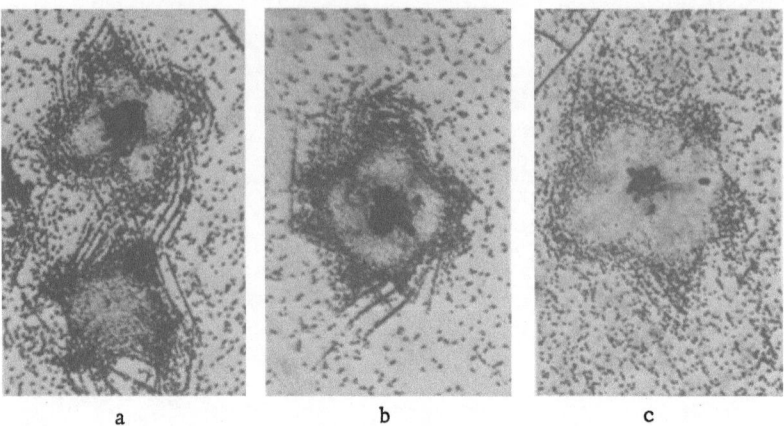

a   b   c

Fig. 5. Deformation rosettes near graphite formed in an ingot of
transformer steel during slow cooling. a, b) × 600; c) × 500.

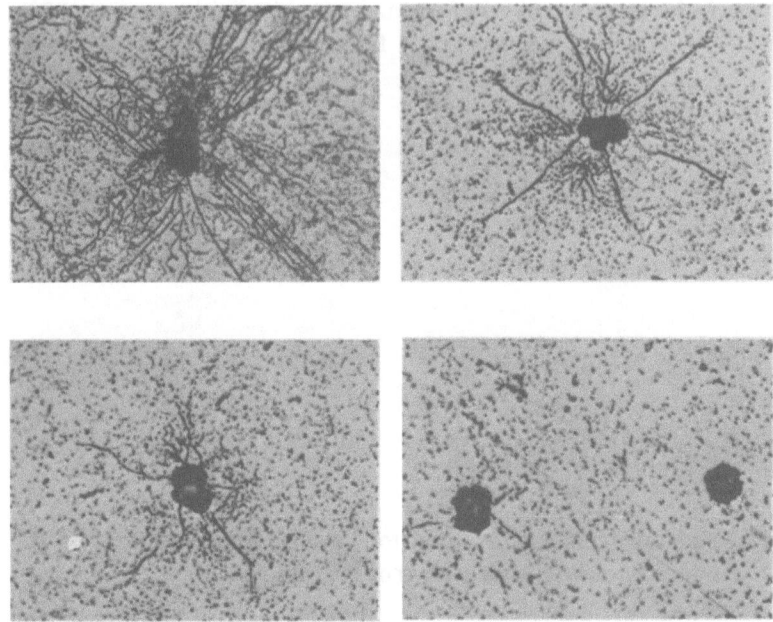

Fig. 6. Effect of thermal cycling on the size and shape of deformation rosettes around graphite (× 500).

The greatest variation of dislocation structure in the rosettes occurs around the points at which the graphite touches the matrix. Here the first recrystallization grains appear. In the early stages these also form a many-petalled rosette (Fig. 3f). With the development of recrystallization, the subgranular structure in the rosettes is removed (Fig. 2c). This type of structure is similar to those usually observed in graphitized steels and malleable cast iron. The presence in the same samples of rosettes with dislocation structures corresponding to different stages in the formation of subgranular structure, right up to the stage of complete removal, is a result of the nonuniform deformation of the matrix during the growth of the graphite and the nonuniform nature of recovery, polygonization, and recrystallization.

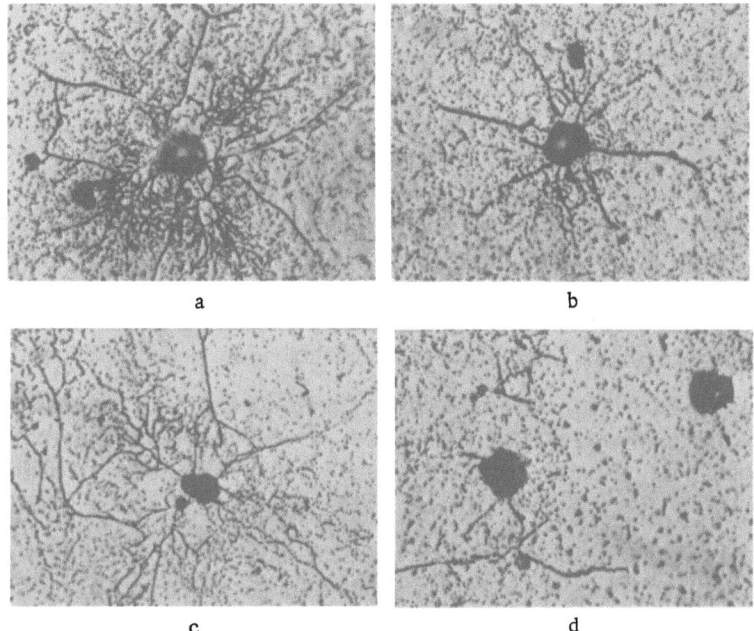

Fig. 7. Structure of samples subjected to thermal cycling and
cooled after graphitization in the furnace (a, c) and in water (b, d).
a, b) 2 cycles; c, d) 3 cycles (× 600).

Considerable deformations around nonmetallic inclusions (and graphite) may also arise
on cooling as a result of the difference in the thermal-expansion coefficients of the inclusions
and the matrix. The rosettes so formed are sometimes similar in shape to those arising at
points subjected to a concentrated load [10]. Nonmetallic inclusions are retained better than
graphite on electropolishing (Fig. 4); frequently, local corrosion appears along their contours,
apparently as a result of the increased deformation created by cooling. On annealing, the
rosettes around the nonmetallic inclusions are removed more rapidly than those around the
graphite.

The strains arising around the graphite as a result of graphite formation and cooling, respectively, are additive. Thus, Fig. 5 shows the deformation rosettes around graphite inclusions
formed on slow cooling in the middle of a transformer-steel ingot. The rosettes have radial
structure, and the dislocation distribution within them is more weakly expressed than in annealed samples. Around the contour of the rosettes we see slip lines evidently formed at low
temperatures on cooling.

Deformation rosettes around the graphite are also observed in samples subjected to
thermal cycling of the following type (after annealing): dissolution of the graphite by holding at
900°C for 0.5 h followed by graphitization at 650°C for 5 h, with subsequent cooling either in
the furnace or in water. The dislocation structure of the rosettes after 1 or 2 cycles may be
seen in Fig. 6a. After 3-5 cycles the rosettes around the graphite are much weaker (Fig. 6b, c)
and after 10 cycles they do not appear at all (Fig. 6d), owing to the increase in the size and
surface of the pores during thermal cycling and the intensification of graphitization on their
surface [11]. Changing the cooling rate after the annealing of the thermally cycled samples
showed that the rosettes were larger in the slowly cooled samples (Fig. 7a, c) than in the water-
cooled material (Fig. 7b, d), owing to the additional strains arising on precipitation of the

graphite (in the course of slow cooling). Water-cooling does not change the dislocation structure of the rosettes very much.

It is difficult to separate the strains arising on graphitization and cooling, respectively. If the rosettes around the graphite only arose during the cooling period (owing to the difference in the linear-expansion coefficients of the graphite and the ferrite), it would be hard to explain the different types of granular and subgranular structures in the rosettes. Even on slow cooling (below 750°C) only the initial stages of polygonization [12] can occur in strained iron silicide. Hence the appearance of deformation rosettes around the graphite must in our case be associated not only with strains caused by cooling, but also with those due to the growth of the graphite resulting from the chemical potential gradient; the latter strains are due to difficulties of accommodation associated with the difference in the specific volumes of the phases [13]. Later investigations should enable us to determine the contribution of each of these.

It follows from the foregoing data that during graphite growth, and on cooling, the matrix may experience considerable anisotropic plastic deformation. The shape and size of the deformation rosettes are determined by the nature and conditions of growth of the graphite, by the properties of the ferrite matrix, and by the softening processes associated with annealing.

In structure and composition, the graphite differs sharply from the iron phases. The formation of graphite in the iron phases is well known to be accompanied by a considerable change in specific volume. The growth of the graphite is associated with the diffusion transport of carbon in solid solution from the dissolving cementite to the graphite and with the removal of matrix atoms from the surface resulting from the inflow of vacancies, slip, and the creeping of dislocations. The motion of carbon and matrix atoms as well as vacancies and dislocations is caused and controlled by the chemical potential gradient. The change in the specific volume of the phases and the existence of the chemical potential gradient during the graphitization of cementite constitute the main reason for the development of stresses and strains in the graphite and the surrounding matrix.

The dislocation structures indicated show that, during the growth of the graphite, the matrix is strained anisotropically, the deformation being localized around the graphite (it is hard to reveal this in the graphite itself). This indicates that, in addition to acts of diffusion, an important part is played by acts of plastic deformation resulting from the activation of sources at the interphase boundary and in the actual ferrite.

By the shape and size of the rosettes created around the graphite we may estimate the nature of the deformation of the ferrite crystals. In a rosette with four rays (see Fig. 3a), these are drawn out along the <110> direction, which corresponds to possible slip planes {110} and {112} [6]. In the dislocation rosettes obtained on the (110) face the rays are arranged along the <111> directions (slip planes {110}, {112}, and {123}). The dislocation rosette formed around graphite illustrated in Fig. 3b is of this type. In this rosette we see four directions of preferential deformation along the <111> directions. Analysis of the rosettes with well-expressed radial or petal structure (see Figs. 2, 3, and 5) shows that the directions of preferential deformation in the ferrite crystals during the growth of graphite are the <110> and <111>; the slip planes are mainly the {110} and {112}. The possible slip of dislocations in several systems simultaneously in the ferrite crystals (in 18 planes at least, not counting planes of the {123} type [6]) complicates the shape of the rosettes. Hence, the shape of the rosette is not so clearly expressed in ferrite crystals as in crystals of the alkali metals, which only have six slip planes. This is also observed on subjecting the crystals to indentor impressions [6, 7].

The shape and structure of the rosettes changes considerably as a result of the superimposition of recovery, polygonization, and recrystallization processes taking place on annealing. These processes reduce the dislocation density in the rosettes and lead to a redistribution of the dislocations and the formation of granular and subgranular structure. From the manner in which the dislocation density changes in the rosettes, the development of subgranular structure, and the appearance of the first recrystallized grains, we may conclude that the places of greatest stress concentration during the growth of the graphite represent the interphase surface. No numerical calculation of the contact stresses was carried out, but from the shape and size of the rosettes we may suppose the stresses are very large and are responsible for the anisotropic plastic deformation of the matrix around the graphite.

The rosettes appearing around the graphite are removed by polygonization and recrystallization. This indicates that the hardening effect created in the matrix during graphite growth and cooling is eliminated in parallel with these softening processes.

The accommodation strain of the matrix associated with the growth of the graphite may have a considerable effect on the development of graphitization. The redistribution and interaction of the dislocations increases the vacancy concentration. The building up of the dislocations, with the formation of grain boundaries and sub-boundaries, creates paths for preferential diffusion flows to and from the graphitization front. Local plastic deformations may affect the shape and size of the growing graphite inclusions, both as a result of the directional displacement of dislocations from the interphase surface (by slipping and creeping under the influence of the chemical potential gradient arising on graphitization) and as a result of the emergence of dislocations at the interphase surface when these are redistributed in the course of softening.

Literature Cited

1. Baranov, A. A., Izv. Akad. Nauk SSSR, 3:123 (1964).
2. Markuszewicz, M., Prace. Inst. Min. Hutniczych., p. 8 (1956).
3. Hurry, E. D., J. Iron Steel Inst., p. 241 (March, 1951).
4. Shestak, B., Czech. J. Phys., 9:399 (1959).
5. Pickering, H. W., Acta Met., 13(4):437 (1965).
6. Kushnir, I. P., Mikhailova, L. K., and Osip'yan, Yu. A., Kristallografiya, 10(1):87 (1965).
7. Shestak, B., Czech. J. Phys., 11(6):444 (1961).
8. Meyer, K., and Gragert, E., Phys. Stat. Sol., 3(11):2005 (1963).
9. Stepanova, V. M., Gumanova, N. A., and Predvoditelev, A. A., Kristallografiya, 10(2):219 (1965).
10. Molotilov, B. V., and Golikov, I. I., Stal', 1:62 (1964).
11. Bunin, K. P., et al., In: Physico-Chemical Bases of the Metallurgical Process. Metallurgizdat, Moscow (1964).
12. Hibbard, W. R., and Dunn, K. J., In: Creep and Recovery [Russian translation], pp. 62-91. Metallurgizdat, Moscow (1961).
13. Higgins, G. T., and Jeminson, G. V., J. Iron Steel Inst., p. 146 (Feb., 1965).

GROWTH AND STRUCTURE OF CEMENTITE CRYSTALS

V. I. Novik and Yu. N. Taran

Dnepropetrovsk Institute of Ferrous Metallurgy

Primary cementite crystals are described in the literature as monolithic plates formed by the growth of plane dendrites [1, 2]. In this paper we shall indicate the microstructure of the plates, the study of which gives a good idea of the mechanism underlying the growth of cementite.

Cementite crystals were grown in alloys of hypereutectic composition containing 4.5% C, 0.03% Si, 0.004% S, 0.01% P, and 5.0% Mn produced on a base of Armco iron and graphite. Manganese was added in order to prevent graphite formation. Owing to the all-around slow cooling of the ingots, large single crystals of primary cementite were obtained in their central shrinkage cavities (Fig. 1). Both extracted crystals and sections of crystals in samples cut from ingots were subjected to examination, revealing the layer and block structure of the cementite. For this purpose we treated the samples in a phosphorochromic electrolyte (90% H_2PO_4 + 10% CrO_3) heated to 80 or 90°C at a current density of 120 A/dm^2 for 20-40 sec, with subsequent etching in the same electrolyte (current density 10 A/dm^2 for 30 sec).

In the cross sections of the cementite plates we observed microrelief in the form of two systems of parallel lines (Fig. 2a). The lines of one system were always parallel to the side faces of the plates and reflected the layer structure of the cementite. The picture of layer-like growth is clearly seen from Fig. 2b (left half of the picture corresponds to the new layer growing on the plate). The lines of the second system were arranged at an angle to those of the first. The size of the angle was determined by the position of the intersecting plane. The second system of lines also appeared in the longitudinal sections of the crystals (Fig. 2c) and was orientationally related to their serrated habit. This system of lines reflected the specific block structure of the cementite. Taking account of the microstructure of the cementite as indicated by microscope and x-ray examination, the structure of part of a branch is shown schematically in Fig. 3. The section is cut by a number of different planes. The lines of microrelief establishing the boundaries of blocks and layers are represented by hatching.

Fig. 1. Crystals of primary cementite in the shrinkage cavity of an ingot (full scale).

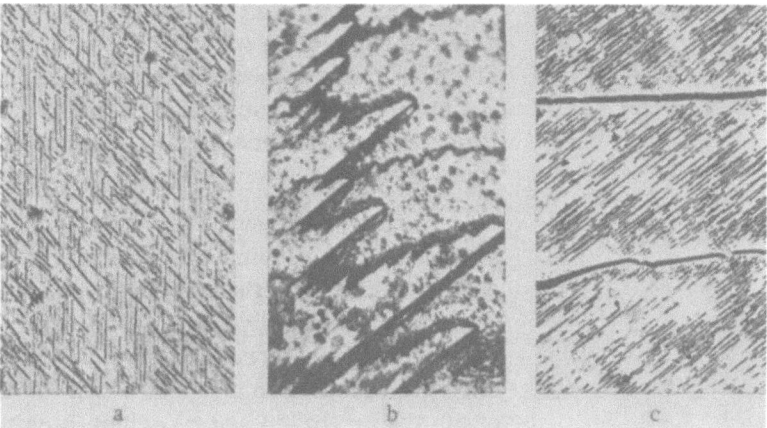

Fig. 2. Microrelief revealed in the transverse (a, × 600) and longitudinal (c, × 250) sections; b) fracture pattern of a cementite crystal, × 28.

Fig. 3. Arrangement of a section of a
cementite crystal.

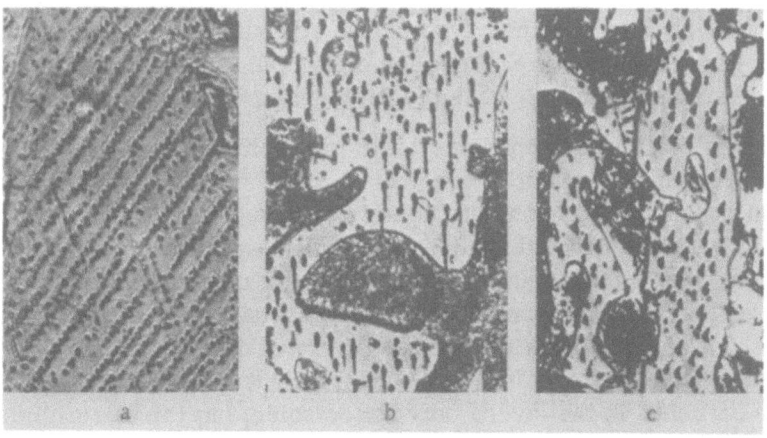

Fig. 4. Etch figures in cementite (a, × 800; b, c, × 600).

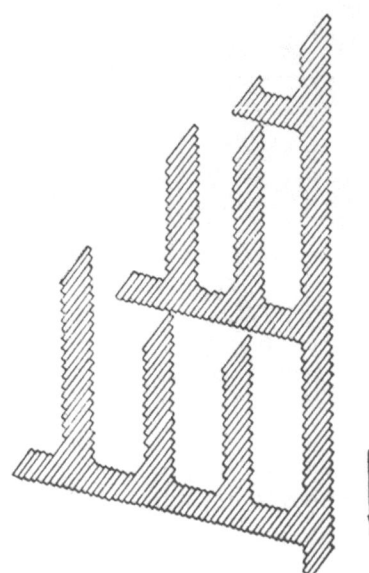

Fig. 5. Arrangement of a dendritic growth element in a cementite plate.

In order to explain the nature of the block boundaries in connection with the concept of the dislocation origin of the etch pits in cementite [3], it was of particular interest to study the distribution of the latter. Electron-microscope data relating to steel [4] indicates the existence of dislocations in cementite.

In order to reveal the etch figures we used thermal etching in a vacuum of 10^{-3} mm Hg at 650°C for 20 min, in addition to the earlier-proposed method of etching in a stream of hydrogen [3]. Before etching, the samples were electrolytically polished in a phosphorochromic electrolyte at 80-90°C with a current density of 30 A/dm^2 for 60 sec. Vacuum etching gives the most stable results and produces sharper characteristics of the shape of the etch figures. The density and distribution of the figures are independent of the method of revelation. The formation of etch figures in cementite, both under vacuum thermal etching and on etching in a stream of hydrogen, is associated with processes of local oxidation and decarburization. This is indicated by the worsening in the etchability of cementite with increasing vacuum. For a residual pressure of 10^{-5} mm Hg, etch figures are not revealed at all.

In the crystals of primary cementite the etch figures are mainly arranged along the block boundaries, the latter frequently appearing as chains of etch pits as shown in Fig. 4 (vacuum etching). The block disorientation determined from the pit density is 1-2'. For calculation purposes the Burgers vector was taken as equal to the average interplane distance of the cementite lattice.

The layer and block structure of the eutectic cementite in alloys of hypoeutectic composition (3.1% C, 0.6% Si, 0.02% S, 0.01% P) is indicated by the shape and arrangement of the etch figures developed. Both the shape and size of the figures vary with etching time. With the conditions of sample treatment indicated above, equiaxial etch pits are revealed. For longer etching periods (20-50 min for etching in vacuum), the pits in the majority of the cementite grains take the form of "pins" or "setsquares" (Fig. 4b, c), oriented similarly for any one crystal. The etch figures obtained reflect the symmetry of the cementite crystals. The figures in the form of pins reflect the layer structure of the crystals; the setsquares represent not only the layer but also the specific block substructure of the cementite. In the cementite of hypoeutectic cast iron there are also systems of etch-pit chains, which evidently correspond to block boundaries.

The fractographic study of cementite crystals in which the crystallization of the upper layers has been interrupted as a result of the loss of liquid from a developing shrinkage pore (natural decantation) enables us to formulate some views regarding the growth of the surface layers of the crystals. The relief on the surface of a cleavage to some extent characterizes the growth of the inner layers of the plate. The cleavage of cementite crystals takes place along the (001) plane, i.e., along the cementite plate. Microanalysis of the surface of crystals and their cleavages showed that, in addition to the layers of dendritic form, relief in the form of a system of parallel bands having serrated boundaries occurred very frequently, as in Fig. 2b. The direction of the branches in the layers may not coincide, but the texture of all the branches and layers in the crystal has a single direction. This may be seen in the matching of the serrated contours of the various systems of branches in the crystal, and is confirmed by the microstructure of the cementite as revealed by etching. The existence of serrated contours in the cementite branches, the "teeth" of these being oriented in the same way over the whole crystal and coinciding with the directions of the sub-boundaries of the crystals revealed by electrolytic treatment, indicates that the formation of the cementite branches takes place by way of the successive growth of blocks. The formation of each layer represents the development of a specific plane dendrite (Fig. 5). As a cementite plate grows, the dendrites find themselves in different states of development; their branches may grow together or intersect one another. The directions of these branches are determined by local conditions of crystallization, but the growth of the substructural elements of all the branches of a given crystal (blocks) takes place in a single direction. The crystal is thus characterized by a sharply expressed texture (see Fig. 2b, c).

The formation of the specific block structure of the crystal takes place in the course of continuous layer-like growth; it is due to the crystallographic nature of the cementite and the influence of impurity on the crystallization process.

Representative of the growth of cementite is the formation of a cementite layer in the course of the eutectic transformation in white cast iron, where the cementite plays the part of the leading phase during the crystallization of the ledeburite eutectic. It may be shown by the stereometric microanalysis of ledeburite columns, using mutually perpendicular sections and repolishing of the samples, that in the initial stages of the eutectic transformation the formation of the new layer of cementite represents the growth of cementite blocks.

Literature Cited

1. Bunin, K. P., Ivantsov, G. I., and Malinochka, Ya. N., Structure of Cast Iron. Mashgiz, Moscow (1952).
2. Hillert, M., and Stenhaüser, H., The Structure of White Cast Iron, Jernkontorets Ann., No. 7 (1960).
3. Tkachenko, F. K., Izv. Akad. Nauk SSSR, No. 1 (1964).
4. Keh, A. S., Acta Met., 11:9 (1963).

DISLOCATION STRUCTURE OF CRYSTALS WITH THE DIAMOND STRUCTURE GROWN IN THE (100) DIRECTION

L. S. Milevskii and V. D. Khvostikova

A. A. Baikov Moscow Institute of Metallurgy

The study of the dislocation structure of crystals with the diamond lattice, started in [1,2], became possible as a result of the adoption of methods of growing silicon single crystals with rectilinear dislocations oriented along the growth axis. The use of the optical-polarization method for the analysis of stresses in silicon [3] made it possible to study the microstresses created by single dislocations. Comparison of experimental data [1, 2] with theoretical [4, 5] led to the deciphering of the [110] dislocation structure of the crystals, in which simple dislocations of two types were observed. These were, first, a 60° dislocation with a (111) slip plane, the Burgers vector $a/2$ of which made an angle of 60° with the [110] dislocation axis, and, secondly, a sessile edge dislocation with a (100) slip plane, the Burgers vector $a/2$ [110] of which was perpendicular to the [110] axis.

Screw dislocations in silicon, oriented in the [110] direction, were studied by the decoration method in [6]. Thus, simple dislocations of three types, the atomic structure of which was considered by Hornstra [7], were observed experimentally in silicon single crystals. In addition to the simple dislocations, Hornstra also considered the atomic structure of certain more complex dislocations, in particular edge dislocations with axis oriented in the [100] direction and Burgers vector $a/2$ [110].

Single dislocations of this type have not so far been studied, owing to the difficulty of obtaining silicon single crystals with dislocations oriented along the growth axis.

In this paper we shall present certain results obtained from a study of single-edge dislocations oriented along the [001] direction.

Silicon single crystals were grown by the Czochralski method under conditions normally resulting in dislocation-free samples. For this purpose seeds oriented in the [001] with random dislocations were used. In order to study the dislocation structure of the crystals we used the methods of chemical etching, the decoration of the dislocations with copper, and the optical-polarization method of studying microstresses [1, 2]. Photographs were taken directly from the screen of an electron-optical converter operating in conjunction with an MP-3 polarization

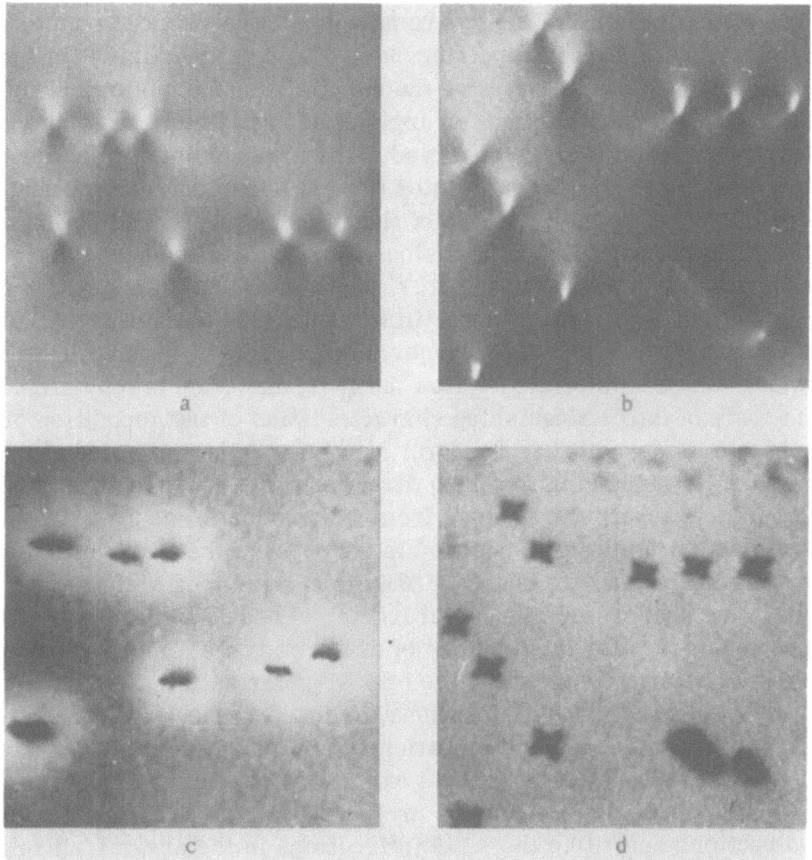

Fig. 1. Dislocations oriented along the axis of observation. a) Double-refraction (birefringence) rosettes observed in crossed Nicols on [001] dislocations; plane of polarization parallel to the frames of the figures; b) nature of the deposition of copper on observing along the [001] dislocation axis; c) birefringence rosettes of sessile edge dislocations oriented in the [110] direction, showing two 60° dislocations; d) deposit on sessile edge dislocations oriented along the [110] direction.

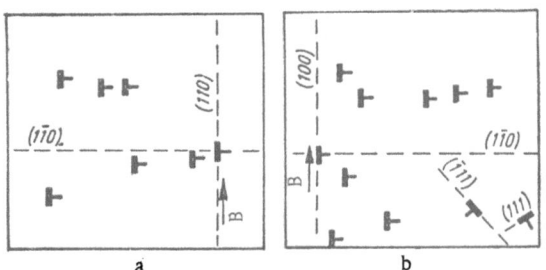

Fig. 2. Arrangement of dislocation structure. a) Edge dislocations in a crystal, [110] type; b) sessile edge dislocations in a crystal, [110] type (in the lower right-hand corner are two 60° dislocations).

microscope. The orientation of the plates studied (cut perpendicular to the crystal growth axis) was checked by an x-ray method with respect to reflections from the (001) and (111) planes.

Figures 1a and b compare results obtained on neighboring plates by the optical polarization method (Fig. 1a) and the method of decoration (Fig. 1b). On looking along the axis of the dislocations, i.e., in the [001] direction, around each dislocation we notice a characteristic double-refraction rosette [2,5], consisting of six petals if the slip plane is parallel to the plane of oscillation of the Nicols. The large petals of the rosette are drawn out along the (110) slip planes, which in Fig. 1a are parallel to the vertical frame. The slip plane is a plane of symmetry of the field of birefringence. From the "coloring" of the petals of the rosettes we see that all the dislocations have the same signs and are arranged in parallel slip planes. On rotating the slip plane through an angle of 45° relative to the plane of polarization, characteristic four-petal rosettes were observed. The arrangement of the dislocations in the sample is shown in Fig. 2a. The Burgers vector $a/2$ [1$\bar{1}$0] denoted by an arrow lies in the plane of the figure at a right angle to the line of the dislocation proceeding in the [001] direction. Dislocations of this type are slipping dislocations and easily change their direction on passing through the crystal. We see from Fig. 1b that, on observing along the axis of the dislocations, there is a slight deviation from rectilinearity. This is especially noticeable for the dislocation situated in the lower left-hand corner, the focusing of which differs from the others. However, the deviation of the dislocations from the [001] direction is only slight and does not interfere with the study of the distinguishing characteristics of the deposition of copper during decoration. The copper is deposited in the form of "strips," the wide side of which is drawn out perpendicular to the slip plane in the [110] direction. The light region surrounding each dislocation in Fig. 1b corresponds to regions from which copper has been removed as a result of deposition. The density of copper deposition is different in the transverse section of the "strip" and especially large near the center of the dislocation, and also in the compressed and drawn-out regions. The sharply anisotropic character of the deposition appears for large copper concentrations and differs for dislocations of different types. For comparison, Fig. 1c, d show sessile edge dislocations observed by the optical-polarization (Fig. 1c) and the copper-decoration (Fig. 1d) methods. These dislocations, oriented in the [110] direction, were studied earlier in [2]. The arrangement of the dislocation structure is shown in Fig. 2b. We see from Fig. 1d that in this case the deposited particles are drawn out in two directions, which on observation along the axis of the dislocation are projected into the form of intersecting fragments of lines. These directions constitute the <110> directions in the crystal. It follows from a comparison of Figs. 1b and 1c that the growth of the deposited particles takes place in the <110> directions intersecting the line of the dislocation but not lying in its slip plane. In fact, only in one of the [1$\bar{1}$0] directions, namely, that parallel to the Burgers vector of the dislocation, is copper deposition practically absent. This direction lies in the plane of the diagram and passes parallel to the vertical frame in Fig. 1b, d.

Further study of the orientation of deposited particles with respect to the axis of the dislocation, as well as a comparison of the nature of copper deposition on dislocations of different types, may prove useful for analyzing the dislocation structure of crystals with dislocations varying in direction and with dislocation loops.

The experimentally observed dislocations oriented in the [100] direction described in the foregoing constitute slipping edge dislocations in silicon crystals; these have a Burgers vector $a/2$ [110] perpendicular to the axis of the dislocation.

Literature Cited

1. Indenbom, V. L., Nikitenko, V. I., and Milevskii, L. S., Dokl. Akad. Nauk SSSR, 141(6): 1360 (1961).
2. Indembom, V. L., Nikitenko, V. I., and Milevskii, L. S., Fiz. Tverd. Tela, 4(1): 231 (1962).
3. Zvyagin, V. I., and Vavilov, V. S., Pribory i Tekhn. Éksperim., 3: 86 (1956).
4. Bullough, R., Phys. Rev., 110(3): 620 (1958).
5. Indembom, V. L., and Tomilovskii, G. E., Kristallografiya, 2(1): 190 (1957).
6. Dash, W. C., J. Appl. Phys., 31: 2275 (1960).
7. Hornstra, J., J. Phys. Chem. Sol., 5(1, 2): 129 (1958).

DISLOCATION REACTION IN CRYSTALS WITH THE DIAMOND STRUCTURE GROWN IN THE (100) DIRECTION

L. S. Milevskii and V. D. Khvostikova

A. A. Baikov Moscow Institute of Metallurgy

The use of the optical-polarization method for studying the dislocation structure of silicon crystals enables the nature and origin of dislocations in lattices with the diamond structure to be studied in detail. It was shown in [1-3] for silicon single crystals that the direction of a dislocation line and the magnitude and sign of its Burgers vector could easily be determined by means of the photoelasticity method, so that dislocations of various signs [4] and types [2, 3] could be clearly distinguished.

One of the possible ways of forming sessile edge dislocations as a result of a dislocation reaction between two "60°" dislocations slipping in the {111} plane was considered; the moment of combination of the 60° dislocations could nevertheless not be illustrated, since no closely adjacent pairs with suitable Burgers vectors remained in the crystal [2]. The reactions were apparently completed in a part of the crystal unsuitable for observation, close to the seed. However, the use of the photoelasticity method for studying dislocation reactions is without doubt extremely promising, since it enables the parameters characterizing the dislocations before and after the reaction to be clearly established. In one of the previous papers [3] we were able to observe dislocations in silicon oriented along the [001] direction. Dislocations of this type proved to be edge dislocations with slip plane (110) and Burgers vector $a/2$ [1$\bar{1}$0], perpendicular to the [001] dislocation axis.

The dislocations observed had considerable mobility and were able to move gradually in the slip plane, apparently under the influence of thermal stresses arising during the growth of the crystal.

The fairly high mobility of dislocations of this type also makes it possible to study the interaction of individual dislocations situated in intersecting slip planes and to study dislocation reactions between them.

The present investigation was carried out on silicon samples grown by the Czochralski method in the [100] direction. The work was carried out by the optical-polarization and decorating methods on neighboring plates cut perpendicular to the crystal growth axis. Photographs were taken from the screen of an electron-optical converter associated with an MP-3 microscope.

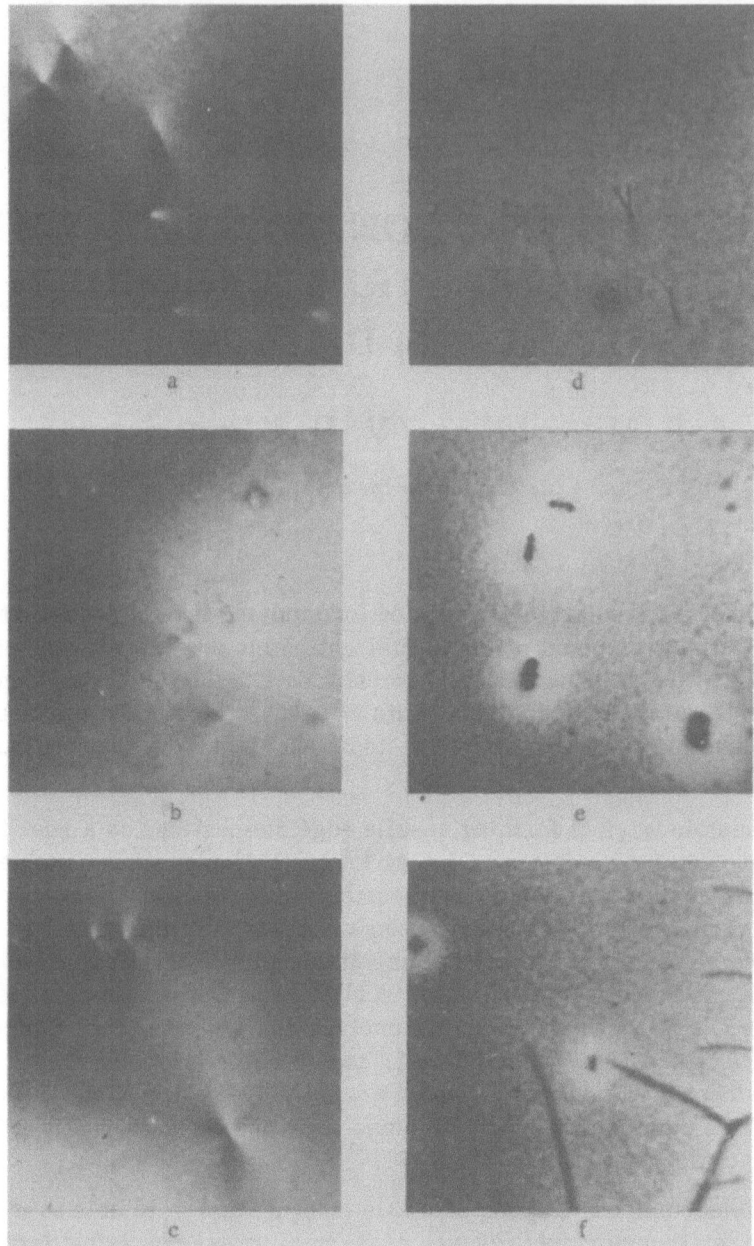

Fig. 1. Dislocations oriented along the [001] growth axis, × 45.
a) Birefringence rosettes; b) the same after a slight displacement of
the dislocations; c) the same after the reaction has taken place; d) slip
established by decoration; e,f) character of deposition before and after
the reaction.

By way of example, let us consider the section of crystal shown in Fig. 1a, b. The micrographs were obtained with crossed Nicols, the planes of polarization being parallel to the frames of the figure. In the upper part (Fig. 1a, b) are dislocations with slip planes parallel to the vertical frame of the figure; in the lower part are dislocations with slip planes of the horizontal type.

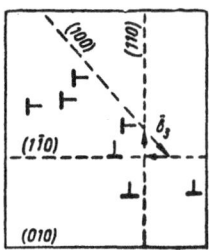

Fig. 2. Arrangement of dislocation structure in the crystal of Fig. 1a. Broken arrow indicates the Burgers vector b_3 of the new dislocation.

The large petals of the birefringence rosettes are drawn out along the slip planes. The arrangement of the dislocation structure of the region in question is indicated in Fig. 2. The broken lines show the (110), ($1\bar{1}0$), and (100) planes and the arrows indicate the Burgers vectors of the dislocations $a/2$ [$1\bar{1}0$] and $a/2$ [110]. In the central part (Fig. 1a) there is a clearly visible pair of dislocations situated quite close together and slipping, intersecting the {110} slip planes at right angles; the upper of these has an indistinct rosette owing to the deviation of the dislocation line from the direction of observation (as a result of slip). We see from a comparison of Fig. 1a, b that the interaction of the elastic fields leads to a change in the mutual arrangement of the dislocations in neighboring plates. The dislocations in the upper part of Fig. 1b are gradually grouped, occupying more stable positions, and then quite quickly built up into a "vertical row" over a section of crystal several centimeters long. The dislocations situated in the lower part of Fig. 1a slip from left to right parallel to the horizontal frame of Fig. 1a. Figure 1b also clearly shows the approach of the pair of dislocations situated in the center. The rather smaller dimensions of the birefringence rosette in Fig. 1b are due to the small thickness of the silicon plate. Further slip leads to the meeting of the dislocation pair under consideration, and the dislocations then enter into reaction.

The Burgers vector of the resultant dislocation equals the sum of the Burgers vectors of the original ones, namely

$$a/2\,[110] + a/2\,[1\bar{1}0] = a\,[010].$$

The reaction is possible, but in the isotropic approximation takes place without change of energy in the system. Figure 1c shows the same region of the crystal after completion of the addition reaction (plate cut at a distance of several millimeters down the ingot). Instead of the two dislocations of Fig. 1a, b we see the resultant dislocation of Fig. 1c, now forming the characteristic four-petal rosette. The (100) slip plane of the new dislocation is a plane of antisymmetry of the birefringence field. This plane is arranged at an angle of 45° to the planes of polarization of the Nicols, which as before are parallel to the frames of the picture. Thus, the result of the addition reaction is a large edge dislocation of a new type with a (100) slip plane, the [010] Burgers vector of which is perpendicular to the [001] axis of the dislocation. The dislocations in the lower part of Fig. 1a, b continue slipping from left to right, as may be seen in Fig. 1c. The resultant dislocation, also a slipping one, moves diagonally into the lower right-hand corner of Fig. 1c. The moment of addition of the dislocations is established by the decoration method in Fig. 1d. The plate in this picture, intermediate with respect to the plates in Fig. 1c and 1b, is inclined at ~20° to the axis of observation. The dislocations are projected into sections of lines, the focus of these being different in depth. Halfway up Fig. 1d, slightly right of center, we see the instant of fusion between the two original dislocations and the appearance of the new one. The two decorating lines come together and unite into one. In order to verify the principles mentioned in [3] regarding the anisotropic character of copper deposition on decorating, and also in order to check the stability of the new dislocation with respect to "splitting up again," we studied several plates distributed along the ingot by the method of decoration.

Figures 1e, f compare the character of copper deposition on the dislocations before and after the addition reaction. We see from Fig. 1e that on observing along the axis of the dislocations the deposited particles are drawn out perpendicular to the slip planes of each dislocation (region next to the lower right-hand corner of the micrograph of Fig. 1b, × 60). After the addition reaction the slip plane of the new dislocation proceeds diagonally with respect to the [110] and [1$\bar{1}$0] directions. According to the conclusions of [3], the deposited particles may now grow freely in two perpendicular <110> directions, since not one of these lies in the slip plane of the dislocation. We do in fact see in Fig. 1f that deposition on the dislocations with Burgers vector $a/2$ [110] takes place in one [1$\bar{1}$0] direction, and on the dislocation with Burgers vector a [010] in two mutually perpendicular directions, forming a "cross." Two dislocations in Fig. 1f have passed into a region of the crystal containing dislocations from other slip systems, which may be seen in the same photograph. The copper concentration in Fig. 1f is rather lower (and hence the size of the deposited particles smaller) than in Fig. 1e.

Thus, the particles actually deposited during the decoration of dislocations in silicon with copper grow in the <110> directions not lying in the slip plane of the dislocation. Dislocations with slip plane (100), Burgers vector a [010], and an orientation along the [001] direction constitute stable slipping edge dislocations in silicon.

Literature Cited

1. Indenbom, V. L., Nikitenko, V. I., and Milevskii, L. S., Dokl. Akad. Nauk SSSR, 141(6): 1360 (1961).
2. Indenbom, V. L., Nikitenko, V. I., and Milevskii, L. S., Fiz. Tverd. Tela, 4(1): 231 (1962).
3. Khvostikova, V. D., and Milevskii, L. S., This collection, p. 209.
4. Milevskii, L. S., Fiz. Tverd. Tela, 4(7):1978 (1962).

X-RAY DIFFRACTION STUDY OF THE DISLOCATION STRUCTURE OF SILICON CRYSTALS

V. F. Miuskov and L. S. Milevskii

A. A. Baikov Moscow Institute of Metallurgy

Institute of Crystallography of the Academy of Sciences of the USSR

X-ray diffraction (xrd) rosettes of single dislocations perpendicular to the surface of the crystal section were first observed by Bonse [1] by means of a two-crystal spectrometer. The rosettes calculated by Bonse for various types of perpendicular dislocations were principally four- or six-petal types with alternating "color." Bonse was only able to obtain black-and-white spots, from which it was impossible to make any detailed judgement of the structure of the rosettes or to compare them with those calculated on the basis of a simplified semiempirical model.

Recently, Danil'chuk and Smorodina [2] were able to obtain xrd rosettes of perpendicular dislocations more complex than those of Bonse by the Borman method with a divergent primary beam (1-3°) of CuK$_\alpha$ radiation, using an anode voltage of 34 kV. The rosettes obtained differed considerably from those of Bonse and (as the authors affirm) had an external similarity with the rosettes obtained by the photoelastic method (pe rosettes).

The topographic resolution in [2] reached 7-10μ, which at the present time cannot be regarded as high. In addition to this, the presence of unfiltered K$_{\alpha_2}$ radiation and the short-wave edge of the continuous spectrum led to an increase in background and lowered the quality of the contrast. This prevented the authors from resolving the details of the xrd rosettes with adequate reliability.

The first attempts to study single dislocations by the xrd method were undertaken on samples of silicon optically examined by Nikitenko; however, the great dislocation density in these and the presence of uncontrolled impurities preventing sufficiently clear results from being obtained.

The principal problem of the present investigation was to study details in the structure of the xrd rosettes and to compare the resultant data with the dislocation structure analyzed by the photoelastic method on the same crystal. We used the camera of [5] with a resolution of 1 μ, which is an order higher than that used in [2], and a primary-beam divergence of less than 2'. This eliminated one of the components of the doublet (the K$_{\alpha_2}$), and also improved the resolution of the camera. In selecting the crystals, attention was paid to choosing those with a fair-ly low dislocation density such that the fields of individual dislocations were not distorted by mutual interaction [8].

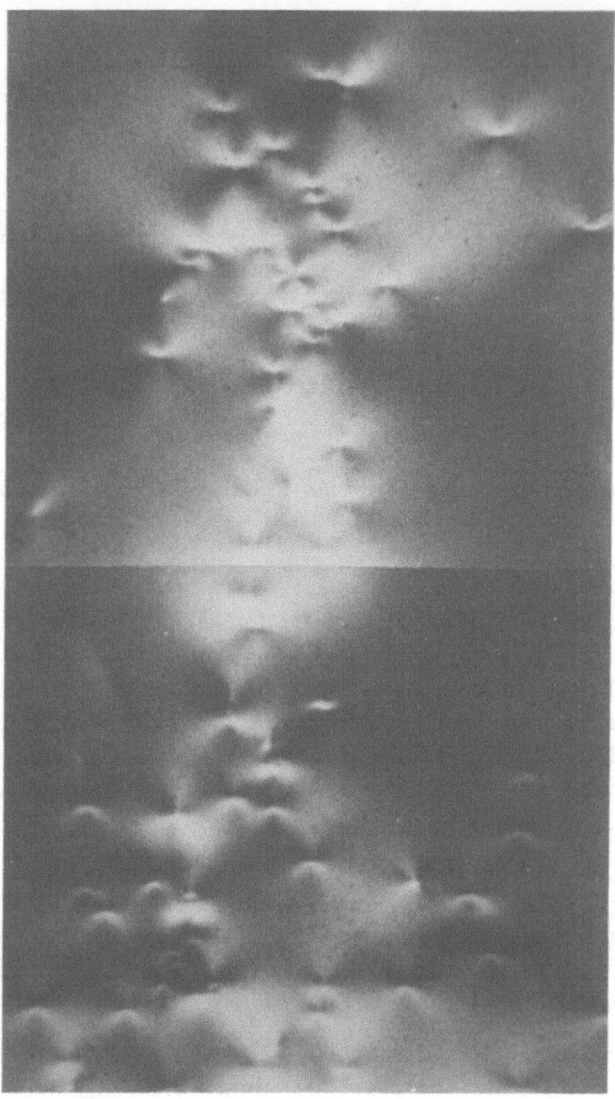

Fig. 1. Photoelastic stress rosettes obtained from a silicon sample. × 24.

Dislocation Structure of a Silicon Crystal

In [3, 4], Indenbom, Nikitenko, and Milevskii demonstrated the possibility of carrying out an optical-polarization analysis of the dislocation structure of silicon crystals. In crystals grown in the [110] direction they observed sessile edge dislocations with Burgers vector $\frac{1}{2}$[110] and slip plane (100), and also 60° dislocations with Burgers vector $\frac{1}{2}$[011] and slip plane ($\bar{1}1\bar{1}$). Later, Milevskii studied silicon crystals [6] oriented along the [110] direction, observing positive and negative edge dislocations arranged in parallel slip planes and studying the creep of these under the influence of dissolved gold [7]. On heating the crystal to 1250°C, creeping took place perpendicularly to the slip plane in accordance with the sign of each dislocation, in the sense of increasing the dimensions of the odd (superfluous) atomic half-plane. For present purposes were therefore selected crystals with dislocations of different signs, the study of which was started in [6, 7]. Since the photoelastic method may be used to carry out a complete analysis of the dislocation structure of crystals containing dislocations passing along the axis of observation, it is convenient for these purposes to use specially grown crystals in which the

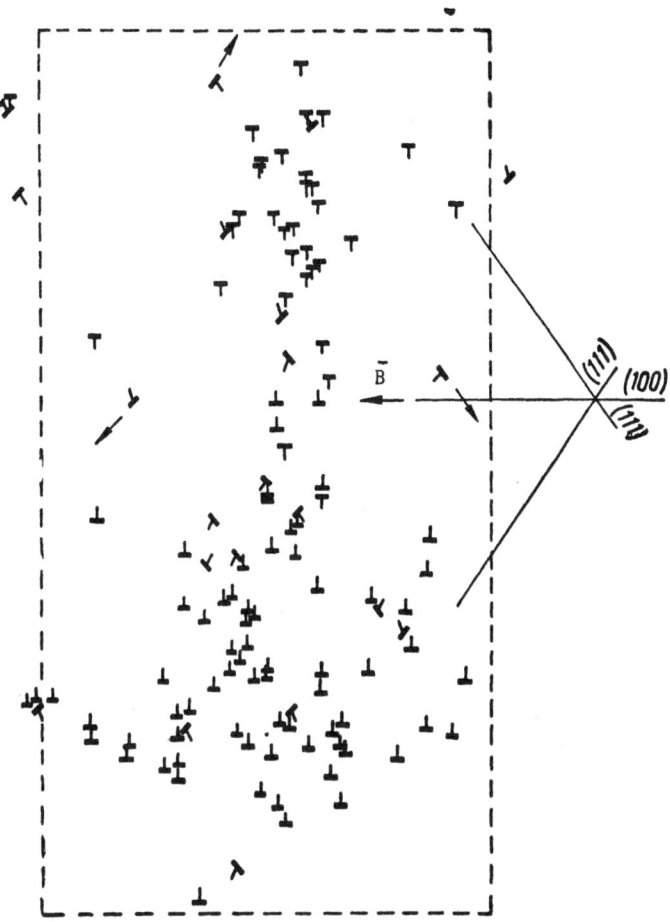

Fig. 2. Arrangement of the dislocation structure of a silicon sample (schematic). ⊥ – positive, τ – negative edge dislocations, ↘ – positive, ↗ – negative 60° dislocations. Arrows show the direction of motion of the 60° dislocations.

dislocations are arranged preferentially along the growth axis, for example, the [110]. Figure 1 shows a region of a silicon crystal examined by means of a polarization microscope (with Nicols). The photograph was taken from the screen of an electron-optical image converter. For the majority of the dislocations the plane of polarization is situated at an angle of 45° to the slip plane. The region in question contains both positive and negative dislocations (upper part) of the edge type with Burgers vectors ±½ [110] and slip planes (001). Four-petal rosettes correspond to these dislocations. In the negative edge dislocations the dark petals are situated underneath on the side of the superfluous atomic half-plane and the light ones on top. The positive dislocations, on the other hand, have the light petals underneath. In addition to the sessile edge dislocations, the crystal contains several 60° dislocations in the (1$\bar{1}$1) and (1$\bar{1}\bar{1}$) slip planes.

The arrangement of the dislocation structure of the crystal is shown in Fig. 2, from which we see the mutual disposition of about 80 edge and 16 60°-dislocations.

X-Ray Diffraction Examination

Some xrd rosettes were obtained for the same crystal as that used for the pe rosettes in Fig. 1. These are shown in Fig. 3. In the scheme of Fig. 2, the area represented in Fig. 3 is enclosed by a broken line. The xrd sample was cut absolutely perpendicular to the growth axis,

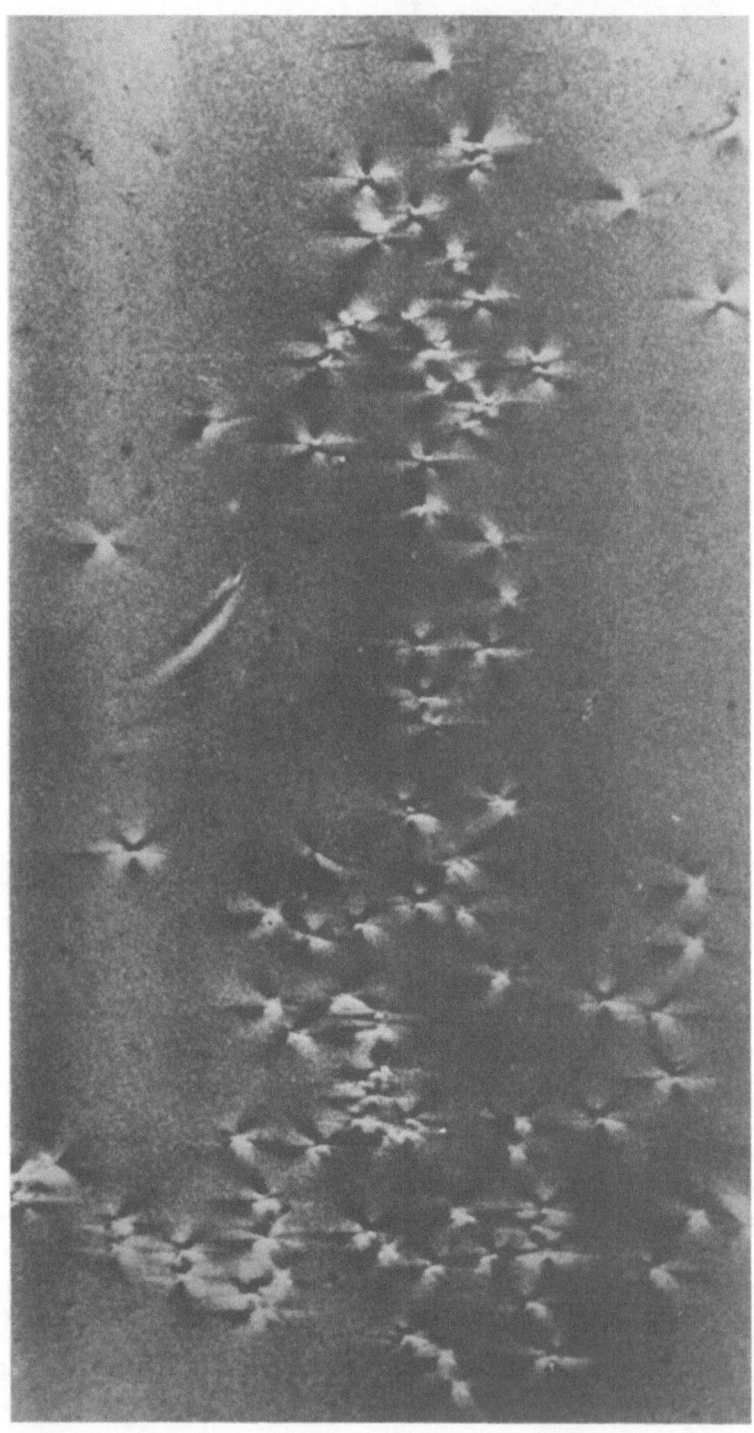

Fig. 3. X-ray diffraction rosettes from the same crystal. Neighboring sample, × 25.

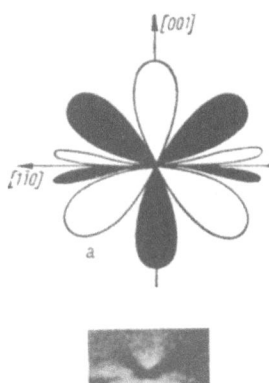

Fig. 4. Arrangement of the petals of an xrd rosette around a positive edge dislocation of the [110] type. This is one of the rosettes of fairly regular form.

the [110]. The thickness of the plate was 2.1 mm. We used the $\bar{2}20$ reflection in copper radiation ($\mu t = 30.5$). In order to obtain optimum resolution, the x-ray picture was taken not by continuous scanning but with a step-by-step movement of the crystal and film through the width of the diffraction image of the stationary crystal, i.e., by 0.70 mm. In other words, we took successive pictures of one section of the crystal after another, so that the photograph in Fig. 3 shows vertical bands where neighboring sections join. With this kind of arrangement for photographing comparatively thick crystals, some dislocations may be absent from the chart if they lie outside the volume embraced by diffraction. However, on comparing the mutual disposition of the xrd rosettes with the dislocation structure of the part of the crystal studied (Fig. 2) as plotted by the photoelastic method, we find that these agree completely. We see from Fig. 3 that the majority of the xrd rosettes have an irregular form. Experience showed that the shape of these changed considerably with the angle between the plane of the crystal cut and the vertical plane of the camera.

By rotating the crystal through 2-3° around the horizontal axis perpendicular to the median line of the primary beam, we may greatly alter the shape of the xrd rosettes of the dislocations in question. Thus, the shape of the rosettes depends on the orientation of the dislocations relative to the direction of the incident beam, so that we may obtain a rosette of regular shape by slightly altering this orientation. It is not possible to bring all dislocations into the correct position at the same time, especially dislocations of opposite signs. We see from the scheme of Fig. 2 that dislocations of opposite signs are concentrated in different parts of the crystal, the positive at the bottom and the negative at the top, considerable macrostresses being formed in each case. In addition to this, dislocations of opposite signs are capable of creeping in opposite directions as a result of interaction with vacancies during the cooling of the crystal in the course of growth [6]. Any curving of the dislocation lines makes it more difficult to obtain xrd rosettes of a single type. Figure 3 also shows a severely inclined dislocation with complex contrast, the formation of which was discussed in [5].

Considering one of the xrd rosettes of regular shape in Fig. 4, we see that this consists of 10 petals of alternating black and white color. We employed double photographic printing of the x-ray negative, so that the distribution of contrast in our figures corresponds to that of the x-ray diffraction picture. For this reason the coloring of the petals of the xrd rosettes in Fig. 3 does not correspond to that observed in [2], where single printing from the original x-ray negative was employed. In our case, the black petals of the xrd rosettes correspond to a maximum intensity of the x-ray diffraction radiation.

The xrd rosette is shown schematically in Fig. 4. The scheme is based on an analysis of a large number of xrd rosettes obtained from dislocations separated by a distance greater than three times the width of an xrd rosette.

Despite the considerable complexity of the xrd rosettes, these may be used for unequivocally determining the sign of the dislocation, i.e., its Burgers vector, since the slip plane is a plane of antisymmetry of the xrd rosette and the plane perpendicular to this is a plane of symmetry of the x-ray diffraction field. We must note the considerable difference between the xrd and pe rosettes, which occurs despite the fact that elastic lattice distortions due to the micro-stresses of single dislocations play a leading part in forming each of these.

The black-and-white coloring of the pe rosettes appears, as indicated in [3], as a result of the superposition of the field of microstresses formed by each individual dislocation and the total macroscopic stress field formed by all the dislocations in the crystal, the color reversing with the sign of the macrostresses. Lines of equal transmission (or extinction) intensity in the pe rosette correspond to lines of equal tangential stresses of a single dislocation, diminishing hyperbolically on passing away from the center of the dislocation. The rosettes obtained by the photoelastic method consist of four petals of equal size when the plane of polarization of one of the crossed Nicols makes an angle of 45° with the slip plane of the dislocation (see Fig. 1), or of six petals when these planes are parallel. In the latter case the larger petals are drawn out along the slip plane.

As indicated earlier, the xrd rosettes consist of 10 petals, which for strict orientation of the dislocation line relative to the direction of the incident x-ray beam include 6 large and 4 small petals, the latter being drawn out along the slip plane of Fig. 4. The large petals, three black and three white, are arranged symmetrically around the center of the dislocation with alternating colors. For the slightest deviation of orientation, or under the influence of the elastic fields of neighboring dislocations, there is a sharp change in all the x-ray pictures, some petals intensifying and others vanishing altogether.

The rosettes in the upper part of Fig. 3 correspond to negative dislocations, the pair of black petals being at the bottom, i.e., in the region of the superfluous atomic half-plane. Positive dislocations are represented by the rosettes in the lower part of Fig. 3, the pair of black petals being raised above the slip plane, i.e., in the region of the superfluous atomic half-plane. Thus, the xrd rosette is directly related to the nature of the elastic distortions of the crystal lattice near the center of the dislocation and contains additional information both of the structure of the field of distortions and of the mechanism underlying the formation of the diffraction image. In contrast to the photoelastic method, which is sensitive to the distribution of tangential stresses, the x-ray diffraction method is clearly sensitive to changes in lattice parameter due to the hydrostatic compression and expansion of the stress field. The method of xrd rosettes here employed may play a considerable part both in refining the structure of the stress field of single dislocations and in constructing a theory for the formation of the diffraction image of a defect in an almost-ideal crystal.

Conclusions

By means of high-resolution x-ray diffraction topography we have established the shapes of the rosettes associated with the distortion fields of dislocations oriented along the [110] direction in silicon crystals.

The slip plane of the dislocation is a plane of antisymmetry of the x-ray diffraction rosette; this enables the sign of the Burgers vector to be determined unequivocally.

The x-ray diffraction rosettes of the fields of distortion around [110] dislocations obtained by reflection from the ($\bar{1}$10) plane consist of 10 petals of alternating black and white color.

Literature Cited

1.   Bonse, U., Z. Phys., 153:278-96 (1958).
2.   Danil'chuk, I. N., and Smorodina, T. A., Fiz. Tverd. Tela, 7(4):1245 (1965).
3.   Indenbom, V. L., Nikitenko, V. I., and Milevskii, L. S., Dokl. Akad. Nauk SSSR, 141(6):
      1360 (1963).

4. Indenbom, V. L., Nikitenko, V. I., and Milevskii, L. S., Fiz. Tverd. Tela, 4:232 (1962).
5. Miuskov, V. F., In: Growth of Crystals, Vol. 5. Izd. Akad. Nauk SSSR (1965). [English translation: Consultants Bureau, New York (1968).] See also Kristallografiya, 2:254 (1963).
6. Milevskii, L. S., Fiz. Tverd. Tela, 5(7):1878 (1962).
7. Milevskii, L. S., Fiz. Tverd. Tela, 5(9):2447 (1962).
8. Lang, A. R., J. Appl. Phys., 30(11):1748 (1959).

MACROMOSAIC SUBSTRUCTURE OF MOLYBDENUM, TUNGSTEN, AND TANTALUM SINGLE CRYSTALS OBTAINED BY NONCRUCIBLE ELECTRON-BEAM ZONE MELTING

V. O. Esin and T. V. Ushkova

Institute of Metal Physics of the Academy of Sciences of the USSR

X-ray diffraction microscopy methods are widely used in the study of single crystals. These methods provide good information regarding the distribution of imperfections on the surface of [1-3] and inside [4, 5] a single crystal, but are inconvenient in that they require special equipment. The method of Schulz, for example, which is most suitable for studying massive, nontransparent metal crystals, requires a sharp-focus x-ray source of appreciable power.

In order to obtain x-ray microdiffraction pictures in the present investigation, we used an ordinary (not sharp-focus) standard x-ray tube of the BSV-2 type (focal spot size 1×14 mm) with a molybdenum anode.

Tungsten, molybdenum, and tantalum single crystals were grown by noncrucible zone melting with electron-beam heating in vacuum (10^{-6} mm Hg); these had the form of a cylinder some 3 mm in diameter and 60-120 mm long. Usually the single crystals were grown from the melt (liquid zone passing along the sample at velocities between 0.1 and 3.5 mm/min), but some of them were obtained by the migration of grain boundaries in the solid phase behind a moving heated zone, the temperature of which lay a little below the melting point.

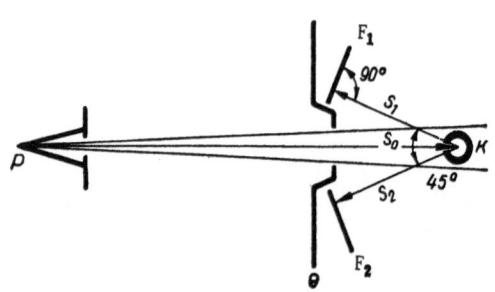

Fig. 1. Geometry for taking the x-ray topographs (top view).

The x-ray topographs were taken in the following way. The crystal K was placed in a diverging white x-ray beam in such a way as to be completely irradiated (over the whole length and diameter) thereby. Back-reflection Laue reflections were recorded on the x-ray film F. In order to increase the resolving power of the method, the distance R from the tube focus to the sample was increased to 420 mm. The distance D from the film to the sample was taken sufficiently large (120 mm) for reflections with different hk*l* values not to fall on each other, but not so large as to increase the exposure

220

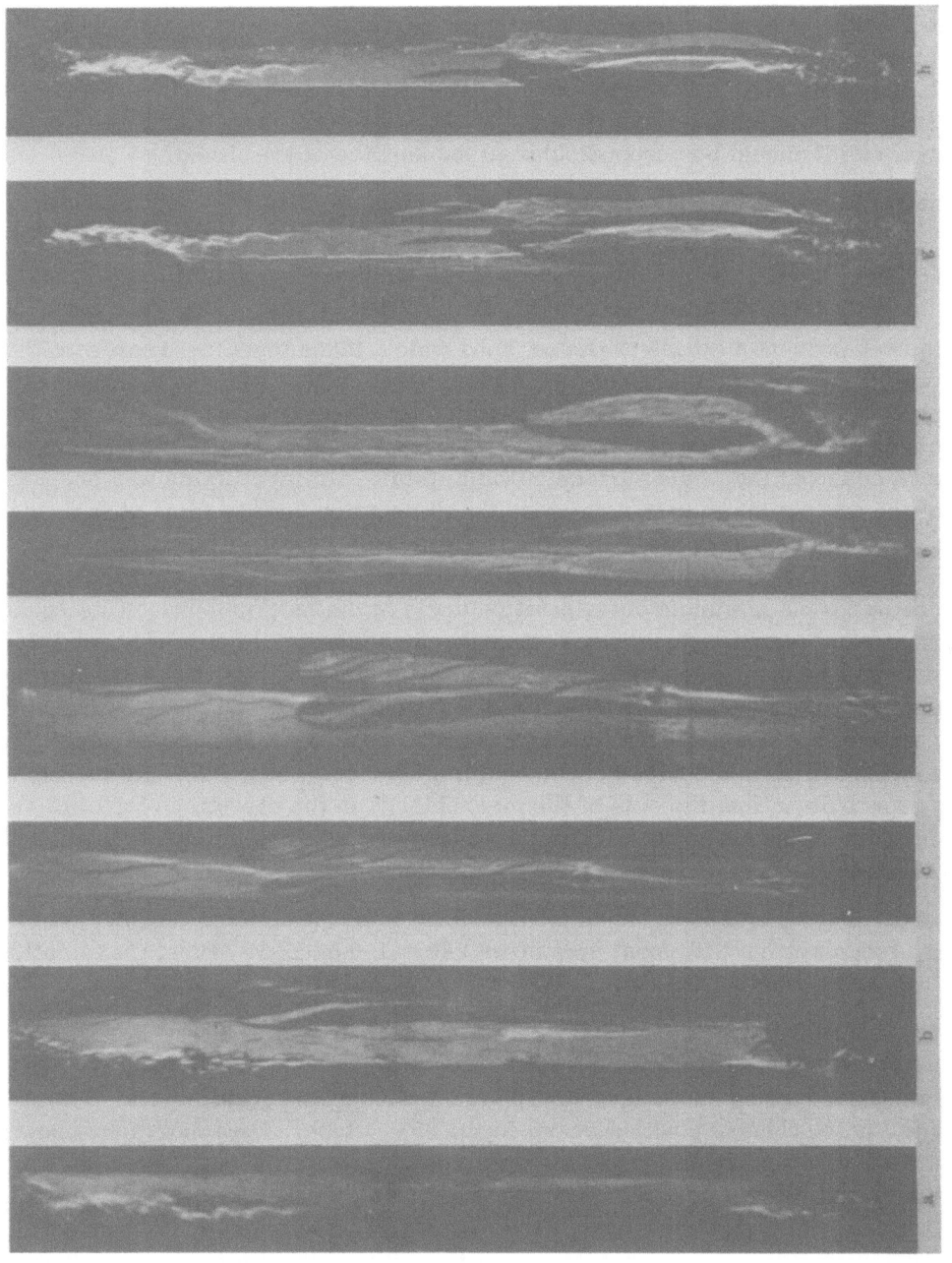

Fig. 2. Series of topographs of the surface of a molybdenum single crystal, × 1.6.

time unduly. For tube voltages of 18 kV and currents of 18 mA, the exposure was about 2 h. Photographs were taken in a beam sliding over the surface of the tube anode, i.e., near the edge of the anode shadow. This made it possible to obtain a linear projection of the focus with dimensions 1×0.37 mm, and gave an angular resolution (for the geometry used in taking the topographs) of 8' in the vertical and 3' in the horizontal direction.

Four successive exposures, rotating the sample in steps of 90° around the cylindrical axis, made it possible to obtain a topographical representation of the elements of macromosaic structure over the whole surface of the crystal. Photographs were obtained simultaneously on two flat films F_1 and F_2 (Fig. 1), arranged on opposite sides of the primary beam. These were arranged in such a way that the diffracted rays S_1 and S_2 diverging from the crystal with an angular interval of 45° should be perpendicular to the surface of the film.

This kind of geometry is convenient, since it enables us, as a first approximation, simply to estimate the disorientation angles of the macroblocks from the formula $\zeta = w/2D$ [6], where ζ is the disorientation angle of the macroblocks, D is the distance from the sample to the film, and w is the width of the white (or black) bands on the topographs.

Figures 2a-h present a series of photographs which, taken together, represent the topography of the whole surface of a molybdenum single crystal grown at a rate of 3.5 mm/min by zone melting with a single pass. The crystal contains two large subregions differing in orientation from the principal crystal by 1°25' and 35', respectively. The boundaries of these subregions may be traced over the whole surface of the crystal. A microsection was prepared on the surface of the sample in order to determine the dislocation density from etch figures. The site of the microsection appears clearly on the topograph (Fig. 2f, g) as a region with poorly resolved defects, arising from mechanical treatment (machining). From the topographs shown we may also estimate the minimum disorientation angle of the macroblocks. This equals approximately 4', i.e., a quantity which may be compared with a visible resolution on the plate (about 1') and with the limiting accuracy of the focusing tube (3'). Figure 3 shows topographs of a tantalum single crystal grown at a velocity of 1.0 mm/min with one pass of the molten zone. These also give topographical information regarding the position of the boundaries of the fragments on the whole cylindrical surface of the sample. The sub-boundaries are mainly drawn out along the growth direction; the size of the macroblocks in the direction of growth reaches several centimeters. Each of these subregions, in turn, consist of a group of finer blocks with smaller disorientation, of the order of 3-10'.

For comparison we took topographs from several molybdenum and tungsten single crystals in a sharp-focus system with focal spot sizes $280 \times 100 \mu$ (BSV-7 tube, Cu radiation, anode voltage 25 kV, current 700 μA) and 50 μ (BSV-5 tube, Mo radiation, anode voltage 27 kV, current 200 μA).

There were no marked differences from the topographs obtained with the ordinary (not sharp-focus) tube. Figure 4a shows a topograph from a tungsten single crystal obtained with a BSV-2 tube and Fig. 4b the same with a sharp-focus BSV-7 tube. The topographs show substructures with closed boundaries in the form of a peculiar network, in addition to coarse fragments. The individual cells of this lattice or network have sizes from 0.4 to 1 mm and a regular hexagonal shape. A substructure of this shape and size was observed in tungsten single crystals by Nakayama et al. [7], and similar networks with much smaller cells (0.2-0.4 μ) relating to elements of finer substructure were recently found in [8]. Thus the method reveals the shape and size of fragments of microsubstructure over the whole surface of the sample and enables us to estimate their disorientation.

The macrostructure observed in the single crystals may be correlated with the corresponding growth. In growing single crystals by zone melting, stable conditions must be maintained

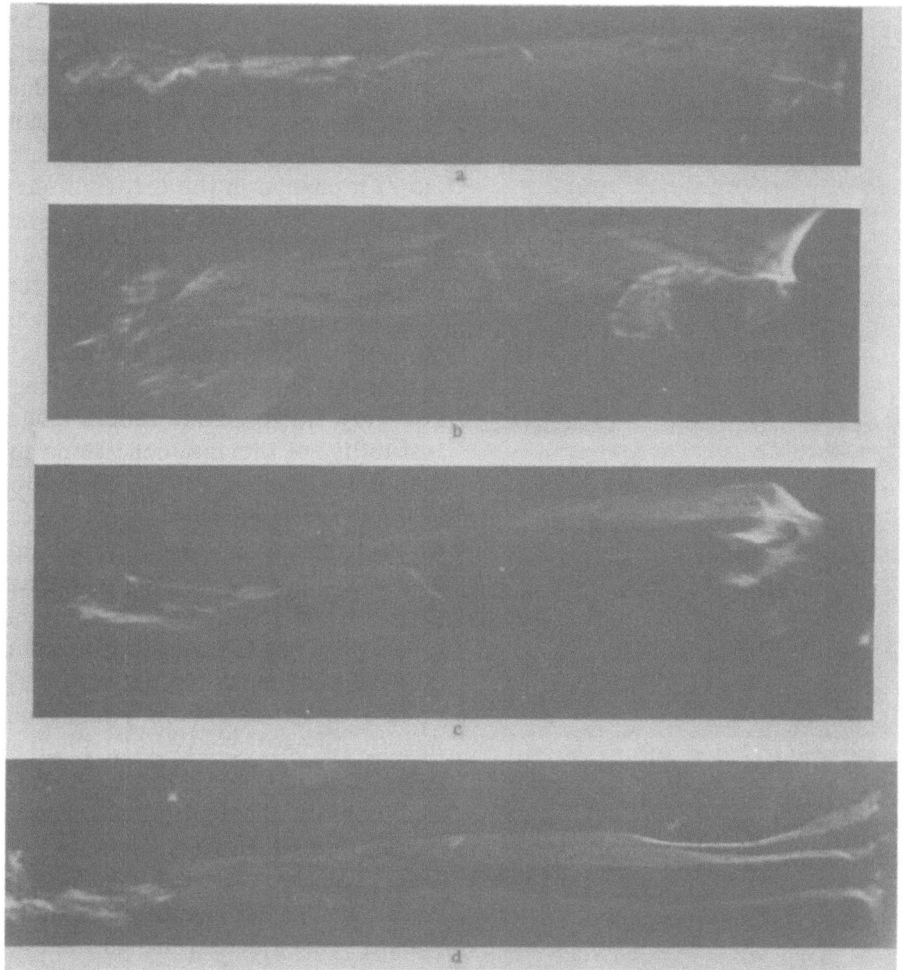

Fig. 3. Series of topographs of a tantalum single crystal.

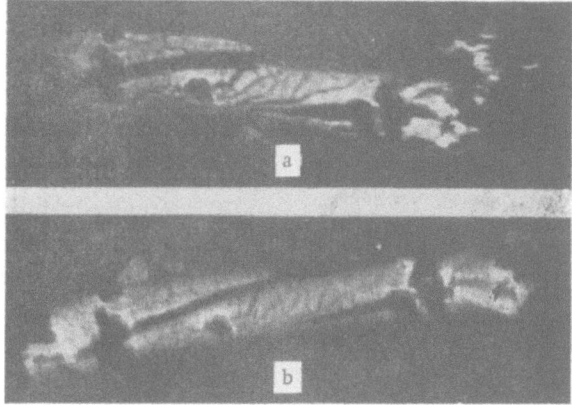

Fig. 4. Topographs from a tungsten single crystal, × 2.

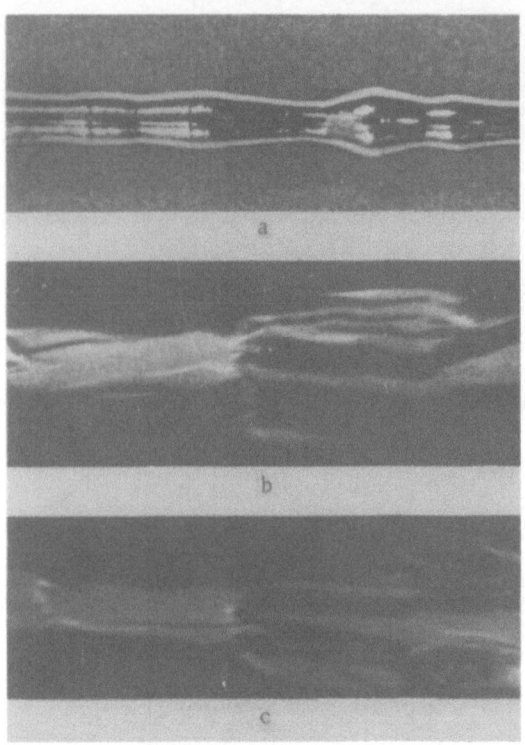

Fig. 5. External shape of a molybdenum single crystal (a) and topographs from the same part of the crystal (b, c). × 2.

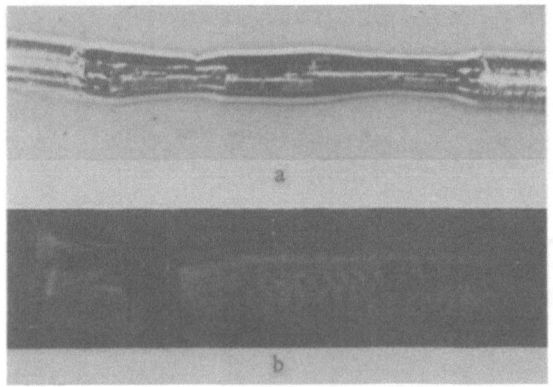

Fig. 6. External shape of a tantalum single crystal (a) and its topograph (b). × 2.

throughout the whole crystallization process: constant heater velocity, constant zone temperature, and constant zone size. Random deviations in the velocity or thermal conditions of the zone lead to a change in its radius, which also produces a change in the external shape of the sample (bulging and necking). Fluctuations in growth rate due to an unreliable set of thermal conditions of inconstancy of the other conditions of zone melting produce severe changes in the perfection of the crystal structure of the resultant single crystals. In fact, in regions of severely varying zone radius (characterizing instability of thermal conditions in the zone and thus a deviation of the solidification rate from the average zone velocity) the crystal grows in a very imperfect way. On re-establishing stabilized conditions, the crystal again starts growing without serious distortions. In places corresponding to changes in crystallization conditions, coarse fragments with disorientations up to 1° may be seen on a background of a comparatively perfect crystal. Figures 5a, b, and c present photographs of the external shape of a molybdenum single crystal (grown at a rate of 3 mm/min with two zone passes) at a necking point, together with topographs from the same region. The disorientation of the fragments resulting from fluctuations in the crystallization conditions reaches 40'. The same may be seen in tantalum single crystals. Figures 6a and b show photographs of the external form of a tantalum crystal grown at 0.1 mm/min with 8 zone passes, together with the corresponding topograph. In the initial region, instability of the conditions led to severe fragmentation of the crystal, the fragments being disoriented by 20' relative to one another.

If fluctuations in growth rate due to instability of the conditions of zone melting are insignificant or absent, severe changes in the perfection of the crystal structure of the resultant single crystals do not occur; crystal growth is quite perfect. Figures 7a and b show photographs of the external shape of a molybdenum single crystal grown at a rate of 1.8 mm/min with one zone pass, together with its topograph. Careful maintenance of the thermal conditions of melting (also reflected in the external form of the crystal) result in quite perfect growth. Only in the initial region is there a small subregion, disoriented by 15' with respect to the main part of the crystal. Scratches introduced onto the surface of the single crystal are of course shown up in the topograph as white, nonreflecting regions.

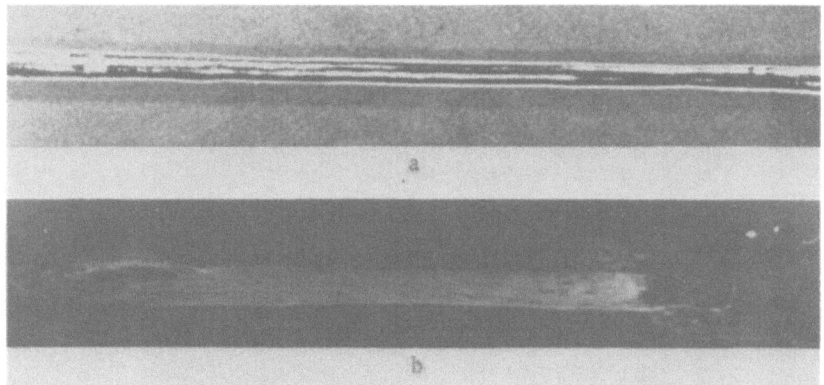

Fig. 7. External shape of a molybdenum single crystal (a) and its topograph (b).

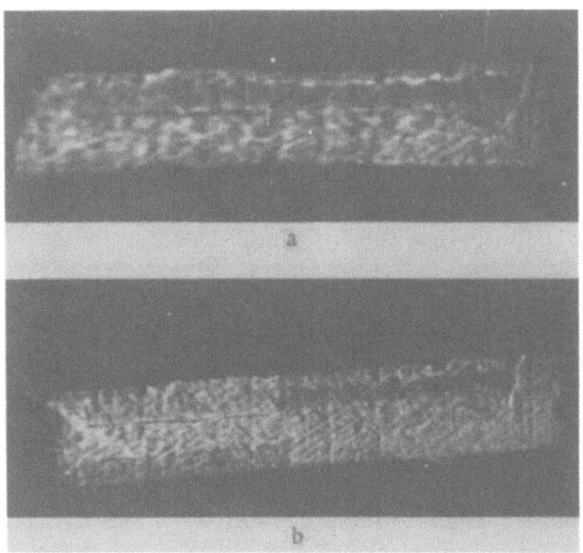

Fig. 8. Topographs from recrystallized molybdenum, ×2. a) Topograph obtained with a standard BSV-2 tube; b) the same with a sharp-focus BSV-7 tube.

It may be noted that the widespread method of determining the perfection of single crystals from the shape and intensity of Laue spots is not really very objective. Owing to the limited diameter of the x-ray beam, a Laue spot only gives information regarding the perfection of a crystal in some localized region, and this may not correspond to the true perfection of the whole crystal. In order to obtain reliable information regarding finer block structure (10^{-3} to 10^{-5}) from the intensity distribution curves in a Laue spot (e.g., [9]), we must eliminate the possibility of capturing a severely disoriented region of the crystal. For this purpose we must first obtain a topograph from the crystal and select the point for the photograph accordingly.

By this method we took topographs of a molybdenum single crystal obtained by zone passes below the melting point at 0.3 mm/min (recrystallization method). The crystals were in the

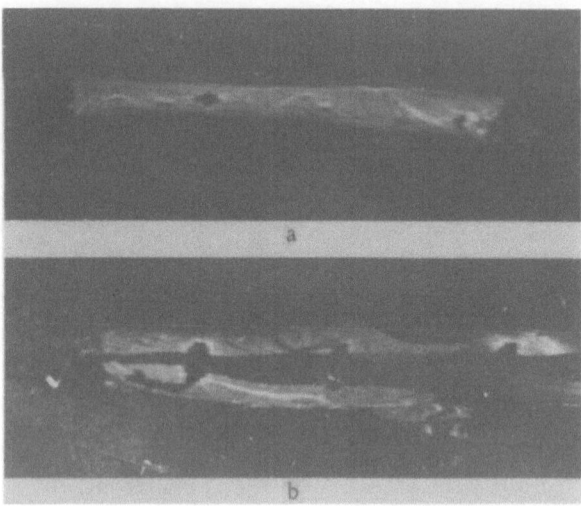

Fig. 9. Topographs from a tungsten single crystal W
(ρ 300°K/ρ 4.2°K = 12,000), ×2.

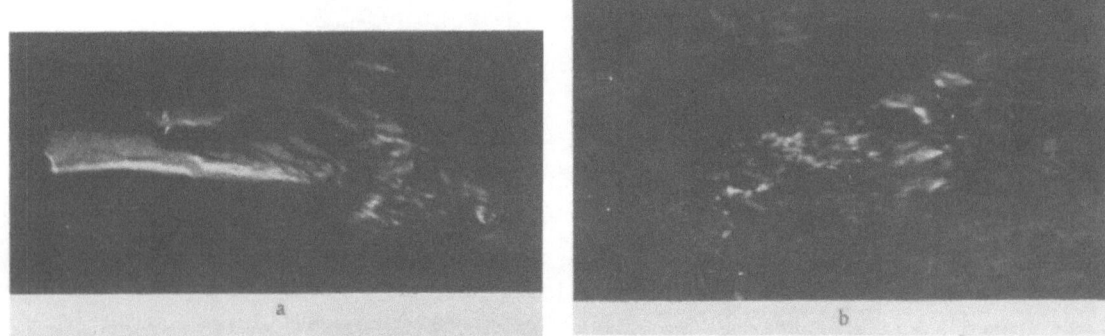

Fig. 10. Topographs from a tungsten single crystal W (ρ 300°K/ρ 4.2°K = 19,000), × 2.

form of thin plates 5 × 25 mm in size and 0.2 mm thick. We see from the topograph of the
crystal surface (Fig. 8a) that the molybdenum single crystal obtained by the method of recrystallization is more perfect than those obtained by zone melting. The disorientation of the fragments of first-order substructure is so insignificant that the shape of the Laue reflection on
the topographs exactly reproduces the shape of the sample; the image of their boundaries is
diffused owing to the smallness of their disorientation, comparable with the resolving power of
the method. We might expect that, on using a sharp focus (large resolution), the elements of
substructure with slight disorientation would appear more clearly. Figure 8b shows a topograph of the surface of the same molybdenum crystal obtained with a sharp-focus BSV-7 tube
(using copper radiation). Small fragments with insignificant disorientations are clearly visible
in this. Evidently photographs taken with an ordinary BSV-2 tube are inadequate for the study
of such substructures, and a sharp-focus source and use of special photographic material [2]
is essential.

    In order to determine the degree of purity of the various single crystals studied we meas-
ured their electrical resistance at room temperature and at the temperature of liquid helium
(liquid hydrogen for tantalum, since it becomes superconducting at helium temperatures) and

calculated the ratio $\rho\,300°K/\rho\,4.2°K$. On comparing this characteristic with the topographs indicating the perfection of the crystal, we found that pure single crystals (with high values of $\rho\,300°K/\rho\,4.2°K$) were as a rule imperfect; their topographs revealed splitting into subgrains of considerable disorientation. For example, Figs. 9a, b and 10a, b show topographs of tungsten single crystals with resistance ratios of 12,000 and 19,000, respectively. Figures 9a and 10a represent topographs of the tungsten single crystals from their most perfect regions, while Figs. 9b and 10b illustrate the least perfect. We see from these topographs that the tungsten single crystal with resistance ratio 19,000 is by no means perfect; it is split into coarse fragments, the disorientation of which reaches 2°. The tungsten single crystal with resistance ratio 12,000 is more perfect; the maximum disorientation of its fragments is 40'. The same phenomenon holds for single crystals of tantalum and molybdenum. In a tantalum single crystal with an electrical resistance ratio of 190 the maximum disorientation angle is 40'; in one with a ratio of 26 the maximum disorientation angle between the macroblocks is 20'. A molybdenum single crystal with a ratio of 5000 has a maximum disorientation angle of the fragments equal to 43', while a crystal with ratio 980 has maximum disorientation angle 15'. These data suggest that the characteristic $\rho\,300°K/\rho\,4.2°K$ is an indicator of the purity of single crystals, but is insensitive to the presence of impurities of the macromosaic type. This is apparently because the coarse fragments in the single crystals have no influence on their electrical resistance, the mean free path of the electron being shorter than the dimensions of the fragments. The boundaries of the coarse defects also give no contribution to the electrical resistance.

In conclusion we may note that in metallic single crystals grown by electron–beam zone melting the detectable elements of substructure (of the first order) have a considerable disorientation (over 3–8'), so that taking topographs with a fine focus changes the picture of subboundary distribution very little. Thus we may quickly and simply obtain information on the macroscopic perfection of metallic single crystals obtained by zone melting with the help of an ordinary standard BSV-2 x-ray tube.

Literature Cited

1. Bačkovsky, I. M., J. Phys. Radium, Ser. VII, 9 : 11 (1938).
2. Schulz, L. G., Trans. AIME, 200 : 1082 (1954).
3. Barrett, C. S., Trans. AIME, 161 :15 (1945).
4. Lang, A. R., Acta Met., 5 : 7 (1957).
5. Newkirk, D. V., and Wernik, D. H., In: Direct Observation of Imperfections in Crystals [Russian translation], pp. 37-78. Metallurgizdat, Moscow (1964).
6. Saito Shozo, J. Phys. Soc. Jap., 17 : 8 (1962).
7. Newkirk, D. V., and Wernik, D. H., In: Direct Observation of Imperfections in Crystals [Russian translation], pp. 332-348. Metallurgizdat, Moscow (1964).
8. Popov, N. M., Savitskii, E. M., and Tsarev, G. L., Dokl. Akad. Nauk SSSR, 162 : 1 (1965).
9. Esin, V. O., and Kralina, A. A., Fiz. Met. i Metalloved., 13(4) : 577-586 (1962).

FINE STRUCTURE OF TUNGSTEN SINGLE CRYSTALS

E. M. Savitskii and G. L. Tsarev

*A. A. Baikov Institute of Metallurgy of the Academy of Sciences
of the USSR*

Tungsten single crystals were studied by light and electron (transmission) microscopy on N. M. Popov's 400-kV apparatus [1]. The objects for study were two tungsten single crystals 4 mm in diameter and 200 mm long, grown by electron-beam zone melting (two passes). The impurity content of the single crystals was C = 0.025% (determined by combustion), O_2 and N_2, 0.001% each (determined by vacuum melting), Mo, Fe, and Si, 0.001% each (spectral analysis). The ratio k = $\rho\,300°K/\rho\,4.2°K$ for the single crystals was 1100. The crystallographic orientation of the longitudinal axis of one single crystal coincided with the [001] and the other with the [011]. The substructure was revealed by electropolishing and electrolytic etching in 2% NaOH. The thin films for transmission microscopy were prepared from the single crystals by electropolishing in NaOH, using a method similar to that given in [2]. In order to study the dislocation structure created by plastic deformation, before thinning, the single-crystal samples were subjected to slight double bending (20°). The secondary phases were indexed and the crystallographic directions determined by means of microdiffraction.

The substructure of the tungsten single crystals revealed by etching consists of subgrains of three orders of magnitude (Figs. 1 and 2). A subgrain of the first order (A, Fig. 1) has dimensions of 0.2-0.5 mm in the transverse section with respect to the growth direction and 1.5-2 mm in the longitudinal direction. The dimensions of the subgrains of the second order (Fig. 1, B) are 0.05-0.1 mm in the transverse section and 0.5-1 mm in the longitudinal direction. Subgrains of the third order are observed at a magnification of about 1000 and appear similar to elements of cellular growth structure (Fig. 2). The cell size in the transverse section is 10-18 μ. The disorientation angles, calculated from the distance between the dislocations in the sub-boundaries, equal 17'30" for subgrains of the first order, 2' for subgrains of the second order, and 0'30" for subgrains of the third order. On studying thin films prepared from undeformed single crystals in the electron microscope, we find growth cells still smaller than those of Fig. 2. In Fig. 3, at point C, we see cells 0.2-0.5 μ in size; these may be regarded as the smallest elements of substructure in tungsten single crystals: subgrains of the fourth order. It should be noted that these cells are observed very frequently, and their images are produced by diffraction contrast, and not by etching effects. At the cell boundaries are segregations (fine precipitates of the carbide W_2C); these were identified by means of microdiffraction. There were also coarse carbide precipitates in the structure of the single crystals. Figure 3 shows a microdiffraction picture taken from an inclusion (in the photograph surrounded by a light spot, the image of the selector diaphragm), in which reflections 1, 2, and 3 correspond to

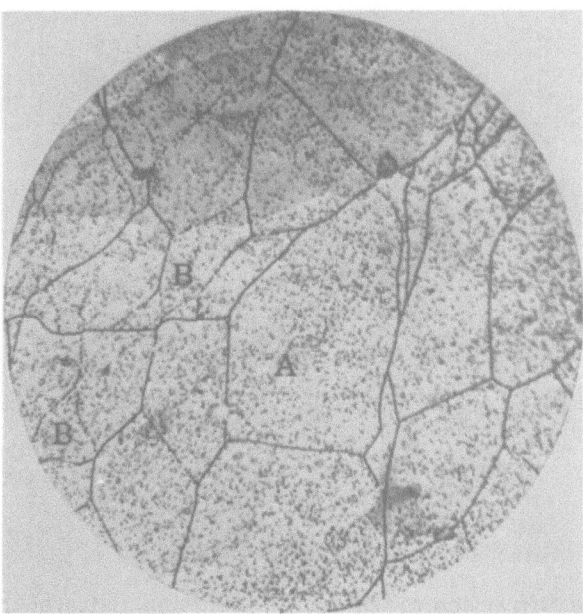

Fig. 1. Substructure of a transverse section of a
tungsten single crystal, × 50. Electropolishing and
electrolytic etching. The (001) plane.

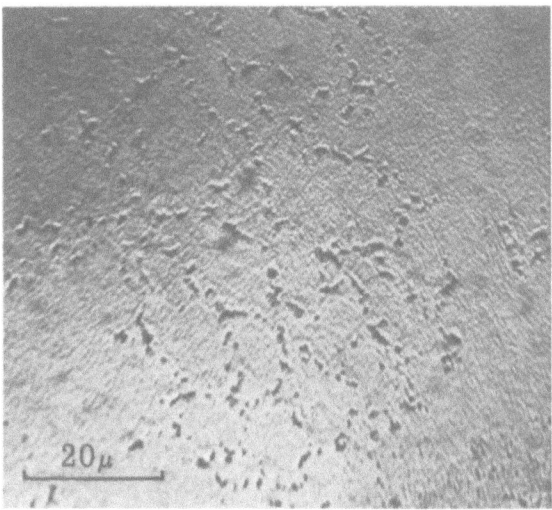

Fig. 2. Cellular growth structure in a tungsten single
crystal, × 50. Electropolishing and electrolytic etching in $CuSO_4 + NH_4OH$. The (011) plane.

the (110), (200), and (110) plane of tungsten and reflections 4, 5, 6, and 7 to the (0002), ($10\bar{1}2$),
($10\bar{1}4$), and (0004) planes of the carbide W_2C, respectively. The size of individual carbides
reaches 0.5-2.5 μ, the dimensions in the growth direction being 1.5-2.5 times greater than in
the transverse direction. All the individual coarse carbides gave point electron-diffraction
pictures; together with the fact of their being drawn out in the growth direction, this indicates

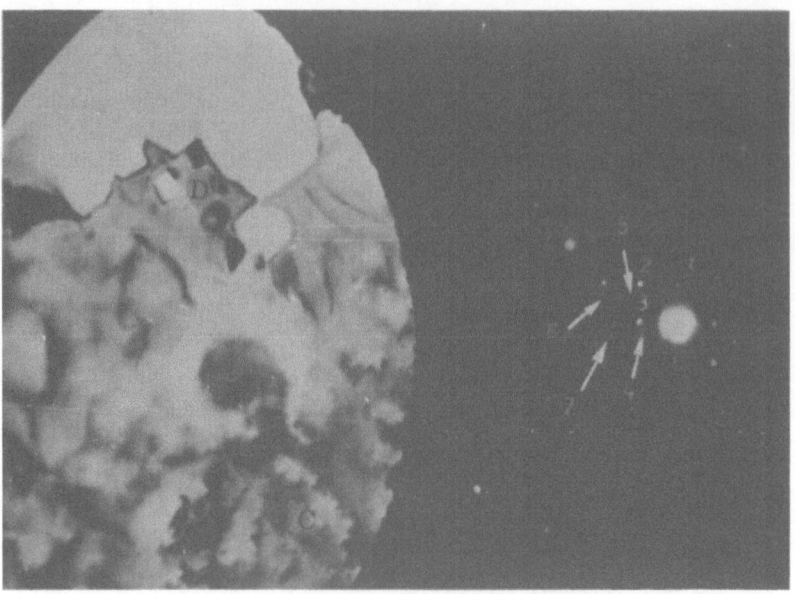

Fig. 3. Microdiffraction picture of an inclusion and growth cell
in a tungsten single crystal. Plane of the film (001).

that the carbides are formed during solidification, and not in the course of precipitation from
the solid phase. The solubility of carbon in
tungsten, even at the temperature of the solidus,
is clearly far below 0.02%, as considered in
later data [3]. Oxides and nitrides were not observed in the structure of the tungsten single
crystals.

The dislocation structure of samples
worked by bending is shown in Figs. 4 and 5.
Dislocations of the D type are screw dislocations, since their lines are parallel to the [110],
the projection of the Burgers vector $a/2$ [111].
The screw character is also indicated by the
zigzag black-and-white contrast of certain dislocations of this type. Dislocations of the C[?]
type are edge dislocations. We cannot regard
these as screw dislocations in another system
with Burgers vector $a/2$, since, in this case,
they should lie in the (110) plane, which is perpendicular to the (001) plane of the film. The
dislocations visible in the photographs are arranged in inclined planes (101) and ($\bar{1}$01). The
screw dislocations contain many steps and
supersteps. At large magnification, some edge
dislocations are resolved as dipoles of width
350-380 Å (G in Fig. 5). The dipoles are unstable and decompose into separate loops (J in

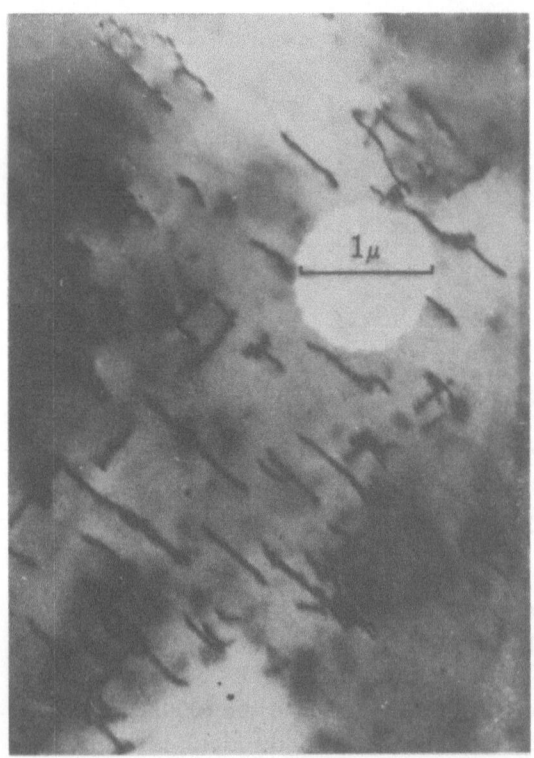

Fig. 4. Screw and edge dislocations in slip
lines. Plane of the film (001).

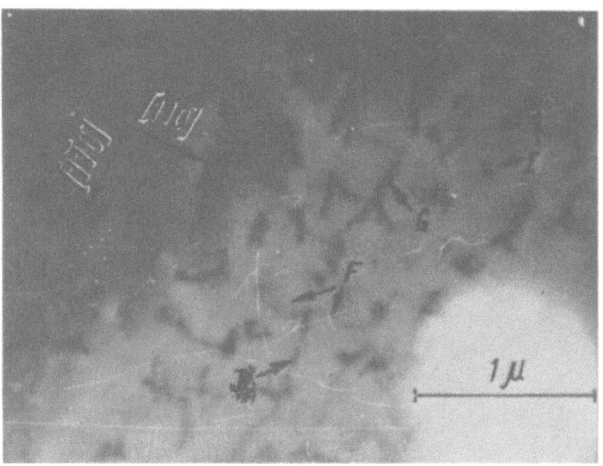

Fig. 5. Dipoles and loops in worked tungsten single crystals. Plane of film (001). Light spot is the image of the scaling diaphragm.

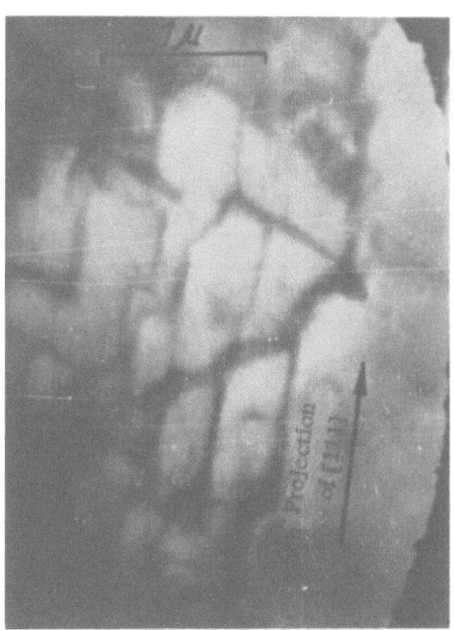

Fig. 6. Motion of dislocations under the influence of the electron beam. Plane of film (001).

Fig. 5). Some loops, for example I in Fig. 5, clearly originate in the passage of dislocations through precipitates. The origin of the edge dislocations and dipoles must be linked with the interaction of screw dislocations with growth dislocations and small precipitates, for the following reason. Figure 6 shows the motion of screw dislocations under the influence of the electron beam. We see that a screw dislocation pulls edge "tails" along after itself; these exhibit a dipole character. The distance between the "tails" is 0.5 μ, so that if we calculate the density of the pinning sites from which the edge dislocations are drawn we obtain $4 \cdot 10^8$ cm^{-2}. This figure agrees closely with the density of growth dislocations revealed by etching, i.e., $2 \cdot 10^7$ cm^{-2}. In [4] the local stopping of the line of a mobile dislocation, leading to the appearance of a dipole, is explained by the formation of long steps as the dislocations pass around obstacles by transverse slip. Supersteps are not capable of moving conservatively and pin the dislocation. On the subsequent return of the dislocation to its original slip plane (double transverse slip) the superstep closes and an independent loop is formed. The same mechanism operates in tungsten single crystals also, but the formation of the supersteps must be linked not with the union of separate steps, as proposed in [4], but with the pinning of the dislocation line on intersecting growth dislocations penetrating the slip plane and the bending of the dislocation line on subsequent slip. In this case we must suppose that the growth dislocations in tungsten single crystals are firmly pinned and are only intersected with difficulty by moving dislocations. The strong pinning of the growth dislocations is confirmed by the following data. If we anneal tungsten at high temperature, hardly any reduction in the dislocation density occurs. If, however, annealing is

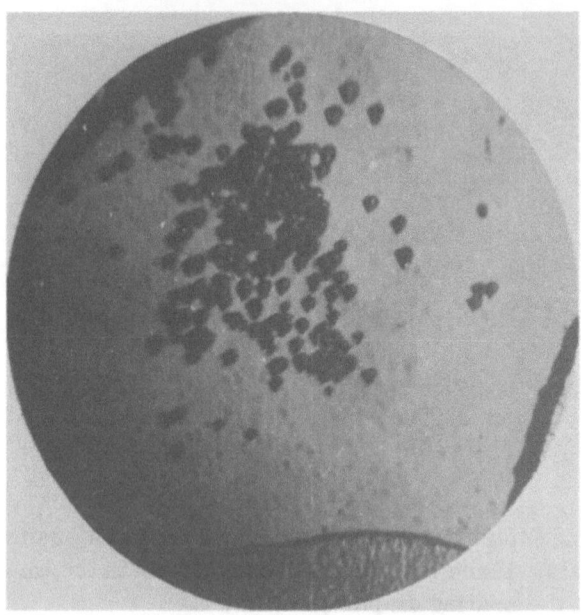

Fig. 7. Growth dislocations in polycrystalline tungsten subjected to decarburization, × 1000.

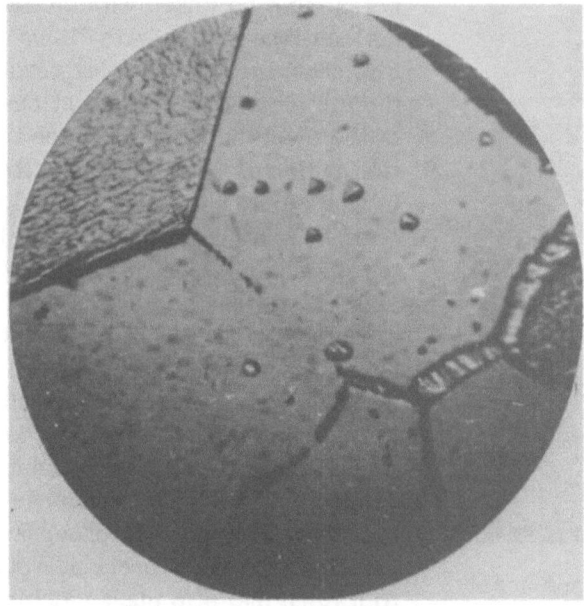

Fig. 8. Same as Fig. 7, but after prolonged decarburization, × 1000.

carried out in the presence of oxygen, the dislocations are unpinned as a result of decarburization and are annihilated. Figure 7 shows the microstructure of polycrystalline tungsten heated
for a long time at approximately 2800° in a weakly oxidizing atmosphere. The parts near the
grain boundaries, in which decarburization has taken place, are free from growth dislocations.
For a sufficiently prolonged holding period, almost all the dislocations are unpinned and annihilated (Fig. 8). Thus we must conclude that the growth dislocations in tungsten are firmly

pinned by W_2C carbide precipitates. An important consequence is that the density of growth dislocations in tungsten is directly connected with the carbon content. The substructure of tungsten single crystals is also very sensitive to carbon content. In single crystals containing 0.0012% of carbon the subgrains are three or four times coarser than in those containing 0.025%. The disorientation of the first-order blocks is 7-9' instead of 17' and that of the second-order ones 30" instead of 1'. It should be noted that the mechanical properties of tungsten single crystals, especially the characteristics of ductility, also greatly depend on carbon content [5].

Conclusions

The substructure of tungsten single crystals consists of subgrains of four orders of magnitude.

Tungsten single crystals grown from material of commercial purity contain a second phase, W_2C.

The dislocation structure of worked single crystals contains edge dipoles, loops, supersteps, and edge dislocations formed by the interaction of mobile dislocations with growth dislocations and precipitates.

The growth dislocations in tungsten are decorated and pinned by W_2C precipitates.

The substructure of tungsten single crystals and the density of growth dislocations are sensitive to carbon content.

The observed motion of the dislocations under the influence of the electron beam indicates that the Peierls—Nabarro force is weak at a temperature of 80-150°.

Literature Cited

1.   Popov, N. M., Izv. Akad. Nauk SSSR, 23(4):436 (1959).
2.   Strat, P., Apparatus for Scientific Research [Russian translation], p. 4 (1961).
3.   Westoren, R. C., and Thompson, V. R., Trans. AIME, 230:931 (1964).
4.   Washburn, J., et al., Phil. Mag., 5:991 (1960).
5.   Savitskii, E. M., and Tsarev, G. L., Izv. Akad. Nauk BelorussSSR, p. 1 (1965).

STUDY OF THE STRUCTURE AND PROPERTIES OF MOLYBDENUM—NIOBIUM ALLOYS IN THE SINGLE-CRYSTAL STATE

E. M. Savitskii, G. S. Burkhanov, and N. N. Bokareva*

A. A. Baikov Moscow Institute of Metallurgy

The main obstacle to the modern technical use of highly-alloyed substitution-type solid solutions based on metals of the VA, and especially VIA, groups is the worsening of the technical ductility. Scientific and industrial experience with refractory metals of the IVA groups shows that the full possibilities of heavily alloying such materials can only be realized if the principal metal and alloying additives are obtained in a state of high purity, especially with respect to interstitial impurities. We therefore thought that it would be interesting to produce single crystals of alloys of refractory metals free from interstitial impurities.

As subject for study we selected the molybdenum—niobium system, containing a continuous series of solid solutions. The two components of the system differ very little in melting point and have bcc lattices with similar parameters and atomic dimensions [1].

Single crystals of molybdenum—niobium solid solutions with specified crystallographic orientation were grown for the first time in an electron-beam zone-melting system [2], using a nitrogen trap and a vacuum of 10^{-4} to 10^{-5} mm Hg. The compositions of the alloys obtained in the single-crystal state were Mo, Mo + 2% Nb, Mo + 4% Nb, Mo + 10% Nb, Mo + 20% Nb, Mo + 35% Nb, Mo + 50% Nb, Mo + 80% Nb, Mo + 90% Nb, Mo + 96% Nb, Mo + 98% Nb, Nb. As original components we used bars of vacuum-melted molybdenum and niobium. The use of a nitrogen trap in the vacuum system prevented the falling of oil vapor and oil-cracking products into the working chamber of the apparatus in the course of melting. The components were melted together directly in the electron-beam zone-melting apparatus. In order to obtain single-crystal alloys of uniform composition, two successive passes of the molten zone were made in opposite directions, the last pass for producing the single crystal being made from a seed. The velocity of the molten zone was between 3 and 6 mm/min, depending on the composition of the alloy.

Chemical analysis showed a uniform distribution of the alloy components along the length of the single crystal. X-ray structural examination showed that alloy single crystals grown

---

*N. P. Khazov, A. E. Tsutskov, and T. S. Stronina took part in the work.

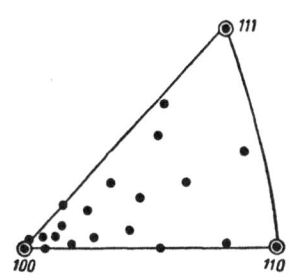

Fig. 1. Orientation of single crystals of molybdenum–niobium alloys in the triangle of the standard stereographic projection.

without any seed had an arbitrary orientation, with a certain tendency toward the [100] corner of the stereographic triangle, as indeed was observed for single crystals of pure refractory metals (Fig. 1).

In obtaining alloy single crystals of specified orientation, we used single crystals of pure molybdenum or niobium as seeds [3], no great change being observed in the proportion of the second component at the contact zone.

Experiments showed that it was quite impossible to obtain single crystals of heavily alloyed material (more than 20% Mo or Nb) without using a seed; this was evidently due to the considerable distortions in the crystal lattice and to the stresses arising in the alloy as a result of these. Only by using a seed taken from a single crystal of a pure metal or alloy as substrate was it possible to obtain alloy single crystals with a greater proportion of the second component.

The single-crystal nature of the alloys on the molybdenum side was tested by electrolytic etching in a 10% aqueous solution of NaOH; for alloys on the niobium side we used chemical etching in a mixture of nitric and hydrofluoric acids ($HNO_3 : HF = 1 : 3$). Alternating lustrous and mat longitudinal bands were obtained on the surface of the single-crystal samples.

All the alloys of the system under consideration were grown with crystallographic orientations of [100] and [110]. The deviation in orientation from the given direction lay between 1 and 5° on growing from a seed.

Selective etching of the pure metal and alloy single crystals gave etch figures representing the outcrops of dislocations on the surface. The crystallographic facing of the etch figures resulted from the {110} planes (the most closely packed planes in the bcc lattice). The etch figures had different shapes on different crystallographic planes of the single crystal: squares on the cube plane and triangles on the octahedral plane. No etch figures could be obtained on the plane with the densest packing, the (110).

Figure 2 shows the fine structure of a molybdenum single crystal and single crystals of Mo-base solid solutions on the (100) plane. We notice single etch figures in the form of squares and also discontinuous mosaic block boundaries consisting of etch pits. The fine structure of the Mo-base alloy single crystals is characterized by a certain increase in dislocation density, refinement of block structure, and increase in block disorientation as the concentration of the alloying element increases. Laue x-ray photographs taken from alloy single crystals show fragmentation of the spots, which confirms the existence of mosaic structure and an increased degree of block disorientation in alloy single crystals (with increasing proportion of the second component), as indicated in Fig. 3.

Measurement of the hardness (Fig. 4) and specific electrical resistance (Fig. 5) of single crystals of molybdenum–niobium alloys showed that these properties varied as functions of composition in accordance with N. S. Kurnakov's law, as in the case of polycrystalline samples. Figure 4 shows the variation in the hardness of molybdenum–niobium single crystals together with that of similar alloys obtained by vacuum arc melting. Owing to their high degree of freedom from interstitial impurities, the single-crystal samples had a lower hardness and were technologically more tractable than the corresponding alloys obtained in the vacuum arc.

Of great interest is the anisotropy in the mechanical properties of molybdenum and molybdenum–niobium single crystals. Mechanical tests were carried out at room temperature

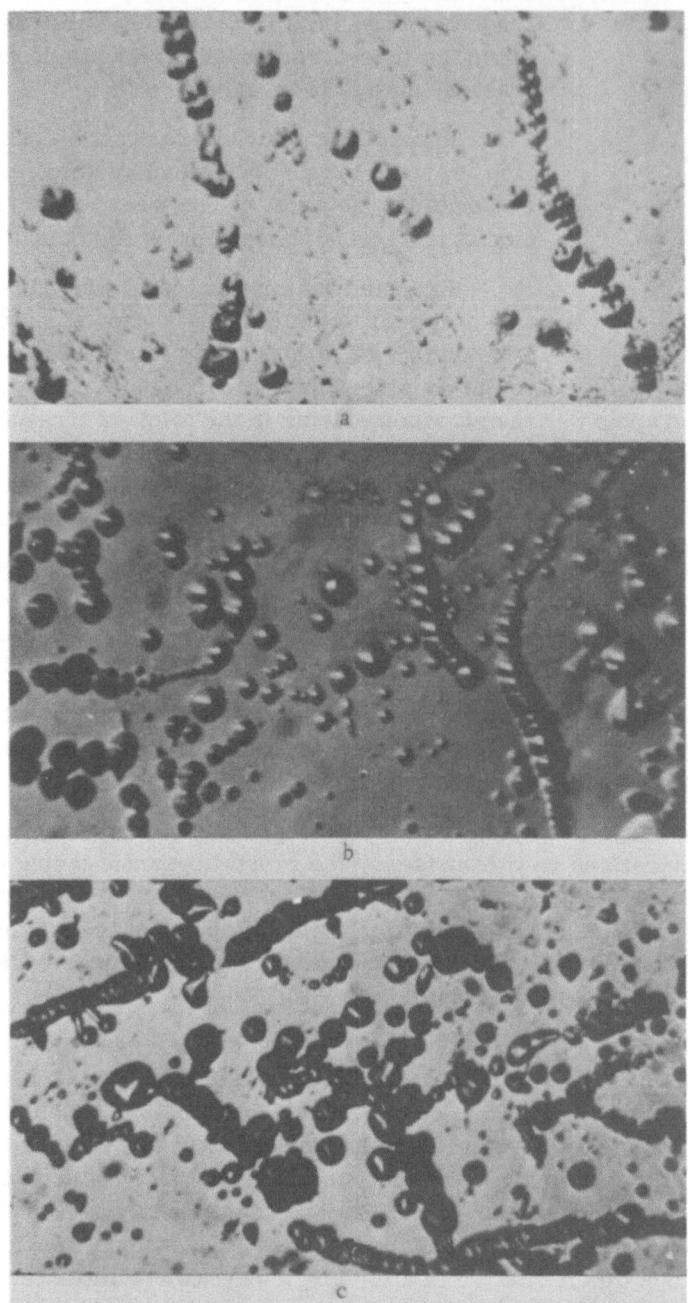

Fig. 2. Substructure of single crystals on the (100) plane, × 500.
Electrolytic etching 5% H_2SO_4, 95% CH_3OH. a) Mo; b) Mo + 4% Nb;
c) Mo + 10% Nb.

on a Gagarin press, with recording of the strain diagram. The samples for testing were prepared by electrolytic etching in a 10% aqueous solution of NaOH in an apparatus constructed in
the Laboratory of Refractory and Rare-Metal Alloys of the Institute of Metallurgy [4]. This
completely removed the work-hardened layer, and the surface of the microsamples prepared
in this way was smooth and lustrous; the samples retained their single-crystal nature.

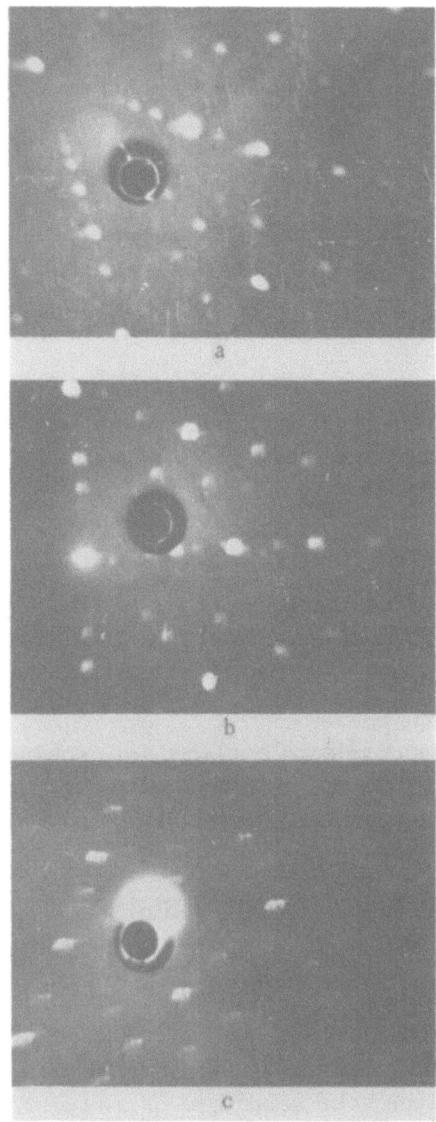

Fig. 3. Laue x-ray photographs taken
in reflection from single crystals of:
a) Mo; b) Mo + 10% Nb; c) Mo + 20% Nb.

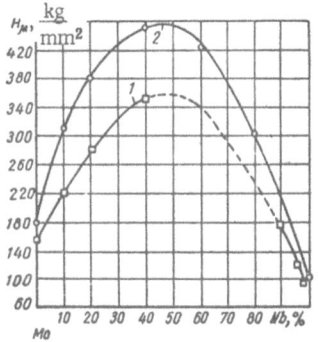

Fig. 4. Hardness of Mo—Nb alloys.
1) Single crystals; 2) after arc melting.

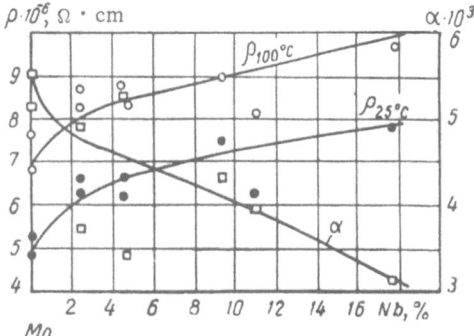

Fig. 5. Specific electrical resistance and
temperature coefficient of electrical resistance of Mo—Nb single crystals.

The results of mechanical tensile tests carried out at room temperature on single crystals of
molybdenum and Mo—Nb alloys (up to 20% Nb) are
shown in Table 1 and Fig. 6.* The mechanical
properties were distinctly anisotropic, the greatest ductility corresponding to single crystals with
orientation [110].

With increasing niobium content the strength of the Mo—Nb alloys becomes greater, while
the ductility falls, although still remaining at a high level. Single crystals of Mo—20% Nb with
[110] orientation showed a neck contraction of up to 40% for an elongation of 13%. In the case
of [100] orientation the neck contraction of the same alloy was 19% for an elongation of 6%. The
difference between the strength and ductility values of single crystals with orientations [100]
and [110] reaches 50%.

It would appear very interesting to study other physical properties (especially emission
properties) for the molybdenum—niobium system as well. We are continuing to engage in

---

* Each point corresponds to two or three samples.

Table 1

Composition	Orienta-tion	σ, kg/mm²	δ, %	ψ, %	H_{μ}, kg/mm²
Mo	100 110	46 29	23 31	47 100	157
Mo+2% Nb	100 110	27	29	74	172
Mo+4% Nb	100 110	38	28	68	180
Mo+20% Nb	100 100	46 59	6 13	19 41	265

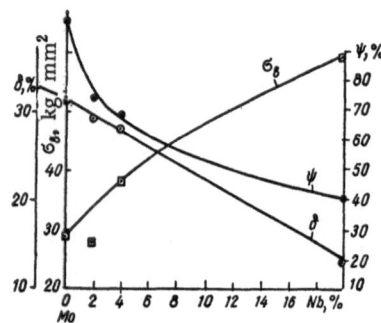

Fig. 6. Mechanical properties of Mo−Nb single crystals at room temperature; orientation [110].

research of this kind for the Mo−Nb system together with other refractory alloys, such as W−Mo, W−Ta, Mo−V, Mo−Re, and W−Re.

Conclusions

We have obtained single crystals of molybdenum−niobium solid solutions over the whole concentration range by electron-beam vacuum zone melting.

We have developed a method for growing oriented single crystals of these alloys by means of seeds taken from the corresponding pure metals.

We have observed mosaic structure in the single-crystal alloys by metallographic and x-ray methods.

We have observed anisotropy in the mechanical properties of molybdenum and molybdenum−niobium single crystals.

We have shown that it is in principle possible to obtain single crystals of a number of refractory alloys in the range corresponding to substitution-type solid solutions.

Literature Cited

1.   Vol, A. E., Structure and Properties of Binary Metallic Systems. Fizmatgiz, Moscow (1959).
2.   Savitskii, E. M., et al., Zavodsk. lab., No. 8 (1961).
3.   Savitskii, E. M., Burkhanov, G. S., Tsarev, G. L., and Bokareva, N. N., Pribory i Tekhn. Éksperim., No. 4 (1965).
4.   Tsarev, G. L., Novokhatskaya, N. I., and Savitskii, E. M., Pribory i Tekhn. Éksperim., No. 1 (1966).

VARIATION OF THE WORK FUNCTION
WITH PARTICLE SIZE

Kh. B. Khokonov and S. N. Zadumkin

Kabardino-Balkarsk State University

The electron work function associated with small particles is of great importance in the solution of a number of theoretical and practical problems involving thermoelectron (thermionic) emission, contact phenomena, photoeffects, etc. As far as we know, however, only one theoretical paper, that of A. P. Maksimenko and V. I. Tverdokhlebov [1], has been devoted to this question. In their calculations, the authors of [1] suppose that the work function φ is exclusively due to forces of interaction between the electron and its mirror image in the conductor, the calculation of φ being carried out at a certain distance x_0 from the metal surface. This does not allow for the work required to remove the electron from the actual metal surface to the distance x_0, although this may contribute an appreciable percentage to the work function [2].

The work function may be calculated as the difference between the energy of the metal after the removal of the electron and that of the un-ionized metal [3, 4]. The energy of a metal object of spherical form with radius r before the removal of an electron (reduced to a single atom) is

$$E = E_0 + \frac{3}{r}\frac{\sigma}{n_V}, \tag{1}$$

where E_0 is the volume part of the energy, σ is the specific surface energy, and n_V is the average number of atoms per unit volume. After the removal of the electron the energy, for one elementary sphere, is

$$E' = E + \left(\frac{\partial E}{\partial z}\right)_R \left(-\frac{1}{N}\right) + \frac{1}{2}\left(\frac{\partial^2 E}{\partial z^2}\right)_R \cdot \frac{1}{N^2} + \ldots + \eta \frac{e^2}{r}\cdot\frac{1}{N}, \tag{2}$$

where R is the radius of the sphere, e is the charge on the electron, z is the number of valence electrons per atom, and η is a coefficient depending on the charge distribution of the ionized metal (of the order of $\frac{1}{2}$). The work function

$$\varphi_r = (E' - E)N. \tag{3}$$

This method of calculating the work function is general and hence allows for the metal-vacuum boundary.

Putting expressions (1) and (2) into (3), we obtain

$$\varphi_r = -\left(\frac{\partial E_0}{\partial z}\right)_R + \frac{1}{2}\left(\frac{\partial^2 E_0}{\partial z^2}\right)_R \frac{1}{N} - \frac{3}{rn_v}\left(\frac{\partial \sigma}{\partial z}\right)_R$$
$$+ \frac{3}{2r}\frac{1}{Nn_v}\left(\frac{\partial^2 \sigma}{\partial z^2}\right)_R + \ldots + \eta\frac{e^2}{r}. \tag{4}$$

The relation between the energies per elementary sphere in the bounded and infinite metal may be put in the form of a series

$$E_0 = E_\infty + \left(\frac{\partial E_\infty}{\partial R}\right)_z (R - R_\infty) + \ldots, \tag{5}$$

where R and R_∞ are the geometric and gravimetric radii of the elementary sphere, in which $R_\infty = R(1 - \delta)$, and δ is the disintegration factor of the metal [5]. In addition to this, the surface energy of a metallic drop depends on the radius [6] in the following way:

$$\sigma = \sigma_\infty\left(1 - \frac{A}{r}\right), \tag{6}$$

where A > 0 is a constant quantity of the order of the interatomic distances, and σ_∞ is the surface energy of the semi-infinite metal. The latter is associated with the work function via the equation [4]:

$$\sigma_\infty \simeq \frac{1}{4}S^{\bullet}n_v z\varphi_\infty, \tag{7}$$

where $S^* = \lambda S$, in which S is a linear parameter of the order of 1 Å, reducing the Thomas–Fermi equation to dimensionless form, and λ is a variational parameter minimizing σ_∞ with due allowance for the exchange correction [7].

Substituting expressions (5)-(7) into (4) and remembering that

$$\varphi_\infty = -\left(\frac{\partial E_\infty}{\partial z}\right)_{R_\infty}, \tag{8}$$

we obtain the following expression for the work function

$$\varphi_r = \varphi_\infty\left(1 + \frac{\alpha}{r} + \frac{\beta}{r^2}\right). \tag{9}$$

Here the quantities α and β have the following meanings:

$$\alpha = \frac{\eta e^2}{\varphi_\infty} - \frac{3}{4}S^{\bullet}\left[1 + \frac{z}{\varphi_\infty}\left(\frac{\partial \varphi_\infty}{\partial z}\right)_{R_\infty}\right] + \frac{R_\infty\delta_0}{\varphi_\infty}\left(\frac{\partial \varphi_\infty}{\partial R}\right)_z, \tag{10}$$

$$\beta = \frac{3}{4}S^{\bullet}A\left[1 + \frac{z}{\varphi_\infty}\left(\frac{\partial \varphi_\infty}{\partial z}\right)_{R_\infty} - \frac{z}{2\varphi_\infty}\left(\frac{\partial^2 \varphi_\infty}{\partial z^2}\right)_{R_\infty}\right]. \tag{11}$$

Relation (9) is valid for dimensions of the metal sphere, such that $r \gtrsim 3bS \simeq 15$ Å.

Calculations of the coefficients α and β, using the Gombash [3] expression for φ_∞, show that $\alpha > 0$ and $\beta > 0$, in which α is of the same order as A, and β is one order smaller.

Thus, expression (9) shows that the work function φ_r increases with diminishing particle size, in qualitative agreement with the experiments of Shuler and Weber [8]. Shuler and Weber's result, to the effect that the electron work function for carbon particles between 100 and 1000 Å in size equals 8.5 eV, however, seems to be too high. The effect of particle size on the electron work function will only arise noticeably for sizes of 50-60 Å or under.

Literature Cited

1. Maksimenko, A. P., and Tverdokhlebov, V. I., Izv. Vysshikh. Uchebn. Zavedenii, 1:84 (1964).
2. Ioffe, A. F., Physics of Semiconductors, p. 280. Moscow-Leningrad (1957).
3. Gombash, P., Statistical Theory of the Atom and Its Application [Russian translation]. Moscow-Leningrad (1951).
4. Zadumkin, S. N., and Egiev, V. G., Fiz. Met. i Metalloved., 21:6 (1966).
5. Rozenberg, G. V., Usp. Fiz. Nauk, 58:487 (1956).
6. Zadumkin, S. N., and Khokonov, Kh. B., Uch. Zap. KBGU, 19:505 (1963).
7. Zadumkin, S. N., Fiz. Met. i Metalloved., 17(3):746 (1964).
8. Shuler, K. E., and Weber, I., J. Chem. Phys., 22:491 (1954).

SECTION III

GENERATION OF CRYSTALLIZATION CENTERS AND THE EFFECT OF HIGH COOLING RATES

GENERATION AND GROWTH OF SOLID-SOLUTION CRYSTALS WITH THE COMPOSITION OF THE ORIGINAL LIQUID

D. S. Kamenetskaya

*Institute of the Physics of Metals and Metal Science of the Central
Scientific-Research Institute of Ferrous Metallurgy*

In considering the crystallization of a two-component liquid, an important question which arises is this: What will be the composition c_2 of a critical nucleus of the solid solution formed in a two-component melt of given composition c_1 (fluctuation-type nucleation)? Earlier [1] we obtained a formula expressing the size of the critical nucleus (i.e., a nucleus such that either its growth or disintegration will lead to a fall in the free energy of the system) as a function of the composition of the liquid, the composition of the nucleus, and the temperature:

$$r_\kappa = \frac{2\sigma v}{f_1(c_1) + (c_2 - c_1)\dfrac{df_1}{dc_1} - f_2(c_2)}, \tag{1}$$

where σ is the surface tension of the crystal–liquid boundary, v is the volume associated with one molecule, $f_1(c_1)$ is the free energy of the liquid of composition c_1 referred to one atom, and $f_2(c_2)$ is the same for the solid solution of composition c_2.

It was found that the critical nucleus of the solid solution arising in a liquid of given composition at a given temperature T may have different compositions and, accordingly, different sizes. For a certain concentration of the nucleus, the curve $r_K(c_2)$ has a minimum. From relation (1) we may find the conditions for the minimum of r_K by setting $dr_K/dc_2 = 0$.

$$\frac{d\sigma}{dc_2} = \frac{\sigma\,(df_1/dc_1 - df_2/dc_2)}{f_1 + (c_2 - c_1)\,df_1/dc_1 - f_2}.$$

If the variation of σ with c_2 may be neglected, the condition for the minimum is the relation $df_1/dc_1 = df_2/dc_2$, i.e., the tangents to the f_1 and f_2 curves must be parallel.

In deriving Eq. (1), and also Eqs. (2)-(4), which will appear shortly, we have not made any simplifying assumptions regarding the way in which f_1 and f_2 depend on temperature and concentration; hence, the main conclusions deduced from these relations are general, for example, that regarding the dependence of r_K on c_2. In order to illustrate this relationship, Fig. 1 contains data relating to $r_K/2\sigma v$ obtained on the approximation of regular solutions for a system consisting of components A and B with the following characteristics: heats of fusion

245

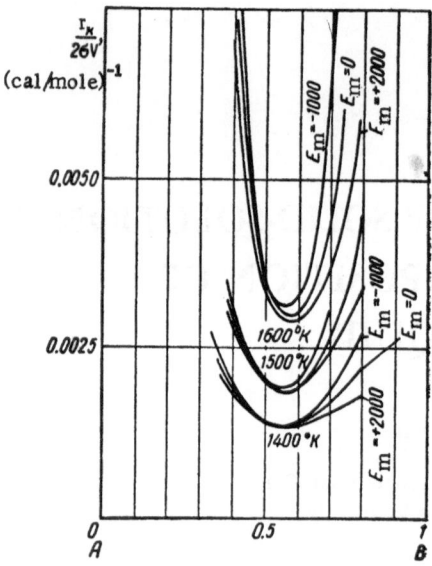

Fig. 1. Radius of critical nucleus as a function of composition for various ΔT and E_m.

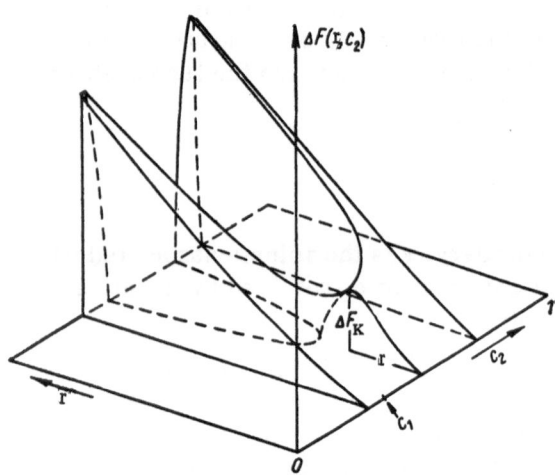

Fig. 2. Schematic form of the surface $\Delta F(r, c_2)$ for a small supercooling.

of the components Q_m^A = 3000, Q_m^B = 4000 cal/mole and melting points T_m^A = 1500, T_m^B = 2000°K. The energies of mixing E_m in the liquid and solid phases are taken as equal to 0, −1000, and +2000 cal/mole. We see from formula (1) and the curves of Fig. 1 that for each supercooling ΔT there is a definite range of possible concentrations of the critical nuclei. For a specific concentration the critical nucleus is the smallest. As supercooling increases the range of possible concentrations of the nucleus widens and the "sharpness" of the minimum is reduced, so that the difference between the size of the critical nucleus of the original composition and the minimum becomes smaller. If the energy characteristics of the alloy did not depend on temperature and concentration, the composition of the minimal nucleus would, for all supercoolings, be equal to the composition of the equilibrium solidus corresponding to the given melt; this condition is satisfied by alloys described on the approximation of regular solutions.

In determining the work of formation of the critical nucleus A_K it was found that this consisted of two parts: the first is associated with the development of the crystal−melt interface and the second with concentration fluctuations:

$$A_\kappa = \frac{1}{3} \sigma S_\kappa + \frac{1}{2} \left(\frac{V_\kappa}{v} \right)^2 (c_2 - c_1)^2 \frac{d^2 F_1}{(dN_1^B)^2}. \quad (2)$$

Here, S_K is the surface (area) and V_K the volume of the critical nucleus, F_1 is the total free energy of the liquid, N_1^B is the total number of atoms of component B in the liquid. Since r_K depends on c_2 for a given c_1 and T, A_K also depends on c_2 for every degree of supercooling.

The concentration of the nucleus for given c_1 corresponding to the minimum value of A_K does not coincide with the concentration of the nucleus of minimum dimensions; it is closer to the concentration of the original liquid.

The quantity A_K is defined as the height of the energy barrier equal to the maximum increment in free energy ΔF on forming a critical nucleus of definite composition. For illustration, Figs. 2 and 3 give a schematic representation of the form of the $\Delta F(r, c_2)$ surface for two different supercoolings $\Delta T_1 < \Delta T_2$.

The $\Delta F(r, c_2)$ surface has a saddle point corresponding to the smallest work of formation of the critical nucleus. This nucleus turns out to be nearer in composition to the liquid than

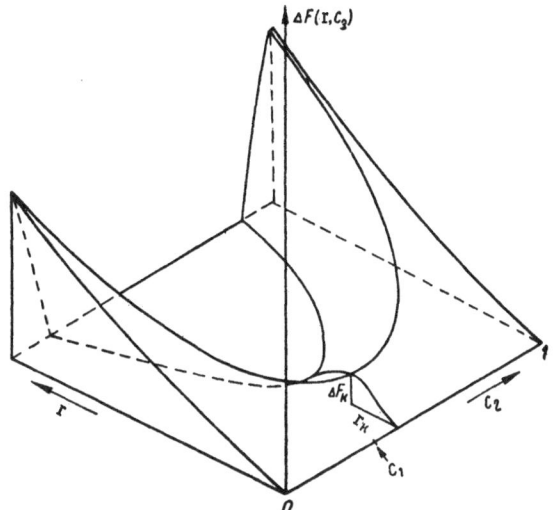

Fig. 3. Same as Fig. 2 but for a large supercooling.

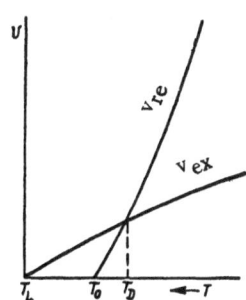

Fig. 4. Rate of transfer of atoms from the
liquid state to the solid (crystalline) as a
function of temperature for various stimuli.

the minimal nucleus. For temperatures above T_0 (the temperature at which the specific free energies of the liquid and solid solutions of the same composition are equal), the generation of crystallization centers of the original composition is impossible; for slight supercoolings relative to T_0 it is possible, but improbable, since A_K is large. For moderate supercoolings the work of formation of a nucleus of the solid solution of the same composition as the liquid approaches the minimal value. Starting from these supercoolings, the generation of a crystal nucleus of the original composition becomes quite probable.

Further growth of the solid-solution crystals is determined by the velocities of the processes underlying the rearrangement of the atoms and the exchange between the solid and liquid phases [2]. The rates of exchange and rearrangement equal

$$U_{ex} = A_{ex} e^{-\frac{U_{ex}}{RT}} [1 - e^{-\frac{\Delta f_{ex}}{RT}}],$$

$$U_{re} = A_{re} e^{-\frac{U_{re}}{RT}} [1 - e^{-\frac{\Delta f_{re}}{RT}}], \tag{3}$$

where

$$\Delta f_{ex} = \left(\frac{df_1}{dc_1} - \frac{df_2}{dc_2}\right) a_2 = (\Delta\mu_A - \Delta\mu_B) a_2, \tag{4}$$

$$\Delta f_{re} = f_1 + (c_2 - c_1)\frac{df_1}{dc_1} - f_2 = (1 - c_2)\Delta\mu_A + c_2\Delta\mu_B,$$

$\Delta\mu_A$ and $\Delta\mu_B$ are the differences in the chemical potentials of the components A and B in the solid and liquid phases, a_2 is the proportion of the substance in the solid phase, U_{ex} is the activation energy of the exchange process, U_{re} is the same for the process of rearrangement. Relations (3) are obtained on the assumption that the liquid and solid phases are homogeneous in composition. Since $U_{re} < U_{ex}$ as a result of the collectivity of the process [3] (this only relates to transformations in the condensed state) and the stimulus Δf_{re} increases with increasing supercooling, we find that with increasing supercooling v_{re} increases more rapidly than v_{ex}. In the case of supercoolings for which $v_{re} \gg v_{ex}$, the solid-solution crystal, once generated, is able to grow without change of concentration. This supercooling must be greater than $T_L - T_0$ (T_L is the temperature of the liquidus). The process of transferring atoms from the liquid phase to the solid at a temperature above T_0 is only possible as a result of exchange between the liquid and the forming crystal (Fig. 4). Starting from a temperature of T_0, the process of rearrangement is also possible; the nature of the v_{re} curve will depend on the rearrangement mechanism. If we are considering metal systems in which the falling part of the growth-rate curve does not occur at large supercoolings, then v_{re} rapidly increases with increasing supercooling, and at a certain value of the latter becomes much larger than v_{ex}. This

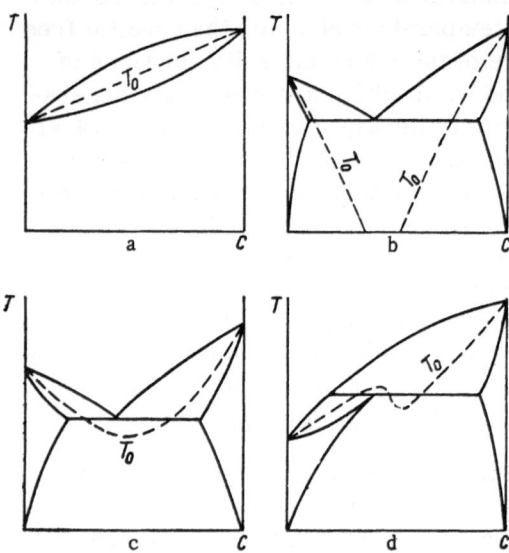

Fig. 5. T_0 curves for systems of various types. a) Continuous series of solid solutions; b) eutectic system consisting of components with different crystal structures; c) the same for isomorphous components; d) peritectic system (components isomorphous).

is attributable to both the faster increase in the energy stimulus for rearrangement as compared with the exchange stimulus on increasing the supercooling, and also the ratio of the kinetic coefficients [$A_{exp}(-U/RT)$], which are larger for the rearrangement process than for exchange [2, 3]. In the temperature range T_0 to T_D, the exchange process is dominant, below T_D rearrangement predominates, and for temperatures $T_B < T_D$ rearrangement takes place so rapidly that exchange is not realized at all. For supercoolings to temperatures below T_B, crystallization of a solid solution of the same composition as the liquid may occur (the so-called "diffusionless" crystallization).

Thus, the upper temperature limit at which crystals of the same composition as the liquid may form and grow is T_0. Both for generation and growth without change of concentration, supercooling relative to T_0 is required; the supercooling is greater (in the case of metallic systems) for generation than for growth. It should be noted that in order to secure growth of the crystals without change of concentration the formation of a nucleus of the same concentration is not obligatory; a nucleus of a different composition may be overgrown by layers of variable concentration until the composition of the outer layer becomes equal to that of the liquid; for a high enough supercooling this crystallization center may grow in the "diffusionless" manner (without change of concentration).

If the supercooling $T_L - T_0$ of the given liquid is small, it may fairly easily be achieved, and in this liquid crystallization may occur without change of concentration. If, however, $T_L - T_0$ is large, realization is more difficult. For $T_0 < 0$, "diffusionless" crystallization is entirely impossible. In systems forming a continuous series of solid solutions, "diffusionless" crystallization should occur comparatively easily. In systems with limited solubility in the solid state, of the eutectic (or peritectic) type, formed from components with identical crystal structure with an appreciable solubility in the solid phase, crystallization without change of composition may occur with the formation of a continuous series of solid solutions, metastable in the middle of the concentration range. The T_0 curves for the various systems have the form indicated in Fig. 5. In systems, the components of which have different crystal structures or very low solubility in the solid phase in the middle of the concentration range, the formation of a solid solution of the original composition is impossible. In alloys for which T_0 is very low or negative, high cooling rates and the associated severe supercoolings should lead to an amorphous state or to the formation of new phases.

Let us turn to what is known from experimental data. What (cooling) rates are required for the suppression of exchange between the phases? Experience shows that separating diffusion may be suppressed both in alloys having a stable single-phase solid solution [4] and, in a number of cases, in alloys consisting of two phases in the equilibrium state [5-9]. The cooling rates necessary for this are tens or hundreds of thousands of degrees per second. In Au−Cu alloys a solid solution with no signs of segregation is obtained for a cooling rate of 61,600 deg/sec, in Ni−Cu alloys for 97,000 [4], in alloys of Al with Mn, Cr, and V for 25,000 [6], and in Ag−Cu alloys for about 100,000 deg/sec [9]. For small cooling rates separating diffusion is not suppressed and conditions are close to equilibrium at the phase boundary [10].

The value of the supercooling required for crystallization of a solid solution of the same composition as the liquid (for a low cooling rate of the order of deg/sec) depends on the nature of the alloy and its concentration. For a Cu−50% Ni alloy we require a supercooling of 110° [11]; for Sb−5% Bi, 70°; and for Na− 0.5 at.% Hg, 8.5° [12]. In the middle of the concentration range greater supercoolings are required than in the neighborhood of one of the components in order to have "diffusionless" crystallization; in systems forming a continuous series of solid solutions smaller supercoolings are needed than in systems of the eutectic type.

At large cooling rates two factors have a part to play: the reduction in the mobility of the atoms, and the supercooling of the liquid, which occurs in view of the fact that at large cooling rates the possibility of crystals forming on active particles existing in the melt cannot be realized. The development of supercooling for large cooling rates was established in [6]. For cooling rates of 25,000 deg/sec, the supercooling for Al was about 200° and for Al with various additives, as follows: Mn, 310°; Cr, 300°; Si, 170°; Ti, 100°; Mg, 115°. The reduction in mobility is shown by the fact that in a number of cases an amorphous ("vitreous") state is obtained; the amorphous state is obtained by quenching from the liquid state in alloys of Te with 10-30% Ge [13], alloys of Te with Al, In, Ga, Si [14] over a certain range of concentrations, and an alloy of Au with 25. at.% Si [15], in some alloys the amorphous state being preserved for a long time (some months in Te−Ge), whereas in other (Au−Si) rapid disintegration occurs. On quenching an Ag−Ge alloy from the liquid state, a metastable electronic compound develops [16]. In this system a solid solution cannot be obtained owing to the different structures of Ag and Ge, whereas Ag and Cu, which have identical lattices, give a continuous series of solid solutions on quenching from the liquid state (metastable in the middle of the concentration range) [9]. The development of considerable supercoolings for high cooling rates plays a fundamental part in the process of "diffusionless" crystallization, since it ensures an adequate supercooling of the liquid below T_0. If the cooling rate is insufficient for complete "diffusionless" crystallization, a mixed structure is obtained: the initial regions correspond to a solid solution of the original composition, and these are surrounded by rims of inhomogeneous composition or dendrites [10, 11, 17].

"Diffusionless" crystallization occurs not only on cooling a liquid, but also on crystallization of an amorphous solution obtained by chemical deposition. An amorphous nickel−phosphorus alloy (8.5-15% P) was obtained in [18] by chemical deposition (precipitation). On heating, this amorphous alloy, with a hexagonal type of packing, passed "diffusionlessly" into a heavily supersaturated solid solution of phosphorus in nickel (hcp) of the same concentration, which then decomposed with the formation of Ni_3P.

Thus, experience shows that for alloys in which "diffusionless" crystallization is possible it does in fact occur for adequate supercoolings, these being achieved either on account of the purity of the melt or on account of high cooling rates. In cases in which "diffusionless" crystallization is impossible, high cooling rates or supercooling lead to the development of metastable phases or to an amorphous state.

Literature Cited

1. Kamenetskaya, D. S., Kristallografiya, Vol. 12, No. 1 (1967).
2. Aptekar', I. L., and Kamenetskaya, D. S., In: Problems of Metal Science and the Physics of Metals, Vol. 8, p. 205. Metallurgiya, Moscow (1964). See also: Fiz. Met. i Metalloved., 14(3):358 (1962).

3. Roitburd, A. L., In: Transactions of the Central Scientific-Research Institute of Ferrous Metallurgy, Vol. 4, p. 56. Metallurgizdat, Moscow (1960); see also Kristallografiya, 7(2) : 291 (1962).
4. Olsen, W. T., and Hultgren, R., J. Met., 188 : 1323 (1950).
5. Bochvar, A. A., and Borin, F. A., In: Tr. MITsMZ, Vol. 1. ONTI, Moscow (1933); see also Bochvar, A. A., Metal Science, p. 200. Metallurgizdat, Moscow (1956).
6. Falkenhagen, G., and Hofmann, W., Z. Metallk., 43(3) : 69 (1952).
7. Fridlyander, I. N., Dokl. Akad. Nauk, 104(3) : 429 (1955).
8. Salli, I. V., and Miroshnichenko, I. S., Dokl. Akad. Nauk SSSR, 132 : 1364 (1960).
9. Duwez, P., Willens, R., and Klement, W., J. Appl. Phys., 31(6) : 1136 (1960).
10. Novikov, I. I., Lyuttsau, V. G., and Zolotorevskii, V. S., Fiz. Met. i Metalloved., 16(2) : 241 (1963).
11. Kamenetskaya, D. S., Rakhmanova, É. P., and Spektor, E. Z., Dokl. Akad. Nauk SSSR, 142(3) : 584 (1962); see also Fiz. Met. i Metalloved., 19(4) : 583 (1965).
12. Kamenetskaya, D. S., In: Problems of Metal Science and the Physics of Metals, Vol. 3, p. 371. Metallurgizdat, Moscow (1952).
13. Willens, R. H., J. Appl. Phys., 33(11) : 3269 (1962).
14. Duwez, P., and Willens, R., Trans. Met. Soc. AIME, 227(2) : 362 (1963).
15. Klement, W., Willens, R., and Duwez, P., Nature, 187 : 869 (Sept., 1960).
16. Duwez, P., Willens, R., and Klement, W., J. Appl. Phys., 31(6) : 1137 (1960).
17. Biloni, H., and Chalmers, B., Trans. AIME, 233(2) : 373 (1965).
18. Moiseev, V. P., Izv. Akad. Nauk SSSR, 26(3) : 384 (1962).

CHARACTERISTICS OF CRYSTALLIZATION
AT HIGH COOLING RATES

I. V. Salli and L. P. Limina

Dnepropetrovsk State University

From a study of the formation of structure in alloys at high cooling rates, a large number of new facts emerge, not entirely explicable on the basis of existing theory. The most interesting characteristics of crystallization at high cooling rates, observed by a number of authors, are (1) the phenomenon of supersaturability (capacity to become supersaturated), and (2) the maintenance of the shape characteristic of low cooling rates in the growing crystals.

In this paper we shall be concerned with the first of these characteristics, namely, supersaturability. It is well known that, in the solidification of binary systems at high cooling rates, in some cases supersaturated solid solutions are formed, sometimes even retaining the composition of the original liquid solution. In other systems it is impossible to obtain a supersaturated primary solid solution above some concentration c_m.

The table summarizes published data on the composition of primary solid solutions for systems with unrestricted solubility in the liquid and restricted solubility in the solid state. By analyzing these data we may conclude that among binary systems suffering phase separation in the solid state there are three quite definite groups of alloys. The first group contains alloys of the eutectic type with intermediate phases, in which high cooling rates lead to the formation of primary solid solutions with a concentration of the second component not exceeding the maximum solubility at the temperature of the eutectic, c_m (Al−Zn, etc.). The second group (consisting of a single system, Ag−Cu) represents eutectic alloys with a continuous series of solid solutions at high cooling rates. The third group contains binary alloys with intermediate phases; among these are systems in which high cooling rates do not produce marked changes in the composition of the solid solution (Al−Cu, Sn−Cu). However, in this group of alloys there are a large number of systems giving rise to primary solid solutions with a supersaturation exceeding c_m by a factor of several times (Al−Mn, Al−Mg, etc.).

It should be noted that in studying these groups of alloys a narrow range of concentrations was employed. In order to make up this deficiency we studied alloys of types representing each of the three groups: Al−Zn, Ag−Cu, and Al−Mn, using a wide range of concentrations and cooling rates. We also studied the Cu−Pb system, which undergoes phase separation in the liquid state.

Rapid cooling of the alloys was carried out in an apparatus with a rotating cylinder, as described in [1]. The solubility and phase composition were determined by x-ray diffraction.

Table 1

Alloy	Type of phase diagram	Exist.of intermed. stage	Solubility of components in each other, wt.%		Type of crystal lattice	
			equilibrium	nonequilibrium	first component	second component
Ni—C	Eutectic	Yes	0.6	1.85	β-fcc	α-hexagonal
Co—C	The same	Yes	0.8	1.65	α-hcp β-fcc	The same
Al—Mn	" "	Yes	1.82	14.5	fcc	α-cubic β-cubic γ-tetragonal δ-fcc
Al—Cr	" "	Yes	0.7	5.8	fcc	α-bcc β-hcp
Sn—Sb	Peritectic	Yes	10.5	16.9	β-tetragonal	Rhombohedral
Al—Ti	Eutectic	Yes	0.15	0.32	fcc	α-hcp β-bcc
Al—V	The same	Yes	0.37	~1	fcc	bcc
Al—Fe	" "	Yes	0.05	0.17	fcc	α-bcc γ-fcc
Al—Mg	" "	Yes	17.4	34-38	fcc	hcp
Ag—Cu	" "	No	8.8·(Cu in Ag) 8·(Cu in Ag)	Complete solubility	fcc	fcc
Cu—Pb	Eutectic	No	~0	~4	fcc	fcc
Fe—C	The same	Yes	1.8	~2	α-bcc, γ-fcc	α-hexagonal
Ni—Sn	" "	Yes	42.0 at.%	17 at.%	α-hcp β-fcc	β-tetragonal
Ni—Ge	" "	Yes	0	22 at.%	The same	Cubic (diamond)
Ni—Si	" "	Yes	17.6 at.%	20 at.%	" "	Cubic (diamond)
Ag—Ge	" "	Yes	9.6 at.%	13.5 at.%	fcc	Cubic (diamond)
Tl—Ag	" "	Yes	0	40 at.%	Hexagonal	fcc
Tl—Au	" "	Yes	0	50 at.%	Hexagonal	fcc
Al—Zn	" "	No	82.2	—	fcc	hcp
Al—Cu	" "	No	5.7	—	fcc	fcc
Pb—Cd	" "	No	0.25·(Pb in Cd) 3.3·(Cd in Pb)	—	fcc fcc	hcp hcp
Al—Si	" "	No	11.7	—	fcc	Cubic (diamond)
Sn—Bi	" "	No	21	—	β-tetragonal	Rhombohedral
Al—Ge	" "	No	7.2	—	fcc	Cubic (diamond)

Note: The first element in all the systems given in the table is the base and the second the alloying element.

In addition to the observations made earlier [2, 3] and confirmed in the present investigation, we noted a number of new features.

It was found in the Ag—Cu system that only in the case when the cooling rate exceeded 10^6 deg/sec was it possible to obtain solid solutions of the same composition as in the original liquid. On reducing the cooling rate the same original melt (in the thicker films) crystallized into a heterogeneous mixture consisting of primary solid solutions with a concentration of the second component not exceeding c_m. This transformation is to some extent reminiscent of the martensite transformation, without the maintenance of coherence, in which only the achievement of maximum cooling rate leads to the formation of a supersaturated solid solution with the concentration of the original phase.

In the Al−Zn alloy no cooling rates gave supersaturations above 65 wt.%, which is almost 1.3 times lower than the maximum solubility at the eutectic temperature. Further increasing the concentration of the original melt led to a reduction in the concentration of the solid solution.

Finally, in the Al−Mn system, on increasing the cooling rate and concentration of the original melt, we were able to increase the solubility to 14.5%, which is 1.5 times greater than existing data. However, it is extremely interesting that, in this case, further increasing the concentration leads to a sharp fall in the solubility of Mn in Al.

In the system incorporating intermediate phases, a similar situation is observed with Fe−C alloys. This suggests that the behavior of alloys with intermediate phases may be likened to binary systems without intermediate phases, in which the solubility cannot be exceeded under equilibrium conditions. For this purpose such systems must be regarded as pseudo-binary, i.e., consisting of an elemental component and a metastable intermediate phase. Thus, in Fe−C alloys we have the Fe−Fe_3C systems and in Al−Mn alloys, Al and the "ε" phase.

Thus, none of the tabulated binary alloys suffering phase separation in the solid state (except Ag−Cu) give a primary solid-solution concentration exceeding c_m at high cooling rates.

Let us now consider a fundamental difference of these alloys from the Ag−Cu system. We see from the table that the systems without intermediate phases consist of components differing from one another in crystal structure, while only in the Ag−Cu system, in which, according to the Hume-Rothery rule, a continuous series of stable solid solutions should be formed, is a continuous series of metastable solutions formed on rapid cooling. Evidently, even in a system with intermediate phases, limited solubility in the solid state in the pseudo-binary system is a result of the fact that the crystal structure of the primary solid solution differs sharply from the structure of the intermediate phase.

We also made a study of Cu−Pb alloys. According to the above classification, these belong to the class similar to Ag−Cu. However, in this alloy the atomic radii differ considerably and the structure of the electron shells is different.

X-ray structural analysis shows that films obtained at high cooling rates consist of a solid solution on a Cu base (approximately 4% Pb), together with pure Pb and Cu. The lattice constant of the supersaturated Cu-base solid solution equals 3.641 kxu.

Thus, in this alloy, we failed to obtain a diffusionless transformation, although we were able to increase the solubility. This fact, however, does not justify us in considering that still further raising the cooling rate might lead to a diffusionless transformation, since the supersaturation reached may constitute the maximum solubility of Pb in Cu at high temperatures.

In conclusion, we note that, in some cases, high cooling rates lead to the formation of an amorphous state of the substance [4] or, in other words, the liquid is supercooled to a temperature at which diffusion separation cannot occur, nor can crystallization of the martensite-transformation type take place.

From the point of view of thermodynamics we may draw the following conclusion. For binary systems in which a diffusionless transformation from the liquid to the solid state does not take place at high cooling rates, the free energy of the liquid solution (right up to the amorphous state) at low temperatures is lower than the free energy of a solid solution which has the same composition, but the crystal structure of one of the components.

Literature Cited

1. Salli, I. V., and Limina, L. P., Zavodsk. Lab., 1 : 120-121 (1965).
2. Falkenhagen, G., and Hofman, W., Z. Metallk., 43(3) : 69 (1952).
3. Duwez, P., Willens, R., and Klement, W., J. Appl. Phys., 31(6) : 1136 (1960).
4. Willens, R., J. Appl. Phys., 33 : 3269-70 (1962).
5. Vol, A. E., Structure and Properties of Binary Metal Systems. Fizmatgiz, Moscow (1959).
6. Hansen, M., and Anderko, K., Constitution of Binary Alloys, I, II [Russian translation]. Metallurgizdat, Moscow (1962).

EFFECT OF COOLING RATE DURING THE CRYSTALLIZATION OF SOLID SOLUTIONS ON THE COMPOSITION OF THE AXIAL PARTS OF THE DENDRITE BRANCHES

I. S. Miroshnichenko

Dnepropetrovsk State University

When the cooling rate is not very low, so that diffusion inside the solid phase cannot occur to the full, nonequilibrium crystallization takes place. The parts of the solid phase crystallizing first are impoverished with respect to the dissolved element.* If the crystals of the solid solution grow in the form of dendrites, then the axial parts of the dendrite branches are impoverished with respect to the dissolved element. The minimum concentration of dissolved material remains constant over a zone sometimes constituting a high proportion of the cross section of the dendrite branch, as shown in Fig. 1a [1, 2].

On taking x-ray diffraction photographs of solid solutions with this kind of distribution of the dissolved element, the shape of the interference contour takes the form indicated in Fig. 1b. The sharp maximum L_1 corresponds to a reflection from the heart of the dendrite branches, having a constant composition. The diffuse interference contour is a result of reflection from the peripheral regions of the dendrite branches with an increased concentration of the dissolved substance.

From the position of the maximum we may determine the crystal lattice parameter of the solid solution in the axial parts of the dendrite branches, and from the corresponding calibration curve (Fig. 2) the amount of dissolved substance in them. The position of the boundary of the diffuse region in many cases indicates the limiting content of the dissolved element in the peripheral parts of the dendrite branches.

We studied the effect of cooling rate during crystallization on the composition of the axial parts of the dendrite branches in solid solutions of magnesium in aluminum, and tin and antimony in copper. The cooling rate varied between 10 and 10^6 deg/sec, or more. This was achieved by solidifying the alloys in metal molds, clapping the melt between copper plates [3], or throwing drops of molten metal on to the inner wall of a rotating copper cylinder [4, 5].

Subjects for study included alloys of the following compositions: Al + 9.6% Mg, Al + 13.0% Mg, Al + 16.4% Mg, Cu + 4.5% Sn, and Cu + 2.9% Sb. The amount of the second component

---

*For cases in which the partition coefficient k < 1.

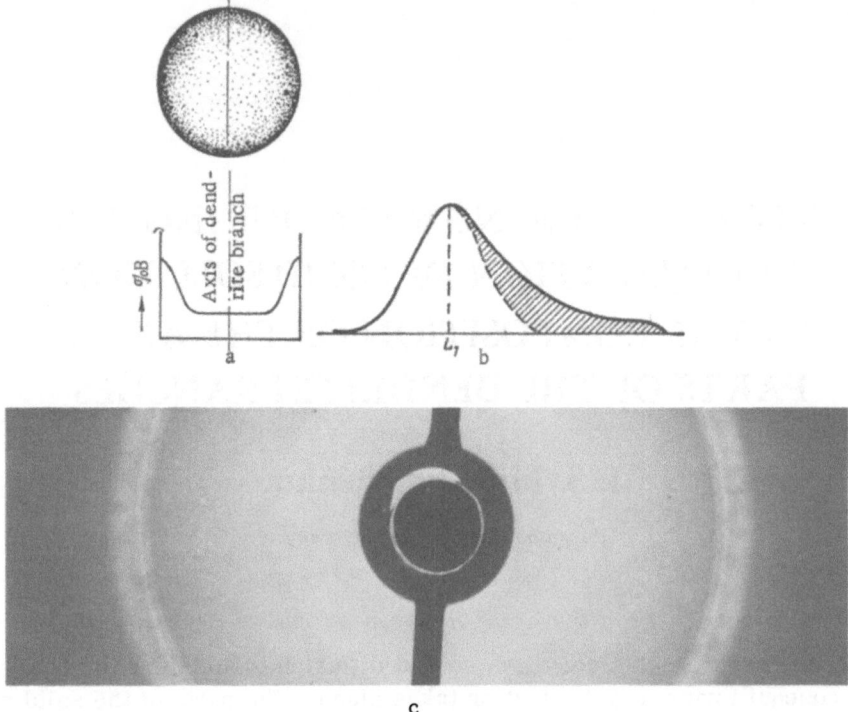

Fig. 1. Distribution of the dissolved element over the cross section of the dendrite cell. a) Schematic representation; b) shape of the interference contour; c) x-ray diffraction photograph of Al−Mg alloy. Back photography, (224) line, Cu radiation, cooling rate 10^4 deg/sec.

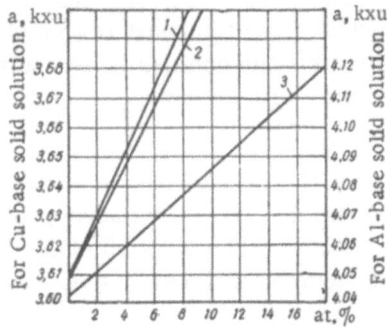

Fig. 2. Crystal lattice parameter of the solid solution as a function of the content of dissolved element. 1) Cu−Sb; 2) Cu−Sn; 3) Al−Mg.

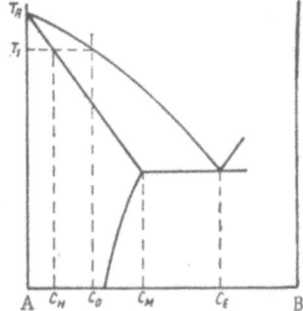

Fig. 3. Type of phase diagram of the alloys studied.

is given in atomic percent. All the alloys belong to the eutectic type with a considerable range of solid solutions (Fig. 3). The limiting solubilities at the eutectic temperatures are Al−Mg−18.9% Mg, Cu−Sn−7.7% Sn, and Cu−Sb−6% Sb.

Results of the Experiments

The compositions of the axial (central) parts of the dendrite branches c_C are shown in Fig. 4 by the shaded region. For comparison, the same graphs show broken lines corresponding

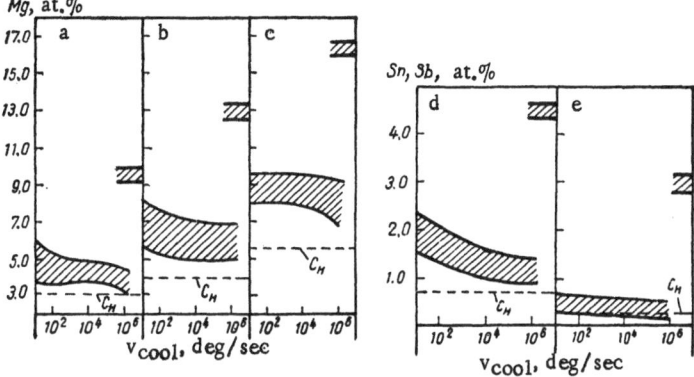

Fig. 4. Composition of the axial parts of the dendrite branches as
a function of cooling rate. a) Al + 9.6% Mg; b) Al + 13.0% Mg;
c) Al + 16.4% Mg; d) Cu + 4.50% Sn; e) Cu + 2.9% Sb.

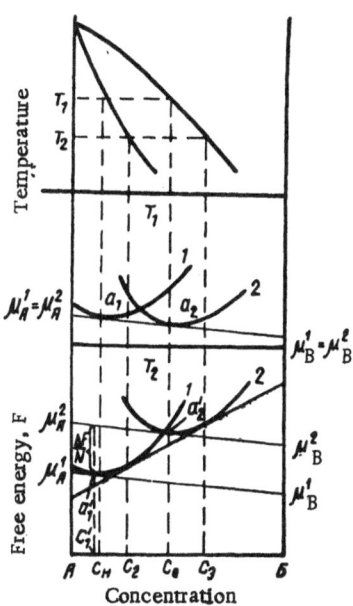

Fig. 5. Arrangement of the free-energy
curves in the absence (T_1) and presence
(T_2) of supercooling. c_1' is the composition of the solid phase growing in liquid
of composition c_0 in the presence of
supercooling at the crystallization front.

to the composition c_H of the equilibrium solidus
of an alloy of the original concentration in the absence of supercooling.

Analysis of the data shows that a change in
the cooling rate from 10 to 10^6 deg/sec has no
great influence on the composition of the axial
parts of the dendrite branches. There is a certain
fall in the content of dissolved material on increasing the cooling rate. The resultant values of c_C
are close to the point c_H of the equilibrium solidus.

The maximum content of dissolved element
in the peripheral parts of the dendrite branches
was no greater than the limiting solubility at the
eutectic temperature in accordance with the equilibrium phase diagram.

On studying the surface of very thin films in
contact with a copper cylinder (cooling rate greater
than 10^6 deg/sec), the x-ray diffraction photographs
showed very fine lines corresponding to reflection
from parts of the alloy having the same composition as the liquid phase c_0. At first these were observed together with lines corresponding to reflection from the impoverished axial parts of the dendrite branches c_C. On further slightly raising the
cooling rate, all the other lines vanished, and only
the reflections obtained from the homogeneous solid
solution with composition of the original liquid remained.

Discussion of Experimental Results

Under practical crystallization conditions, diffusion in the melts is limited. Hence, a
layer of liquid enriched with atoms of the dissolved substance arises at the front of the growing
solid phase. If the crystallization front is plane, the maximum content of the dissolved substance

in the layer is c_0/k. The solid phase thus formed will have the composition of the original liquid, c_0 [6].

Solidification conditions characterized by a fairly high concentration of dissolved atoms in the liquid, a low temperature gradient, and a high crystallization rate, ensure a dendritic form of crystal growth. In this case the crystallization front is not plane, but consists of individual needle-like projections, acting as bases for the formation of dendrite branches. This cut-up form of the crystallization front has the result that the distribution of the dissolved substance in the enriched layer of liquid will differ from that adjacent to a plane front. At the tips of the needles the content of dissolved atoms falls as a result of the development of an additional diffusion flow parallel to the crystallization front. In the intervals between the projections there is a buildup of atoms of the dissolved substance [7, 8].

The change in the concentration of dissolved atoms at the tips of the projections must influence their composition. Here we suppose that the compositions of the liquid and the growing solid phases follow the lines of the liquidus and solidus on the equilibrium phase diagram.*

The influence of the growth rate of the crystals on the intensity with which the impurity atoms pass out from the tip of the projection is complex. On the one hand, an increase in the velocity of the crystallization front should prevent the outflow of atoms from this. On the other hand, a reduction in the radius of curvature of the tip resulting from the increased velocity should strengthen the diffusion flow of dissolved atoms parallel to the crystallization front.

The slight impoverishment of the heart of the dendrite branches with respect to dissolved atoms as cooling rate increases (Fig. 4) is evidently a result of the increasing influence of the reduction in the radius of curvature of the tip of the projection on the outflow of excess atoms.

The composition of the axial parts of the dendrite branches, especially those formed at high cooling rates, is close to the composition of point c_H on the equilibrium solidus. This indicates a considerable outflow of excess atoms from the tip of the projection. The axial parts of the dendrite branches are formed in liquid having a composition close to the original. There is a kind of "intergrowth" of the enriched layer of liquid by the growing crystal.

We may suppose that the temperature of the melt at the tip of the projection is in this case also close to equilibrium, i.e., the supercooling at the crystallization front is negligible. However, the composition of the crystallizing solid phase should clearly not change a great deal when an appreciable supercooling develops at the crystallization front (Fig. 5).

In a liquid of composition c_0 supercooled to a temperature T_2, crystals of composition c_L^1 will possess the greatest growth stimulus. Only for crystals of this composition will $\mu_A^2 - \mu_A^1 = \mu_B^2 - \mu_B^1$, so that the reduction in the free energy per atom of crystallizing material $(\Delta F/N)$ will be greatest [10, 11].

In the range of supercoolings achieved in practice, the displacement of the free-energy curves along the axis of concentrations, and the change in their radius of curvature will be very slight [12]. Hence the concentrations c_L^1 and c_H will be close to one another. Thus, supercooling at the crystallization front should not lead to any severe change in the composition of the crystallizing phase, if the composition of the liquid at the crystallization front remains unaltered.

On increasing the crystallization rate to values of about 10^6 deg/sec, homogeneous solid solutions of the same composition as the liquid phase develop in the systems studied. In Fig. 4

---

* More exactly, they should follow the kinetic liquidus and solidus lines [9].

this may be noted from the sharp change in the composition of the axial parts of the dendrite branches. The final result does not contradict the principles of the diffusionless transformation [11]. Nevertheless, we have insufficient data as yet to judge the mechanism underlying the transfer of atoms from one phase to another.

Literature Cited

1. Kohn, A., and Philibert, J. Met. Treatment, 27(179) : 180 (1960).
2. Malinochka, Ya. N., Liteinoe Proizv., No. 10, p. 28 (1963).
3. Miroshnichenko, I. S., and Salli, I. V., Zavodsk. Lab., 11 : 1398 (1959).
4. Duwez, P., and Willens, R., Trans. Met. Soc. AIME, 227 : 362 (1963).
5. Salli, I. V., and Limina, L. P., Zavodsk. Lab., 1 : 120 (1965).
6. Tiller, W. A., et al., Acta Met., 1 : 428 (July, 1953).
7. Lyubov, B. Ya., and Temkin, D. E., In: Growth of Crystals, Vol. 3, p. 59. Izd. Akad. Nauk SSSR, Moscow (1961). [English translation: Consultants Bureau, New York (1962).]
8. Hurl, D. T., In: Growth Processes and the Growing of Single Crystals [Russian translation], p. 325. IL, Moscow (1963).
9. Borisov, V. T., Dokl. Akad. Nauk SSSR, 142(1) : 69 (1962).
10. Hiller, M., Acta Met., 1(6) : 764 (1953).
11. Aptekar', I. L., and Kamenetskaya, D. S., In: Problems of Metal Science and the Physics of Metals, p. 205. Metallurgiya, Moscow (1964).
12. Kamenetskaya, D. S., In: Problems of Metal Science and the Physics of Metals, p. 113. Metallurgizdat, Moscow (1949).

EFFECT OF ULTRASOUND ON THE RATE OF NUCLEUS FORMATION AND THE SHAPE OF THE CRYSTALLIZATION FRONT IN ORGANIC MATERIALS

O. V. Abramov

Institute of Metallurgy and the Physics of Metals of the Central Scientific-Research Institute of Ferrous Metallurgy

Analysis of the crystallization of metals and organic substances in an ultrasonic field suggests that the mechanism underlying the effect of ultrasound is quite complicated and includes its influence on both the parameters of crystallization and the dispersion of the growing crystals.

Earlier research into the mechanism in question has been devoted to particular aspects of the complex effects of ultrasound, while others have been neglected. In studying the mechanism underlying these effects it is desirable to consider the influence of ultrasound on the rate of nucleus formation and the shape of the crystallization front, as well as dispersion processes, and also to estimate the contribution made by such processes to the formation of fine-grained structure in bars of solidified material.

Experiments were accordingly made on transparent organic substances.

The influence of ultrasound on the rate of forming crystallization centers was estimated from the change in the degree of supercooling necessary for the development of crystallization centers in the substances studied (threshold of metastability) under the influence of ultrasound.

The arrangement of the apparatus used for the experiments is illustrated in Fig. 1.

An ampoule containing the material under examination 4 was placed in the crystallizer (mold) 6 and surrounded by water at a given temperature supplied from the thermostat 8 through a rubber hose 7. The temperature of the water was regulated to an accuracy of 0.1°C by the thermometer 3 mounted in the crystallizer. Ultrasonic vibrations were introduced into the crystallizer through the waveguide 2 linked to a magnetostriction converter 1. The converter was fed from a 10-kW ultrasonic generator.

This apparently inefficient way of introducing the ultrasound into the substance under consideration through a layer of water and the glass of the ampoule was employed in order to eliminate the generation of crystallization centers at the radiating end of the waveguide, which

260

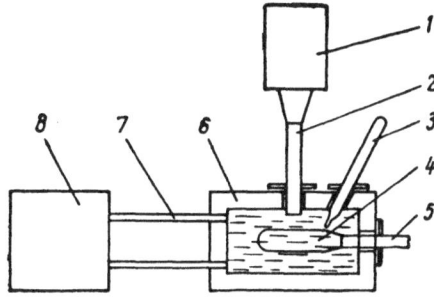

Fig. 1. Arrangement of apparatus for studying the effect of ultrasound on the generation rate of crystallization centers.

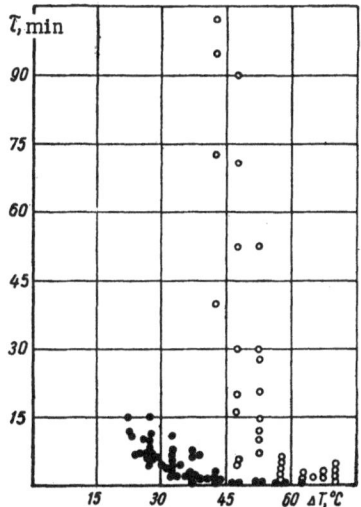

Fig. 2. Influence of ultrasound on the metastability threshold of Betal. O — Control experiments; ● — ultrasonic treatment.

may take place if the vibrations are introduced directly into the melt, and also to minimize the ultrasonic heating of the sample.

As subjects for study we chose Betal, naphthalene, azobenzene, Salol, and thymol. The ampoule containing the molten substance (at a fixed value of superheating) was inserted into the crystallizer by means of the holder 5. Water at a fixed temperature was passed from the thermostat into the crystallizer, i.e., a preselected value of supercooling was assigned to the sample. The waiting time for the appearance of the first crystallization center was then determined. In experiments involving ultrasonic treatment, the ultrasound was introduced after holding the ampoule for 0.5 to 1.5 min at the assigned degree of supercooling. In these experiments also the waiting time for the appearance of the first crystallization center was determined.

All the substances chosen had sharp thresholds of metastability. A reduction of 1-2° in the value of the supercooling greatly increased the waiting time for the appearance of the first crystallization center. Thus, in Betol, on supercooling by 42.5°, the shortest recorded waiting time for the appearance of the first center was 40 min, while on supercooling by 41.5° a 5-h holding period was insufficient to produce crystallization centers (Fig. 2). An analogous relationship between the waiting time τ and the supercooling ΔT was found for the other substances.

The introduction of ultrasound at maximum intensity resulted in the generation of crystallization centers at lower values of supercooling in all these cases, although the rise in the threshold of metastability differed for different substances (Table 1).

Table 1. Effect of Ultrasound on the Threshold of Metastability

Substance	Superheating temp., °C	Melting point, °C	ΔT	$\Delta T_{u.s.}$	$\Delta T - \Delta T_{u.s.}$
Betol	100	92.5	42.6	22.6	20
Naphthalene	90	80	8	2	6
Azobenzene	75	68	16	5	11
Azobenzene	55	44	12	10	2
Salol	65	44	21	10	11
Salol	75	44	38	30	8
Salol	85	44	49	46	3
Pure thymol	55	49.9	49.9	48	1.9
Contaminated thymol	55	48.9	30	20	10
Thymol + SiO_2 and C	55	49.9	38.9	14.4	24.5
Thymol + boron carbide	55	49.9	40.0	34	6

Immediately after introducing the ultrasound, a large number of crystallization centers were generated within the volume of the sample material, the waiting time for the appearance of crystallization centers being greatly reduced.

We also studied the effect of the purity of the sample material on the change in the threshold of metastability produced by the ultrasound. The experiments were made with thymol and Salol. In addition to chemically pure thymol, samples subjected to special purification and other samples containing deliberately introduced insoluble impurities (SiO_2, C, BC) were used. The effect of ultrasound on the metastability threshold of purified thymol was very slight. The introduction of insoluble impurities of SiO_2 and C led to a considerable reduction in the value of supercooling necessary for the generation of crystallization centers. Boron carbide had a negligible influence on the metastability threshold of thymol.

In experiments with Salol we studied the effect of the superheating of the melt in the course of melting on the change in the metastability threshold under the influence of ultrasound. As the superheating temperature increased the insoluble impurities became more and more de-activated, this constituting a peculiar kind of purification of the sample and reducing the threshold of metastability. With increasing purity of the Salol, the influence of ultrasound on the change in the metastability threshold temperature lessened.

The results of these experiments confirm the data of [1, 2], in which it was noted that the introduction of insoluble impurities improves the processing of materials in an ultrasonic field. In [2] we considered a possible mechanism for the activation of impurities; this was based on the effects of viscous friction between the surface of the impurity particles and the melt in an ultrasonic field.

Experiments carried out on the whole series of organic substances showed that ultrasound increased the rate of generation of crystallization centers, the influence being greatest in the presence of insoluble impurities in the test sample.

In the next series of experiments we studied the effect of ultrasound on the rate of generating crystallization centers and on the dispersion of the growing crystals under conditions in which the ultrasonic emitter was introduced directly into the melt. The experiments were made with thymol in the apparatus illustrated in Fig. 3. The sample of thymol was placed in a double-walled test tube 7 with an internal diameter of 14 mm and a length of 100 mm. Water from the thermostats 4 passed through the jacket of the test tube. The provision of two thermostats, the system of taps 3, and the two zones in the jacket of the test tube made it possible to set up a variety of conditions of heating and cooling the melt. Thus the thymol could be melted, superheated, or given some specific temperature to an accuracy of 0.2°C.

The waveguide and radiator 5 were introduced into the test tube from the top. The waveguide was heated by means of a special heater 6 with thermal control to the temperature of the melt so as to prevent crystallization at its cold end. The power introduced into the melt was measured by the method described in [3].

For regulating the temperature a Chromel–Copel thermocouple 2 was introduced into the melt. The emf of the thermocouple was passed to an electronic potentiometer 1 (an ÉPP-09) with a 2-mV scale [1]. In the thermocouple circuit was a compensating source of emf of a sign opposite to that of the thermocouple. The value of the compensating emf was taken in such a way as to enable the temperature scale of current interest to be represented on the 2-mV scale of the ÉPP-09.

The system was observed through an MBS-1 binocular microscope 8 with a magnification of 4 to 84 times.

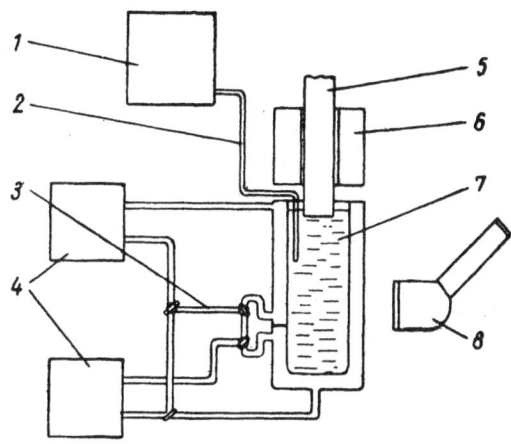

Fig. 3. Arrangement of the apparatus for studying the crystallization of thymol.

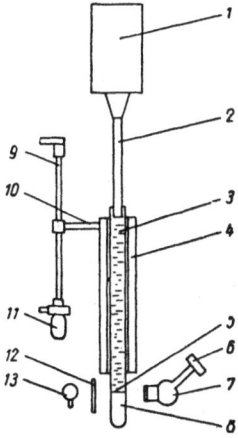

Fig. 4. Apparatus for observing the shape of the front during the directional crystallization of naphthalene.

The melting point of thymol, 48.6°C, was established in preliminary experiments. This indicated that the thymol was slightly contaminated with impurities. No special purification was attempted.

On cooling the melt (without any seed) in the absence of ultrasound, crystallization of the thymol started at 30-32°C. At this temperature, crystals so formed grew at a considerable rate. Crystallization of the whole melt took 20-25 min. Holding the melt at 47°C (supercooling of 1.6°C) for an hour did not lead to the appearance of crystallization centers. On introducing ultrasound into the melt, a cloud of small crystals formed around the radiator, and these very rapidly (fractions of a second) propagated over the whole volume of the test tube. The introduction of ultrasound also led to the development of cavitation phenomena in the melt, as represented by the formation of pulsating cavitation bubbles. A cloud of small crystals formed around the bubbles and propagated over the whole volume of the melt. The cavitation bubbles formed mainly at the radiator—melt boundary and more rarely at the tube walls and thermocouple.

On switching off the ultrasound the crystallites thus formed grew rapidly. The crystallization process ended in this case after 3 to 5 min.

Prolonged treatment of the melt led to its being heated above the melting point. In these experiments the ultrasonic power introduced into the melt was 200 W, which for a radiator diameter of 10 mm gave an intensity of about 160 W/cm^2. An analogous effect of the ultrasound also occurred after reducing the intensity to 2 W/cm^2.

In the next series of experiments we studied the dispersion of the growing particles of ultrasound.

The lower part of the test tube was cooled intensively, while the temperature of the upper part was maintained at several degrees above the melting point. This eliminated the possibility of crystallization centers being formed in the neighborhood of the radiator. A seed was introduced into the lower part of the tube and crystallization commenced at this point. The growth rate of the crystals was 4-5 mm/min. On introducing ultrasound at maximum intensity (I = 160 W/cm^2) cavitation phenomena developed at the crystal—melt interface and crystallites were torn away from the interface and scattered over the whole volume. The crystallization rate increased sharply.

In order to estimate the contributions respectively made by the dispersion and generation of crystals under the influence of ultrasound at maximum intensity to the refinement of the structure, we made the following experiment. In the upper part of the tube the temperature was

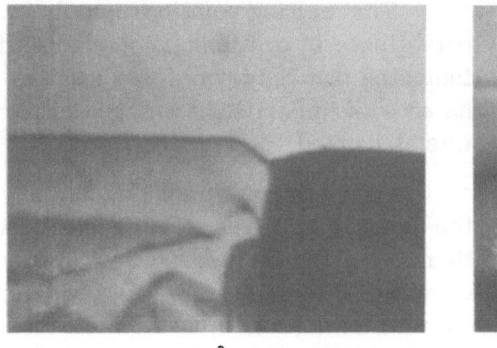 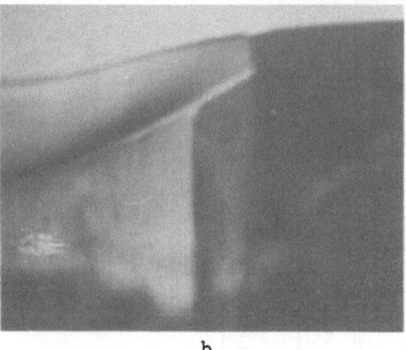

Fig. 5. Shape of crystallization front for a furnace velocity of 45 mm/h. a) In the absence of ultrasound; b) on introducing ultrasound at a power of 20 W/cm^2 (× 120).

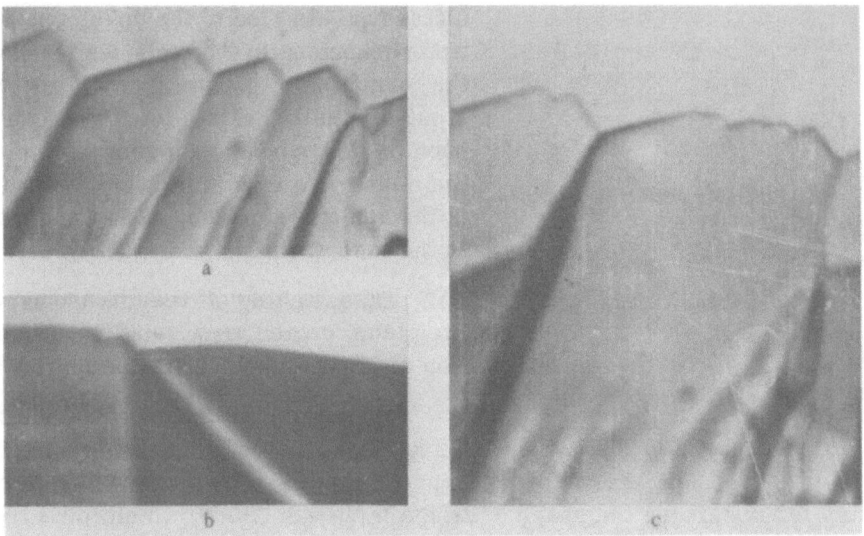

Fig. 6. Shape of crystallization front for a furnace velocity of 75 mm/h. a) In the absence of ultrasound; b) 2 sec after introducing ultrasound at a power of 20 W/cm^2; c) 4 sec after introducing ultrasound at a power of 20 W/cm^2 (× 120).

kept 1.5°C below the melting point, while the lower part was cooled sharply. Ultrasound was introduced while directional crystallization was taking place at the bottom of the tube. Dispersion of the growing crystals was observed at the crystallization front, while nucleation of new centers occurred in the radiator region. The number of crystals split off from the phase boundary was much greater than the number of crystallites developing under the influence of ultrasound in the neighborhood of the radiator. Reducing the temperature at the radiator produced a sharp increase in the rate of forming crystallization centers in the supercooled melt.

The ratio of the contributions of dispersion and generation was also estimated for the case in which thymol was poured into a mold (from above) while ultrasound was introduced from below through an opening in the bottom of the rectangular glass mold, constructed in such

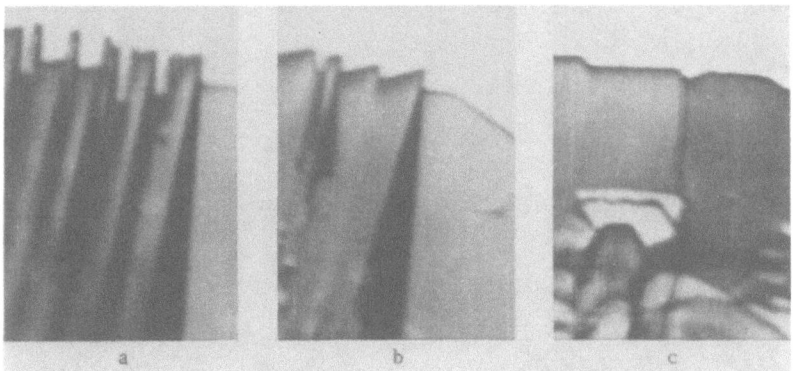

Fig. 7. Shape of crystallization front for a furnace velocity of 95 mm/h.
a) In the absence of ultrasound; b) 2 sec after introducing ultrasound at a
power of 20 W/cm^2; c) 4 sec after introducing ultrasound at a power of
20 W/cm^2 ($\times$ 120).

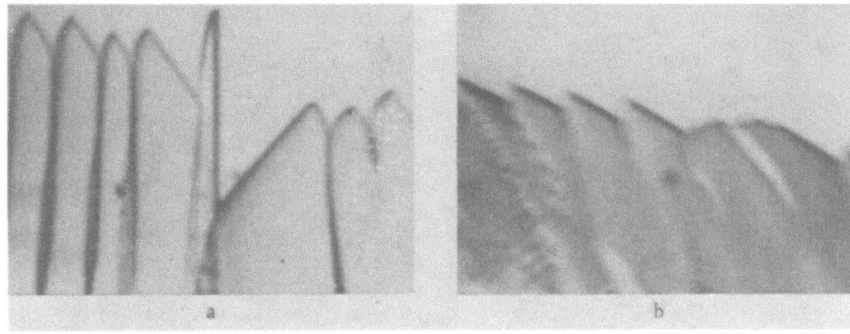

Fig. 8. Shape of crystallization front for a furnace velocity of 130 mm/h.
a) In the absence of ultrasound; b) 5 sec after introducing ultrasound at a
power of 20 W/cm^2 ($\times$ 120).

a way as to enable the crystallization process to be observed. A thymol melt superheated to a
temperature of 54°C was poured into the mold. Crystallization in the absence of ultrasound
started with the generation and growth of crystals near the steel radiator at a temperature of
24-26°C. Gradually crystals were also generated at the walls of the mold. The melt crystallized completely in 30-40 min. In the case in which ultrasound was introduced (the ultrasound
being switched on before pouring the melt), a "cloud" of crystals developed on the radiator at
a temperature of 48.3°C and then spread over the volume of the melt. The crystals falling on
the walls started gradually growing and were dispersed in their turn. Intensive crystal formation continued at the radiator. The whole volume of the melt was filled with crystallites. The
main contribution to the refinement of the structure in this case came from the formation of
crystals at the radiator. The part played by the dispersion of the crystals growing from the
walls was very slight.

These experiments led to the conclusion that, depending on the nature of the substance
and the conditions of crystallization, either the dispersion of the growing crystals or the increase in generation rate may predominate in an ultrasonic field. In both these processes
cavitation phenomena play a leading part.

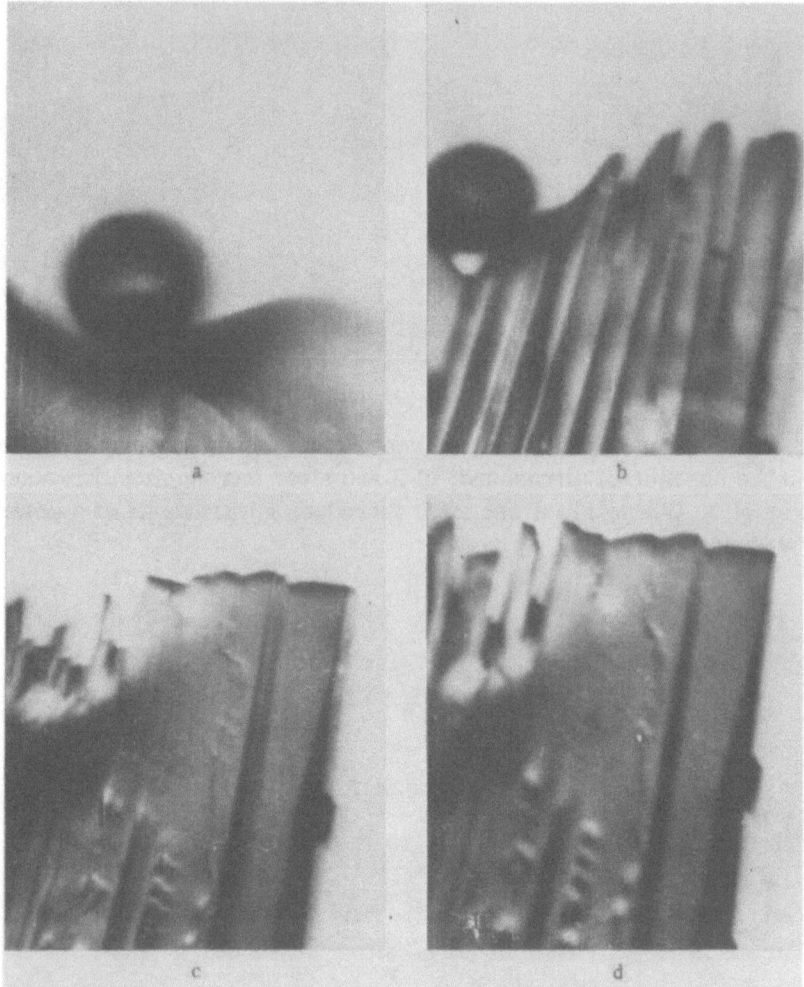

Fig. 9. Shape of crystallization front on introducing ultrasound at a power of 35 W/cm² into the melt, with a furnace velocity of 95 mm/h. a) Cavitation bubble; b) breaking up of the crystallization front by the pulsating cavitation bubble; c, d) growth of crystals at the site of a burst cavitation bubble (c, 2 sec; d, 4 sec after bursting).

On reducing the power of the ultrasound introduced to 2 W/cm², dispersion processes no longer occurred. Ultrasound smoothed the crystallization front, making it more nearly plane.

More detailed study of the effect of ultrasound on the shape of the crystallization front was carried out with naphthalene colored with 0.2 wt.% azobenzene. The different colorings of the components and their solutions enabled us to use the colorimetric method in order to determine the effective partition coefficient of the impurity. The effect of ultrasound on the shape of the crystallization front was compared with the change in the effective partition coefficient.

The experiments were carried out with the apparatus shown schematically in Fig. 4. A test tube containing the solution of azobenzene in naphthalene 3 was placed in the heater 4. Motion of the heater, required in order to initiate motion of the crystallization front, was effected by means of a motor 11 with reducing gear, screw 9, and nut 10. The velocity could be varied smoothly from 5 to 200 mm/h. Elastic vibrations were introduced into the solution through the end of the waveguide 2 linked to the magnetostriction converter 1. The frequency of the elastic vibrations was kept constant and the level of the power introduced was fixed between 20 and 35 W/cm^2.

The crystallization front was observed in polarized light (passing through the polaroid 12 from the lamp 13) by means of an MBS microscope 7 fitted with a photographic attachment 6 ($\times$ 120). The change in the form of the crystallization front moving from bottom (solid phase) to top was photographed.

The experiments showed that for heater velocities up to 60 mm/h the crystallization front was smooth in the absence of ultrasound (Fig. 5a). Increasing the velocity above this led to the development of a broken front (Figs. 6a, 7a, 8a). As the velocity increased the broken state of the front became more pronounced.

The introduction of ultrasound (20 W/cm^2) into the melt smoothed the crystallization front; for furnace velocities of up to 120 mm/h the crystallization front became smooth 2-3 sec after switching the ultrasound on (Figs. 5a, 6b, 6c, 7b, 7c). For a furnace velocity greater than 120 mm/h there was also a certain smoothing of the crystallization front, but the latter remained to some extent broken up. The tips of the projecting crystals became less sharp (Fig. 8b). The effect of ultrasound appeared in its influence on the growth conditions of the crystals rather than in their mechanical fracture. Increasing the power to 35 W/cm^2 led to the appearance of cavitation bubbles in the melt. The pulsating cavitation bubbles formed hollows on the surface of the crystallization front (Fig. 9a, b). When the bubbles burst, there was an intensive growth of needle-shaped crystals out of the hollow on the crystallization front. The crystallization front became broken up (Fig. 9c, d) for all furnace velocities.

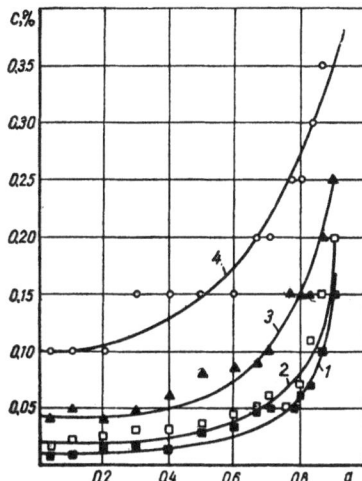

Fig. 10. Concentration curves of the impurity distribution along a purified bar for various furnace velocities.
1) v = 45 mm/h; 2) v = 75 mm/h;
3) v = 95 mm/h; 4) v = 130 mm/h.

On analyzing the various possible reasons for the influence of ultrasound on the shape of the crystallization front, it must be pointed out that ultrasound should give rise to the development of microcurrents at the crystallization front [4] and promote agitation in the liquid zone. These processes (microcurrents and agitation) lead to the breakup of the diffusion layer and to a reduction in its thickness, evening out the concentration of impurity in the liquid zone and increasing the temperature gradient near the front. Thus, our measurements showed that for a furnace velocity of 45 mm/h the temperature gradient in the immediate neighborhood of the front (up to a distance of 0.8 mm) increases from 45 deg/cm in the absence of ultrasound to 80 deg/cm in its presence.

An increase in the temperature gradient and a reduction in the concentration of impurity at the front inescapably entail a reduction in the zone of concentration supercooling and promote the formation of a smooth crystallization front [5]. The smoothing of the front in turn leads to a reduction in the thickness of the diffusion layer and, in addition to this, impedes the capture of impurities by the growing solid-phase crystals and thus reduces the effective partition coefficient of the impurity.

When cavitation phenomena develop in the melt, mechanical fragmentation of the growing crystals takes place and the front becomes broken up. This leads to an increase in the thickness of the diffusion layer and facilitates the capture of impurities by the growing crystals, i.e., increases the effective partition coefficient.

Table 2. Relation Between the Effective Partition Coefficient
and the Shape of the Crystallization Front

v, mm/h	I, W/cm^2	k	Shape of front
45	0	0.075	Smooth
	20	0,065	Smooth
	35	0.08	Slightly broken up
75	0	0.14	Broken up
	20	0.075	Smooth
	35	0.15	Broken up
95	0	0.25	Broken up
	20	0.075	Smooth
	35	0.2	Broken up
130	0	0.5	Severely broken up
	20	0.1	Slightly broken up
	35	0.45	Severely broken up

In order to compare the shape of the crystallization front with the effective partition coefficient, we plotted impurity-distribution curves along the bar of solidified material during directional crystallization with different furnace velocities and ultrasound power levels (Fig. 10). From these curves we calculated the effective partition coefficient k

$$\frac{c}{c_0} = k(1-g)^{k-1}$$

Here, c is the impurity concentration in the given part of the bar, c_0 is the original impurity content, and y is the relative proportion of solidified material in the bar.

It follows from Table 2 that there is a clear correspondence between the shape of the front and the partition coefficient: the less broken up the front, the smaller is the effective partition coefficient. The smoothing of the crystallization front on introducing ultrasound may be used in order to intensify the processes of zone refining.

Thus our experiments show that the introduction of elastic vibrations into the melt increases the rate of generating crystallization centers and changes the shape of the crystallization front, smoothing it for low ultrasound intensities and producing dispersion of the growing particles for high intensities.

Literature Cited

1.   Danilov, V. I., and Chezhemov, G. Kh., In: Problems of Metal Sciences and the Physics of Metals, Vol. 4, pp. 34-39. Metallurgizdat, Moscow (1955).
2.   Abramov, O. V., and Teumin, I. I., Fiz. Met. i Metalloved., 15(5) : 710 (1963).
3.   Abramov, O. V., and Teumin, I. I., In: Use of Ultrasound in the Production of Alloys and Their Heat Treatment, Vol. 4, pp. 14-21. Moscow (1962).
4.   Kubanskii, P. N., Zh. Tekhn. Fiz., 22 : 4 (1952).
5.   Rutter, J., In: Liquid Metals and Their Solidification [Russian translation]. Moscow (1962).